PHENOMENA IN IONIZED GASES
XXII ICPIG

PHENOMENA IN IONIZED GASES

XXII ICPIG

Hoboken, NJ July–August 1995

EDITORS
Kurt H. Becker
The City University of New York (City College)

Wayne E. Carr
Erich E. Kunhardt
Stevens Institute of Technology

AIP CONFERENCE PROCEEDINGS 363

American Institute of Physics   Woodbury, New York

Authorization to photocopy items for internal or personal use, beyond the free copying permitted under the 1978 U.S. Copyright Law (see statement below), is granted by the American Institute of Physics for users registered with the Copyright Clearance Center (CCC) Transactional Reporting Service, provided that the base fee of $6.00 per copy is paid directly to CCC, 222 Rosewood Drive, Danvers, MA 01923. For those organizations that have been granted a photocopy license by CCC, a separate system of payment has been arranged. The fee code for users of the Transactional Reporting Service is: 1-56396-550-X/ 96 /$6.00.

© 1996 American Institute of Physics

Individual readers of this volume and nonprofit libraries, acting for them, are permitted to make fair use of the material in it, such as copying an article for use in teaching or research. Permission is granted to quote from this volume in scientific work with the customary acknowledgment of the source. To reprint a figure, table, or other excerpt requires the consent of one of the original authors and notification to AIP. Republication or systematic or multiple reproduction of any material in this volume is permitted only under license from AIP. Address inquiries to Office of Rights and Permissions, 500 Sunnyside Boulevard, Woodbury, NY 11797-2999; phone 516-576-2268; fax: 516-576-2499; e-mail: rights@aip.org.

L.C. Catalog Card No. 96-83294
ISBN 1-56396-550-X
DOE CONF- 950749

Printed in the United States of America

CONTENTS

Preface vii
Sponsors of the XXII. ICPIG ix
ICPIG Committees x

UV/VUV High Sensitivity Absorption Spectroscopy for Diagnosing Lighting and Processing Plasmas and for Basic Data (Penning Prize Lecture) 1
    J. E. Lawler, M. A. Childs, K. L. Menningen, L. W. Anderson, S. D. Bergeson, and K. L. Mullman

Ion—Plasma Interaction 12
    C. Deutsch, C. Fleurier, D. Gardès, and G. Maynard

Surface-Wave Sustained Plasmas: Toward a Better Understanding of RF and Microwave Discharges 25
    M. Moisan, C. M. Ferreira, J. Hubert, J. Margot, and Z. Zakrzewski

The Present State and Development Trends of Discharges. Physics and Chemistry of Plasmas. 41
    A. A. Ivanov

Fundamental Electron Collision Processes Relevant to Low-Temperature Plasmas 75
    K. H. Becker

Low Frequency Oscillations and Chaos in Plasmas 90
    A. Piel and T. Klinger

H.F. Emission Related to Particle Beams Injected into Ionospheric Plasma from Spacecraft. 105
    Z. Kłos

Different Aspects of Self-Consistent Modeling of Non-Equilibrium Discharges 121
    M. Capitelli, C. Gorse, D. Iasillo, and S. Longo

Formation and Evolution of the Cathode Sheath on the Streamer Arrival 136
    M. Černák

On the Use of the Plasma in III-V Semiconductor Processing 146
    G. Bruno, P. Capezzuto, and M. Losurdo

Breakdown and Discharges in Dense Gases Governed by Runaway Electrons. 156
    L. P. Babich

Magnetic and Electric Probe Diagnostics in Inductive Plasmas 166
    V. Godyak, R. Piejak, and B. Alexandrovich

Numerical and Analytical Kinetic and Fluid Models for RF Discharges 176
    C.-H. Wu, F. Dai, C. Li, F. F. Young, and J. Tsai

Physics of High-Power Electron Beam-Plasma Interaction 192
    V. S. Koidan

Progress in the Understanding of Non-1D Glow Discharges: Experiment and Theory 204
    J. Derouard and L. Pitchford

External Magnetic Field Influence on Properties of High-Power
Laser-Produced Plasma 214
    J. Wołowski, A. Kasperczuk, and T. Pisarczyk
Air Ions and Aerosol Science 224
    H. Tammet
Molecular Beam Studies of Collisional Autoionization Processes 234
    B. Brunetti, S. Falcinelli, and F. Vecchiocattivi
The Propagation of Positive Streamers in Air and at Air/Insulator Surfaces ... 247
    N. L. Allen
Microwave Discharges used as Excitation Sources for Spectro-Chemical
Analysis 257
    A. Gamero
The Bloch–Elwert Structure as a Universal Manifestation of the Rate
of "Soft" Inelastic Collisions in Plasmas: Bremsstrahlung, Electrical
Conductivity, Ionization Loss 267
    V. I. Kogan
Arc Spot Ignition on Cold Electrodes in an Ambient Gas Atmosphere 278
    J. Mentel, R. Bayer, J. Schein, and M. Schumann
Tomography for Discharge Plasmas 289
    N. Iwama
Thin Film Deposition on Internal Walls of Cavities and Complex Hollow
Substrates 299
    L. Soukup, M. Šícha, L. Jastrabík, and M. Novák
Atmospheric Pressure Dielectric Controlled Glow Discharges: Diagnostics
and Modelling 306
    F. Massines, R. B. Gadri, P. Decomps, A. Rabehi, P. Ségur, and C. Mayoux
Capacitively Coupled Radio Frequency Discharges 316
    J. E. Allen
Corona Physics and Diagnostics 332
    R. S. Sigmond
Discharge Physics Issues in the Scaling of High Power CO_2 Lasers 345
    E. Desoppere and C. Leys
Nonlinear Surface Waves in Plasmas 355
    S. M. Vuković, N. B. Aleksić, and D. V. Timotijević
Optical and Probe Diagnostics of RF Discharges 363
    V. Kapička, M. Šícha, and A. Brablec
Cluster Lamp—A New Kind of Light Generation Mechanism 373
    R. Scholl and G. Natour

Author Index 383

PREFACE

The XXII. International Conference on Phenomena in Ionized Gases (ICPIG) was held on the campus of Stevens Institute of Technology in Hoboken, NJ (USA) from July 31 to August 4, 1995. The Conference venue also included a workshop on "Environmental Applications of Plasmas" which was held on July 30, 1995. The Local ICPIG Organizing Committee was chaired by Kurt H. Becker and Erich E. Kunhardt. Wayne E. Carr served as the Conference Secretary. Members of the Local Organizing Committee included representatives from academia, industry and government laboratories. This was the first ICPIG held in the United States. The past Chairman of the ICPIG International Scientific Committee, Arthur H. Guenther from Sandia National Laboratories, Albuquerque (USA) and a native of Hoboken, was Honorary Chairman of the XXII. ICPIG.

The Conference was attended by about 350 participants from 35 countries. There was a significant student attendance and about 50 accompanying persons participated in the social program. Approximately 450 contributed papers were presented in 4 poster sessions. Those contributed papers received before June 15, 1995 were published in the Book of Contributed Papers and distributed to all Conference participants. Post-deadline papers will be published in a separate supplement to the Book of Contributed Papers.

This Book of Invited Papers contains written versions of the invited talks presented at the XXII. ICPIG by 34 invited speakers, of which, however, four did not submit a manuscript. Invited talks consisted of General Invited Talks and Topical Invited Talks. In addition, this book contains a written version of the Penning Prize Lecture delivered by the 1995 recipient of the Penning Prize, James E. Lawler of the University of Wisconsin, Madison, WI (USA). James E. Lawler was the third recipient of the Penning Prize which is sponsored by Philips Lighting, Eindhoven, The Netherlands. The previous Penning Prize recipients were Hans von Engel, United Kingdom (1991) and Yuri Raizer, Russia (1993).

We are grateful for the generous financial support from the sponsors listed on the following page, which made it possible to provide partial or full support to about 100 participants. Special emphasis was put on supporting young scientists, scientists from the Republics of the former Soviet Union, scientists from Eastern Europe and scientists from developing countries. We acknowledge with special thanks the efforts of the staff of the Stevens Office of Student Housing, Dining Services and Center Operations, in particular Deborah Lanza, Kathleen McHugh and Samuel Sheps who were instrumental in making the non-scientific part of the XXII. ICPIG as enjoyable and successful as the scientific part.

Kurt H. Becker
Wayne E. Carr
Erich E. Kunhardt

Sponsors of the XXII. ICPIG

Innovative Science and Technology Office - BMDO

The Greve Foundation

OSRAM Sylvania

Philips Lighting, Eindhoven (Penning Prize)

Phillips Laboratory

Sandia National Laboratory

Los Alamos National Laboratory

Lawrence Livermore National Laboratory

US Air Force - EOARD

International Union of Pure and Applied Physics (IUPAP)

International Union of Radio Science (URSI)

The American Physical Society

XXII. International Conference on Phenomena in Ionized Gases (ICPIG)

ICPIG International Scientific Committee (1993-1995)

A.J. Davies (Chairman)	Great Britain
R. d'Agostino	Italy
A. Bouchoule	France
S. Denus	Poland
K. Günther	Germany
F.J. de Hoog	The Netherlands
H. Kikuchi	Japan
E.E. Kunhardt	USA
B. Milic	Yugoslavia
A.H. Oien	Norway
A.A. Rukhadze	Russia
M. Sicha	Czech Republic

Honorary Chairman of XXII. ICPIG

A.H. Guenther   Sandia National Laboratory

Local Organizing Committee of XXII. ICPIG

Co-Chairmen
K.H. Becker	City College of CUNY
E.E. Kunhardt	Stevens Institute of Technology

Secretary
W.E. Carr	Stevens Institute of Technology

Members
R. Barker	US AFOSR
V. Byszewski	OSRAM Sylvania
A. Chernikov	Stevens Institute of Technology
R.A. Gottscho	AT&T Bell Laboratories
Y. King	Phillips Laboratory XP
M.E. Mauel	Columbia University
K.H. Schoenbach	Old Dominion University
F. Williams	University of Nebraska
R. DeWitt	US Department of Energy

IUPAP SPONSORSHIP

To secure IUPAP sponsorship, the organizers have provided assurances that ICPIG will be conducted in accordance with IUPAP principles as stated in the ICSU Document Universality of Science (6th edition, 1989) regarding the free circulation of scientists for international purposes. In particular, no bona fide scientist will be excluded from participation on the grounds of national origin, nationality, or political considerations unrelated to science.

PENNING PRIZE 1995[*]

James E. Lawler

University of Wisconsin, Madison, WI (USA)

Penning Prize Lecture:

"UV/VUV High Sensitivity Absorption Spectroscopy for Diagnosing Lighting and Processing Plasmas and for Basic Data"

[*] Sponsored by Philips Lighting, Eindhoven, The Netherlands

UV/VUV High Sensitivity Absorption Spectroscopy for Diagnosing Lighting and Processing Plasmas and for Basic Data

J. E. Lawler, M. A. Childs, K. L. Menningen, L. W. Anderson, S. D. Bergeson, and K. L. Mullman

Department of Physics, University of Wisconsin, 1150 University Ave., Madison, WI 53706 USA

Abstract. High sensitivity absorption spectroscopy involves the use of modern diode and CCD (charge coupled device) detector arrays to observe fractional absorptions of ultraviolet (UV) and vacuum ultraviolet (VUV) radiation as small as 0.00001. Stable arc lamps provide a continuum in some experiments, but experiments at very high spectral resolution or at VUV wavelengths require the greater spectral radiance of a synchrotron. Absolute densities of excited atoms, atomic ions, and molecular radicals are measured in both processing and lighting plasmas. Basic spectroscopic data needed for the analysis of astrophysical observations from the Hubble Space Telescope are measured using absorption of Fe^+ in a hollow cathode discharge.

INTRODUCTION

Absorption spectroscopy has long been used to diagnose glow discharge plasmas. Although it lacks the spatial and temporal resolution of laser induced fluorescence, absorption spectroscopy provides absolute densities, a major advantage over other detection schemes. It is also useful for recording the spectra of atomic or molecular species that do not radiate because the excited level predissociates or is collisionally quenched. The measurement of absolute densities is often critical in an analysis of the ionization, power, or chemical balance of a glow discharge plasma.

Early absorption experiments often involved using one glow discharge (line emission source) to diagnose another glow discharge (absorbing sample). This approach is effective so long as the experimentalist has knowledge of the spectral line profiles in both discharges. Laser absorption spectroscopy offers superior spatial and temporal resolution but is not routinely used at far UV or VUV wavelengths, since the high order nonlinear effects required to shift the laser radiation to these short wavelengths usually degrades the amplitude stability of the radiation and limits the sensitivity of the experiment. It is also possible to use an arc lamp or a synchrotron (continuum source) to diagnose a glow discharge (absorbing sample). Experiments of this type now achieve a remarkable sensitivity to very small fractional absorptions because of the use of diode or CCD detector arrays. The use of a detector array provides two key advantages over a single channel, sequentially scanned absorption experiment. First, a sequentially scanned spectrometer maps any noise or drift in the continuum source onto the spectrum and thus limits the sensitivity of the absorption experiment. A multielement array detects all spectral channels simultaneously and largely avoids this source of noise as long as the continuum does not change its spectral shape. Photon shot noise is then the primary limit on sensitivity. Second, a detector array rapidly accumulates good photon statistics in all spectral channels.

The following sections of this paper describe the use of detector arrays for high sensitivity absorption spectroscopy in the study of a Hg-Ar glow discharge lighting plasma, in the study of a H_2-CH_4 glow discharge used for chemical vapor deposition (CVD) of diamond films, and in the determination of atomic oscillator strengths for the analysis of astronomical data.

Hg-Ar GLOW DISCHARGES

Wamsley *et al.* experimented with the use of an intensified CCD array to perform high sensitivity absorption spectroscopy on Hg-Ar glow discharges very similar to the discharge in commercial fluorescent lamps (1). The discharge consisted of a 400 mA axial dc current passed through a 35 mm inside diameter fused silica tube containing 2.5 Torr of Ar and 5.2 mTorr of Hg. Although the positive column of fluorescent lamp discharges is fairly well understood, the negative glow and the Faraday dark space regions near the hot cathode are not so well understood. Cathodic phenomena limit the life of most fluorescent lamps.

Wamsley *et al.* used a Xe arc lamp as a continuum source, a 0.5 m focal length echelle spectrometer, and an image-intensified CCD array (1). Figure 1 is a schematic diagram of their experiment. The use of an echelle grating in high order produces a high resolving power, but requires a premonochromator to separate orders. This experiment achieved sensitivity to fractional absorptions of 0.001 with only 6 sec of integration, as shown in Fig. 2.

This experiment was motivated in part by an earlier single channel, sequentially scanned experiment (2). Phase-sensitive detection was used in the single channel experiment to discriminate against the line emission from the glow discharge and detect the continuum from a Xe arc discharge that had absorption features after transmission through the glow discharge. Wamsley *et al.* measured absorption of a resonance line at 194 nm to determine the absolute density of the ground level of Hg^+ and measured absorption of various lines at visible and near-UV wavelengths to determine the absolute density of excited 6 ^{3}P Hg atoms (2). The measurements were

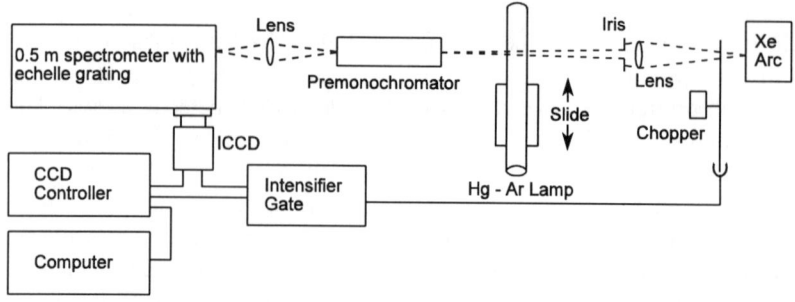

FIGURE 1. Schematic diagram of the experiment used by Wamsley et al. to probe Hg-Ar discharges (1).

made in the negative glow, Faraday dark space, and positive column regions of an operating fluorescent lamp discharge. Wamsley et al. realized that the sensitivity of the single channel, sequentially scanned experiment was limited by shot noise (Poisson statistical noise) at 194 nm and was limited by arc lamp drift or low frequency noise at visible wavelengths, where the arc lamp is brighter. Both of these noise sources were suppressed in a second experiment that utilized a CCD detector array. The detector array made the improved experiment largely insensitive to arc lamp drift or noise and simultaneously accumulated good photon statistics in all

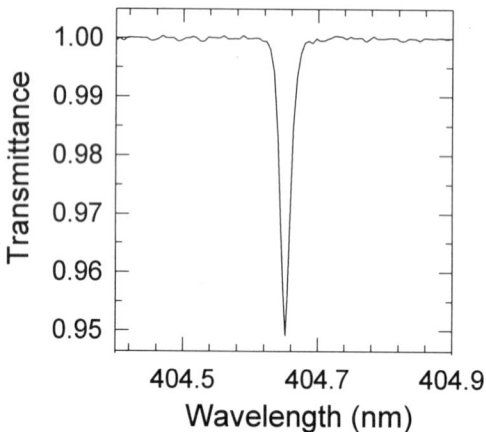

FIGURE 2. Absorption spectrum of the 404.7 nm transition of Hg obtained using the apparatus of Wamsley et al.(1) Note that sensitivity to a fractional absorption as small as 0.001 was achieved.

spectral channels.

Detector arrays usually cannot be "read out" rapidly enough to perform multichannel phase-sensitive detection. It is necessary to use a different technique to discriminate against the line emission from the glow discharge and detect the continuum from the arc discharge with absorption features after transmission through the glow discharge. Wamsley et al. demonstrated a digital subtraction technique to solve this problem. They also explored the relative spectral radiance of glow discharges, arc discharges, and synchrotrons. They found that under many conditions the spectral radiance of a glow discharge, even at the center of emission lines, is comparable to, or less than, the spectral radiance of an arc discharge. This is important because the signal from the glow discharge line emission must be comparable to, or less than, the signal from the arc discharge for the digital subtraction to be practical. The spectral radiance of synchrotron is much greater than either. The basic procedure used by Wamsley et al. involved recording a spectrum of the arc plus glow discharge emission and then a spectrum of the glow discharge by itself. The latter spectrum is subtracted from the former. The difference spectrum is then divided by a (dark count corrected) spectrum of the arc lamp continuum to eliminate effects from pixel-to-pixel variations in the detector array and from any structure in the arc lamp continuum (1).

These experiments used an image intensifier to provide precise electronic gating of the CCD array. Integration times must precisely match if one is to subtract two spectra and detect very small fractional absorptions. Mechanical shutters are not well suited to matching the integration times to part-per-million accuracy, although they are used to block the arc lamp during alternate spectra. Diode arrays provide such gating automatically during read out, but CCD arrays are somewhat less convenient because stored charge is moved across the CCD array during readout. The read out process can smear the spectrum unless precautions are taken. More recently Bergeson et al. have found that gating can be accomplished by using a frame transfer to a shielded part of the CCD array as described below (3). This approach is preferable because an image intensifier is both expensive and fragile.

The choice of a diode or CCD array must be based upon the expected light level of the experiment. CCD arrays are much more sensitive than diode arrays. A CCD array is typically required for an experiment with a spectrometer that has a very high spectral resolving power and a small etendue. A good scientific grade CCD array will have a read noise of 5 to 10 photoelectrons, while a good scientific grade diode array will have a read noise of 1500 to 2000 photoelectrons. A well designed experiment should collect signal photoelectrons rapidly enough to overwhelm dark noise and should collect sufficient signal photoelectrons during readout such that shot noise exceeds the read noise.

The goal of the work by Wamsley et al. was to understand the ionization balance of the cathode region of hot cathode Hg-Ar discharges (2). The results from the absorption spectroscopy experiment using the detector array were encouraging because fractional absorptions of 0.001 were detectable. This represented an order of magnitude improvement over the sensitivity of the single channel, sequentially scanned experiment.

CH_4 - H_2 GLOW DISCHARGES

Childs *et al.* and Menningen *et al.* used high sensitivity absorption spectroscopy to study molecular radicals in CH_4 - H_2 glow discharges used for diamond film CVD (4-6). Glow discharges in molecular gases are widely used for plasma processing, such as (semiconductor) etching, film deposition, and various surface treatments. Measurements of molecular radical densities in these discharges requires high sensitivity because numerous rotational and vibrational levels of the molecule are appreciably populated, effectively reducing the density of absorbers for a particular transition. Growth of diamond and diamond-like carbon films in CH_4 - H_2 glow discharges, although not yet in widespread industrial use, is particularly interesting for study because of its great potential.

Childs *et al.* studied a discharge occuring between a 3 mm diameter tantalum tube, that served as the hollow cathode, and a 50 mm x 12 mm x 0.5 mm strip of silicon wafer, that served as the anode and growth substrate (4). The feed gas, a few percent mole fraction CH_4 in H_2, flowed through the hollow cathode at 40 sccm. A pressure of 30 Torr was maintained and dc currents of 650 mA to 1.0 A were used. The optical apparatus consisted of a Xe arc lamp as a continuum source, a 0.5 m focal length spectrometer, and a diode array. Figure 3 is a schematic diagram of the experiment. The more modest spectral resolving power (without an echelle grating) and the resulting larger light flux enabled Childs *et al.* to use a diode array. Diode arrays, although less sensitive than CCD arrays, offer the advantage that the readout mechansim provides precision gating, as discussed earlier. A narrow band interference filter was used to reduce stray light in the spectrometer. The light shutter used to block the arc lamp continuum during alternate spectra is not shown in Fig. 3. Childs *et al.* (5) achieved sensitivities to fractional absorptions of 0.00001 and measured the absolute densities of several important molecular radicals, including CH_3 and CH, as shown in Fig. 4.

It is widely believed that CH_3 is a key gas phase precursor of diamond films under many conditions. This radical, like many polyatomics, cannot be detected using emission spectroscopy or laser-induced fluorescence because it dissociates when electronically excited. The CH_3 absorption feature at 216 nm is quite broad, about 1.2 nm (see Fig. 4), due to rapid dissociation. This allowed Childs *et al.* to open the entrance slit of the spectrometer to a spectral bandpass of 0.76 nm and achieve very high signal-to-noise ratios in the far UV using only an arc lamp continuum and a diode array. This experiment is sufficiently inexpensive and simple that it may ultimately be useful as a real time, *in situ* diagnostic in a production

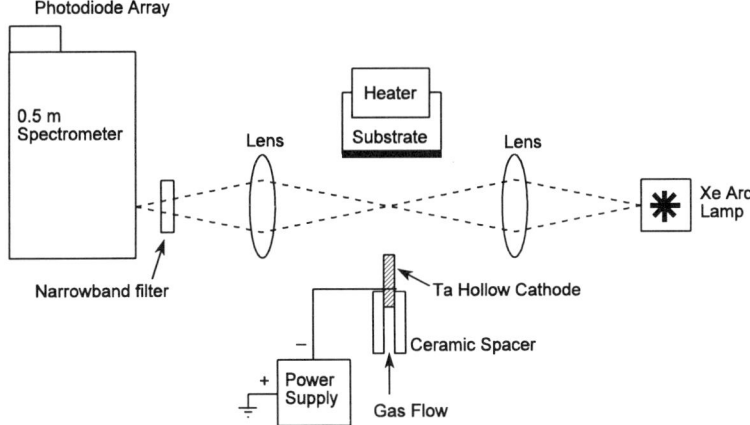

FIGURE 3. Schematic diagram of the apparatus used by Childs *et al.* (4) to obtain absorption spectra in a hollow cathode dc discharge chemical vapor deposition system.

environment. Childs *et al.* obtained spatial maps that are useful in assessing the role of CH_3 in diamond film growth (5).

Measurements of the CH density are very valuable because the ratio of the CH to CH_3 densities, $[CH]/[CH_3]$, is simply related to the hydrogen dissociation ratio, $[H]/[H_2]$. The fast, two body abstraction reactions, $H + CH_3 \rightleftharpoons H_2 + CH_2$ and $H + CH_2 \rightleftharpoons H_2 + CH$, are equilibrated in typical diamond CVD systems at pressures

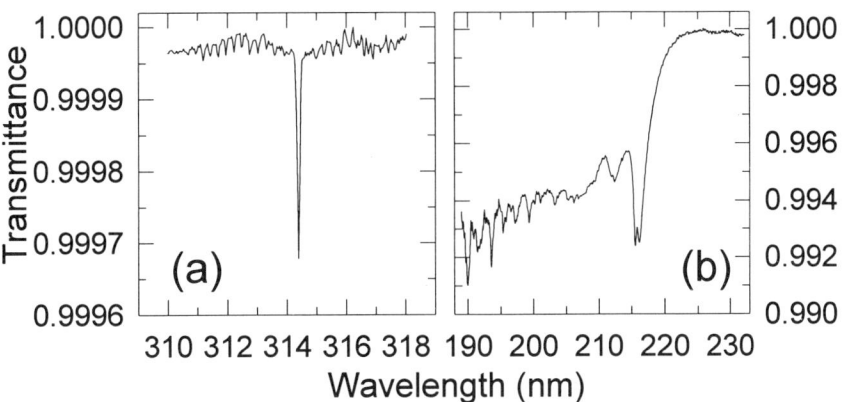

FIGURE 4. Sample absorption spectra taken with the apparatus of Childs *et al.*(5) (a) The CH $A\ ^2\Delta \leftarrow X\ ^2\Pi$ transition near 314 nm. Several R and P branch lines are visible to either side of the line-like Q branch. Note that sensitivity to a fractional absorption of 0.00001 was achieved. (b) The $CH_3\ B\ ^2A_1' \leftarrow X\ ^2A_2''$ feature near 216 nm. The feature between 211 and 213 nm is a weak CH_3 absorption feature. The features below 205 nm are due to C_2H_2.

in the tens of Torr. Their reactions rates are sufficiently large to dominate diffusive transport and thus reach equilibrium at the local gas temperature. Both the rates of these reactions and the equilibrium constants, as well as most of the gas phase chemistry of diamond CVD systems, are well known because of earlier research on combustion systems. These basic ideas were used to determine the hydrogen dissociation ratio in a hot filament deposition system by Childs *et al.* (5) and in a glow discharge deposition system by Menningen *et al* (6). Atomic hydrogen is a key radical in almost all diamond CVD systems. It plays many roles including: (a) driving the gas phase abstraction reactions which initiate the hydrocarbon chemistry as described above, (b) passivating the diamond surface to inhibit reconstruction into graphite, (c) opening growth sites on the passivated surface through surface abstraction reactions, and (d) preferentially etching graphite deposits as they form (7). A reliable, low cost, and fairly simple spectroscopic experiment for monitoring the hydrogen dissociation ratio is quite valuable.

Menningen *et al.* report using the CH and CH_3 densities to determine the hydrogen dissociation ratio in a CH_4 - H_2 glow discharge used to grow diamond films. Menningen *et al.* also thoroughly explored the issue of detector array nonlinearity, which can affect the accuracy of digital subtraction (6).

Absorption spectra of CH as shown in Fig. 4 required sensitivities to fractional absorptions of about 0.00005. These sensitivities were achieved in integration times of less than 30 minutes. Research by Childs *et al.* and by Menningen *et al.* has led to a more quantitative understanding of the gas phase chemistry of diamond CVD.

Fe HOLLOW CATHODE GLOW DISCHARGES

Bergeson *et al.* are pushing the technique of high sensitivity absorption spectroscopy in the UV and VUV to the limits of existing technology (3). The scientific goal of their laboratory astrophysics experiment is to measure accurate oscillator strengths for Fe^+ lines and lines of other atoms and ions in the UV and VUV. Relative absorption oscillator strengths from a common lower level are combined with emission branching fraction measurements to obtain very accurate sets of relative oscillator strengths. This somewhat redundant approach to determine relative oscillator strengths reduces the uncertainty in VUV emission branching fractions due to the difficulties of VUV radiometric calibration. The relative oscillator strengths are normalized using radiative lifetimes measured using laser induced fluorescence. This data is urgently need for the analysis of astrophysical spectra in the UV and VUV recorded by the Goddard High Resolution Spectrograph on the Hubble Space Telescope.

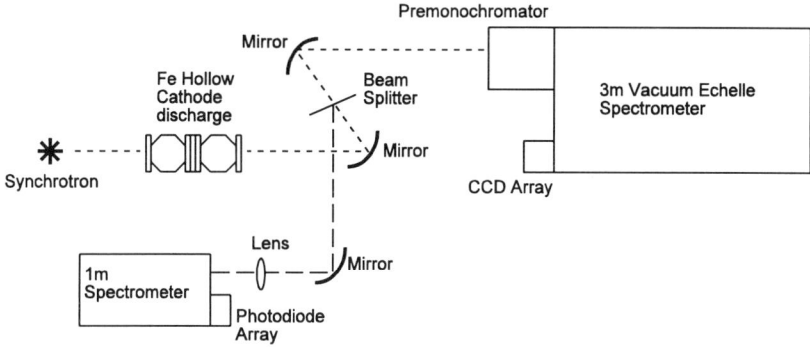

FIGURE 5. Schematic diagram of the apparatus used by Bergeson et al.(3) to measure oscillator strengths of various Fe⁺ transitions.

Bergeson et al. are using an intense Fe hollow cathode discharge (3). The cathode is 1.0 cm in diameter and 13.0 cm long and is lined with Fe. The discharge occurs in 1.2 Torr of Ar with currents up to 2.0 A. The very rich Fe and Fe⁺ spectra require extremely high spectral resolving powers at UV and VUV wavelengths. A 3 m focal length vaccum echelle spectrometer with a spectral resolving power of several hundred thousand is being used. The high spectral resolving powers and resulting diminished light throughput require an advanced VUV sensitive CCD array.

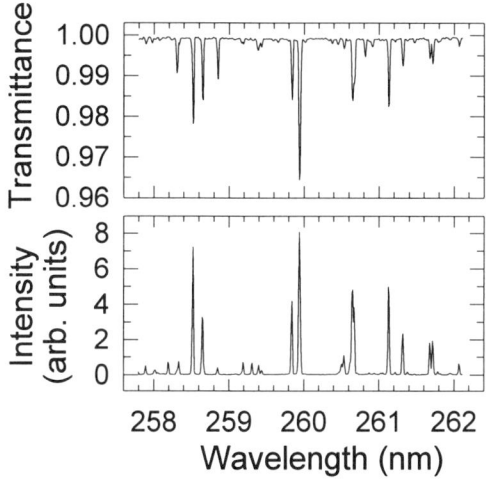

FIGURE 6. Sample Fe⁺ absorption (top graph) and emission (bottom graph) spectra obtained from the reference channel (1 m focal length spectrometer with an entrance slit width of 25 μm) in the apparatus of Bergeson et al.(3)

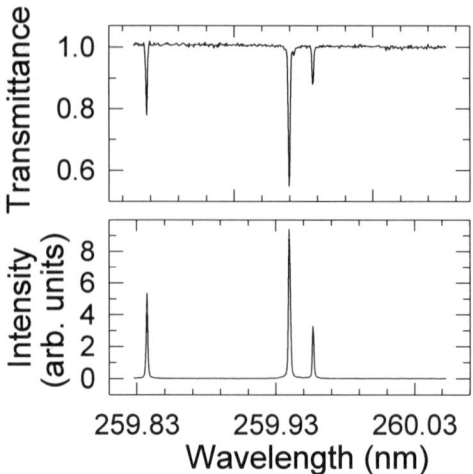

FIGURE 7. Sample Fe⁺ absorption (top graph) and emission (bottom graph) spectra obtained using the 3 m focal length, vacuum echelle spectrometer in the apparatus of Bergeson *et al.* and an entrance slit width of 50 μm (3). The lines only partly resolved near the center of Fig. 6 are completely resolved in this figure.

Synchrotron radiation from an electron storage ring provides the UV and VUV continuum. The 800 Mev, 200 mA storage ring with a 2.083 m magnetic radius provides a spectral radiance more than 1000 times greater than that of an arc lamp at UV and VUV wavelengths. Figure 5 is schematic of this experiment. A reference channel uses a 1 m focal length vacuum spectrometer with an 2400 groove/mm holographic grating and a diode array. This channel is used to monitor a reference line arising from the lower level of interest and chosen so that it can be resolved with the 1 m spectrometer. Error from possible drifts in the Fe hollow cathode discharge are avoided by simultaneously monitoring the reference line and the unknown line absorption.

Figure 6 shows sample emission and absorption spectra which demonstrate the resolving power and signal-to-noise ratio of the reference channel. Figure 7 shows sample emission and absorption spectra taken with the 3 m focal length echelle spectrometer and the CCD detector array. The Fe⁺ lines that are partly resolved near the center of Fig. 6 are completely resolved in Fig. 7. The 3 m focal length vacuum spectrometer with the 300 groove/mm echelle grating is used in 25th order and produces a reciprocal linear dispersion of 0.022 nm/mm. For comparison the 1.0 m focal length spectrometer with the 2400 groove/mm grating used in first order produces a reciprocal linear dispersion of 0.42 nm/mm.

The quality of the data and the broad applicability of this experiment indicate that it will be very important for laboratory astrophysics by producing accurate oscillator strength data. It will also be useful as a diagnostic in a variety of lighting and processing plasmas. Although this experiment cannot be described as simple or inexpensive, it does offer the possibility of measuring absolute atom or ion column densities as small as 10^8 cm^{-2}. Such a column density corresponds to a density of 10^7 cm^{-3} in a 10 cm long plasma. By comparison it is not easy to detect a density of 10^7 cm^{-3} atoms or ions using laser induced fluorescence in a typical glow discharge due to background light from discharge.

SUMMARY

The development of multielement detector arrays such as photodiode arrays and CCD arrays has made it possible to greatly enhance the performance of absorption spectroscopy experiments. The measurement of the absolute densities of small numbers of atoms, ions, or molecular radicals in glow discharges has provided very valuable data for the understanding of lighting plasmas, chemical vapor deposition environments, and astrophysical spectra. The simplicity and relatively low cost of the basic apparatus, together with the high sensitivity achievable in such experiments, indicates that future application of the techniques described above to a diverse range of systems will yield many important and exciting results.

ACKNOWLEDGEMENTS

The experiments described above have been accomplished with the support of the General Electric Company, the Army Research Office, the National Aeornautics and Space Administration, and the Synchrotron Radiation Center in Stoughton, WI, a National Science Foundation supported facility.

REFERENCES

1. R. C. Wamsley, K. Mitsuhashi, and J. E. Lawler, *Rev. Sci. Instrum.* **64**, 45 (1993).
2. R. C. Wamsley, J. E. Lawler, J. H. Ingold, L. Bigio, and V. D. Roberts, *Appl. Phys. Lett.* **57**, 2416 (1990).
3. S. D. Bergeson, K. L. Mullman, and J. E. Lawler, to be published.
4. M. A. Childs, K. L. Menningen, H. Toyoda, L. W. Anderson, and J. E. Lawler, *Europhys. Lett.* **25**, 729 (1994).
5. M. A. Childs, K. L. Menningen, H. Toyoda, Y. Ueda, L. W. Anderson, and J. E. Lawler, *Phys. Lett. A* **194**, 119 (1994).
6. K. L. Menningen, M. A. Childs, H. Toyoda, Y. Ueda, L. W. Anderson, and J. E. Lawler, *Contrib. Plasma Phys.* (1995), accepted for publication.
7. J. C. Angus and C. C. Hayman, *Science* **241**, 913 (1988).

Ion — Plasma Interaction

C. Deutsch, C. Fleurier, D. Gardès and G. Maynard

GDR-CNRS-918 — Orsay and Orléans, France

Abstract. The synchronous firing of dense and strongly ionized plasmas with the time structure of bunched and energetic multicharged ion beams allows to probe for the first time, the long searched enhanced plasma stopping and the enhanced projectile charge at target exit, as well.

INTRODUCTION

The interaction of ion beams with dense and fully ionized plasmas has been recently promoted as a major domain of investigation (1-6). It lies at the borderline of atomic and discharge physics. These topics are of a crucial significance in asserting the feasibility of manipulating intense beams of light or heavy ions toward the goal of compressing hollow microspheres (a few hundred microns in diameter) up to the ignition of the deuterium + tritium fuel delivering α particle and neutron thermonuclear yields.

The basic mechanisms underlying the physics of charged particles stopping in various states of matter have been the object of intense scrutiny since the very early days of quantum mechanics. For instance, the seminal 1913-15 Niels Bohr papers on the atomic model, are mostly dedicated to an impact-parameter formulation of ion stopping. Since then, the electromagnetic coupling between projectiles and target particles has always been the topic of ever increasingly sophisticated approaches. As a result, we now have a very large body of data for the stopping of ions in a neutral target.

However, all these studies were invariably conducted with an electrically neutral target material. Recently, the consideration of fully ionized targets, composed of ions and electrons, has emerged as a novel challenge with major concerns for thermonuclear research, high-energy particle acceleration, and related fields of interest.

For many years ion-plasma interactions were encountered in a variety of situations of technological and engineering concern. A conspicuous example is thus afforded by the intense deuterium ion beams in the 500-800 keV energy range, routinely used for the aditional heating of tokamaks and other magnetically confined plasmas. However, the specific issues associated to the basic physics of ion beam stopping in a strongly ionized plasma medium was never considered a topic of fundamental interest.

This situation changed quite abruptly at the beginning of the 80's with the emerging possibilities of achieving the inertial compression of hollow spheres, a few millimeters in diameter, containing the deuterium + tritium thermonuclear fuel through energetic and intense light ion or heavy ion beams.

Considered intensities are in the Megaamps/cm^2 range for light ions (D^+ or Li^+ at few MeV/nucleon) and in the 10 kiloamps/cm^2 range for heavy ions (B_i^+, U^{n+} at 50 MeV/nucleon). Then, suddenly, the basic physics of ion stopping in dense plasmas got pushed ahead at the forefront of the interest of many physicists, in view of its uncontournable relevance in assessing the basic trends of the driver-pellet interaction.

REDUCTION PRINCIPLE

One of the misconceptions which seems to have prevented physicists to addressing earlier the issues of charged particle stopping in dense plasmas, was the prejudice that an inferno of collective plasma effects should occur and blur out the basic mechanisms associated to the ion beam-plasma interaction itself.

Such expectations are unfortunately fulfilled by the interaction of intense electron beams with dense plasmas, which explains they are considered useless for inertial confinement fusion (ICF). On the other hand, it has been documented recently through a lot of numerical simulations and adequately designed experiments that the above deleterious collective behavior take only a nearly negligible energy toll during the ion-plasma interaction.

This latter can thus be safely reduced to a linear superposition of single ion stopping in a hot medium mostly constituted of free electrons.

Such a fortunate occurrence is indeed straightforwardly based on a very simple argument. Actually, as far as the most intense ion beams are considered for ICF purposes, the mean interior distance remains, at least, two orders of magnitude above the electron screening length in the considered dense plasma targets. The same situation also prevails for the cold target exposed to the incoming ion beam.

These simple considerations allow us to reduce the beam-target interaction to an ion-target one, by neglecting collective aspects in a first approach. In so doing, we have built the contents for the so-called reduction principle. This welcome simplification should nevertheless be taken with some reservations for the case of protons. The corresponding current densities may well range up to MA/cm^2, so that some caution should be exercised in every practical situation dealing with light ion beams.

With these minor restrictions taken into account, one is entitled to make use of conventional wisdom as to which intense beams are likely to appear dilute in the target. Had we considered intense electron beams, the collective phenomena would not have been so easily eliminated. For instance, potential wells can develop in the plasma produced by heating a thin foil, so that incoming projectiles are likely to be trapped and accelerated backward after several bouncing periods.

So, if the ion beams are not submitted to the filamentation instability, which can reduce drastically the mean ion relative distance, the projectiles do not see each other while stopping in dense matter.

At this juncture, it should be noticed that the opposite situation with many ion fragments flying within a few atomic relative distance of each other is very actively investigated at the moment. Such an interest arises from the possibility of using intense cluster ion beams (Au_n^+, C_{60}^+, etc...) for compressing a thermonuclear pellet through a multifragmentation of the cluster ion at the beam-target interface. Leaving aside such an exotic situation, the reduction principle allows for a very efficient simplification of the interaction of intense atomic ion beams with a dense target.

STANDARD STOPPING MODEL (SSM)

A lot of theoretical and numerical investigations have already anticipated the presently ongoing experimental programs detailed below. Amongst many others, the seminal work of T.A. Mehlhorn (3) (Sandia, Albuquerque) and also L. Nardi and Z. Zinamon (3) (Weizmann Institute) was instrumental in suggesting a basic scheme for the understanding of the ion-plasma interaction.

The corresponding argument runs essentially as follows: the standard theoretical framework designed for cold matter stopping may be extended to plasma stopping provided due attention is paid to the projectile effective charge as well as to the increasingly significant free electron contribution with respect to the projectile energy loss. More specifically, these free electrons display an enhanced capability to respond to the incoming electric field. In addition, highly excited bound electrons to the target are also more versatile, and as a result more efficient at taking out the projectile kinetic energy when compared to the least excited ones. These considerations hold as long as the projectile-target electron interaction may be worked out within a Born approximation.

As a result the theoretical task thus essentially consists of supplementing the above picture with adequate quantitative insight for every projectile-plasma pair of practical interest.

The utmost significant feature of the relevant physics is afforded by the markedly enhanced projectile charge state within the target plasma as compared to its cold target homologue. According to an argument going back to J.S. Bell (1953), it is much more difficult for the incoming projectile to get recombined from free electron plane wave states than from orbitals already bound to other target ions. Actually, the respective rates differ at least by three orders of magnitude.

Such an emphasis on a proeminent free electron stopping combined with the enhanced projectile charge leads to expect an enhanced plasma stopping (EPS) compared to cold matter stopping with same electron density in the equivalent target.

The predicted EPS arises from a few simple changes, entirely included in the most straightforward extension to a plasma target, of the well known (Bethe) cold-matter stopping expression. For partially ionized material, free electrons, bound electrons, and plasma ions contribute to energy loss. Ignoring relativistic corrections and the usually very small ion contribution, the stopping power dE/dx is given by

$$\frac{dE}{dx} = \frac{4\pi N_0 e^4 \rho}{A_T m_e V_1^2} Z_{eff}^2 \left[\overline{Z} \ln \Lambda_F + (Z_T - \overline{Z}) \ln \Lambda_B \right]$$

where $E = M V_1^2 / 2$, ρ is the density of the stopping medium, N_0 is Avogadro's number, e is the electron charge, m_e is the electron mass, Z_{eff} is the effective charge of beam ions, V_1 is the projectile velocity, A_T is the target atomic weight, Z_T is the target atomic number, $\overline{Z}$ is the average ionization in target, and Λ_B, Λ_F are the arguments of the Coulomb logarithms for bound and free electrons, respectively.

For high target-electron velocities, Λ_B is given by the familiar Bethe expression

$$\Lambda_B = \frac{2 m_e v_1^2}{I_{av}}$$

where I_{av} is a geometric average of the effective excitation and ionization potential of the bound electrons. The expression for Λ_F is

$$\Lambda_F = \frac{2 m_e v_1^2}{\hbar \omega_p}$$

where ω_p is the plasma frequency. At low velocity, when one has $\Lambda_F < 1$, it must be modified.

Above expression implies the neglect of any collective stopping effect due to the high intensity of the ion beam, in agreement with recent investigations of the target corona instabilities (5). It is the high-temperature limit of more sophisticated estimates for the bound- and free-electron stopping power in the dense target plasma. Moreover, for partially stripped projectiles, an equally significant enhanced stopping also arises from the strongly reduced recombination (6), between incoming ion and free electrons. Such drastic behavior maintains a relatively high Z_{eff}, in contrast to that in cold matter, where the ion projectile can easily pick up bound electrons from target atoms (or ions) located near its trajectory. Thus the EPS physical content rests essentially on the much enhanced response of plasma free electrons, together with highly increased Z_{eff} values compared to nominally equivalent cold target, i.e., ones with the same line-integrated nuclear density (number/cm^2). n_e is the free-electron density target and ℓ the linear ion range within.

Enhanced Plasma Stopping

Bethe-like stopping appears two or three times bigger in a fully ionized plasma than in the equivalent cold gas with the same density of electrons bound to atomic and molecular orbitals. Dramatic illustrations are shown in Figure 1 where we systematically compare for C^{n+}, the respective evolution in cold gas and fully ionized hydrogen of Z_{eff} (left vertical axis) and energy loss (right vertical axis) in terms of penetration depth. A Monte Carlo code, making use of every excitation, ionization and recombination cross section, has been developed (8), together with a stopping calculation to follow dynamically Z_{eff} in terms of projectile velocity V_1. In cold gas, Z_{eff} data in terms of penetration range, and pertaining to several ionization stages, decay monotonically from initial value to a common cold-gas asymptotic limit (Z_1 = projectile atomic number).

In contradistinction, Z_{eff} plasmas rise steadily. For the lightest carbon element, all ionization stages end up in C^{6+} for a maximum range corresponding to the experimental SPQRII setup detailed below. This effect increases with Z_1. As a rule, projectile effective charges in plasma lie much higher (6) than their cold-gas homologous, except very near the end range, where a kind of catastrophic recombination takes place. As a consequence, stopping powers $\sim Z_{eff}^2$ may increase by orders of magnitude. These specific Z_{eff} behaviors in plasma are essentially due to a quasihindered recombination between ion projectile and target electron. The latter has a type of plane-wave behavior that prevents it from being used as a bound orbital. In cold gas, all electron states are bound ones.

Therefore, projectile recombination is made possible by transferring to it a bound electron instead of a target ion. As far as experimental verification of these predictions is concerned, target ionization looks like a much more significant parameter than temperature (7). We thus expect to witness these Z_{eff} plasma behaviors, even at a rather modest target temperature, a few eV, provided the target remains fully ionized.

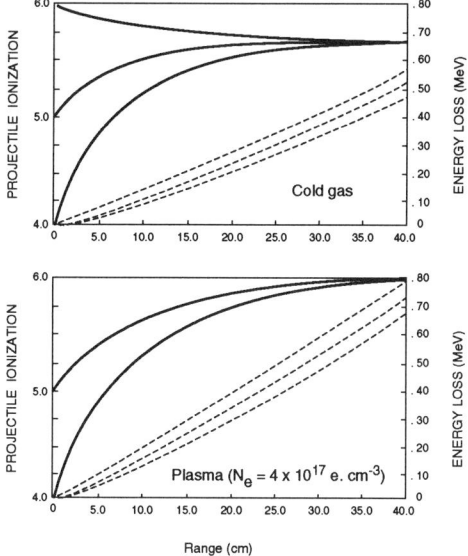

FIGURE 1. Effective charges and energy losses of energetic multicharged ions in cold gas and fully ionized hydrogen. C^{n+} at 2 MeV/a.m.u

Principles of Ion-Beam—Plasma Experiments

The previous line of reasoning suggests a clear path to an experimental verification of above previsions. The reduction principle stated may be implemented to test EPS results. One thus feels entitled to replace a complex interaction of intense ions beams with a given target by that of a dilute beam out of a standard accelerating structure. The plasma target may thus be fired independently of the ion beam, provided the discharge ignition is synchronized with the beam time structure. In this fashion, it appears feasible to emulate a dense, plasma produced, say, by a space-charge-dominated ion beam of inertial-confinement fusion (ICF) interest.

Such an approach builds up a backbone for the socalled SPQRII (Stopping Plasma Quantitatively Reinforced) (9-13)project started at Orsay. One can thus make use of standard accelerating structures delivering dilute pulses of ions in any charge state with any requested kinetic energy. One is left with a practical task of simultaneously performing beam spectrometry and plasma diagnostics. Before going on further, it is appropriate to derive a most useful scaling rule (7),

$$\frac{\Delta E}{E} = \frac{n_e \ell}{E^2} \; .$$

from above equations between maximum energy loss and linear density of target electrons. ℓ denotes penetration depth. It allows us to match projectile energy with target-free-electron density. For a given $n_e\ell$, only a narrow range of E values will provide the largest stopping power which may be probed by matching standard accelerating facilities like Tandem and Linacs with a dense plasma synchronously but independently fired on the beam line.

It should also be appreciated that we developed first the so-called SPQRI concept in order to probe essentially the enhanced projectile ionization in plasma (EPIP), through a laser ablated target plasma.

SPQRII SETUP

An experimental realization of the above scheme for the interaction of ion beams with strongly ionized hydrogen plasma is achieved by Ohmnic heating of a low-pressure hydrogen column. The gas is confined in an alumina tube and inserted on a beam line from a tandem Van de Graaff (see Fig. 2).

The plasma results from the powerful discharge of the capacitor band delivering 5 kJ when operated at 15 kV. Special coupling ports have been developed to insure the connection between the plasma target and the vacuum of the beam line.

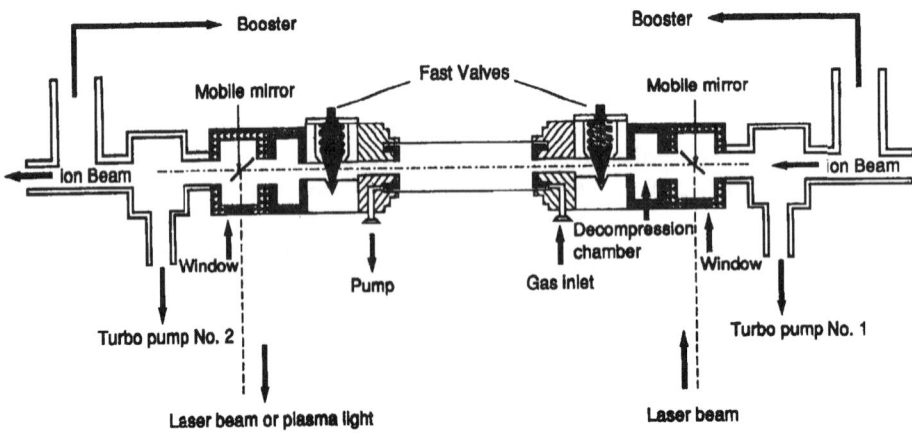

FIGURE 2. Experimental setup. A linear plasma column is confined in an alumina tube connected to the beam line with two fast valves. Plasma diagnostics are realized using a laser beam injected during the plasma shot via two removable mirror systems.

They consist of two fast valves associated with a differential pumping system. They are only opened during the plasma ignition and they reduce significantly the amount of cold gas at both sides of the plasma tube during the energy loss measurement. A variant of this setup devoted to plasma stopping of projectiles that are heavier than Ar has already been presented.

The energy loss of the heavy ions was performed using capacitive phase probes, which deliver an accurate timing signal that corresponds to the beam burst flowing through the detector. The energy-loss measurement results from a time-of-flight (TOF) comparison of the ions before and after interaction with the plasma.

The ion beam has a time microstructure lasting 2 nsec repeated every 800 nsec. The plasma lifetime corresponds to about 100 µs and it can be fired every 2 min.

200 µsec after the opening of the fast valves, a time coincidence between the plasma light and a beam pulse starts the computer-aided measurement-and-control (CAMAC) acquisition system. Time signals delivered by the phase probes are then sequentially registered using a time-digital-converter multistop. The visualization of the time spectrum, and its memory storage are devoted to a special microprocessor controlled with a personal computer. Two Faraday caves enclose the plasma target and the data-acquisition equipment. They insure a shielding from the rf noise generated by the plasma discharge.

Energy-Loss Measurements

Measured stopping quantities can now be compared to their theoretical SSM counterparts. At this point, it is convenient to recall that plasma stopping has to be referenced to cold-gas stopping with the same density of recombined electrons. This fictitious cold-gas equivalent is the true plasma reference. There is no initial cold-gas fill preceding the plasma ignition.

Our numerical codes based on the SSM carefully include the residual bound-electron contribution to discharge stopping. Therefore, a meaningful theory-experiment confrontation is likely to occur.

Measured ΔE follows closely the n_e time profile. As far as accuracy is concerned, it should be disclosed that uncertainty on time-of-flight measurements only amounts to ± 0.5 nsec. A larger fluctuation persists from shot to shot. Maximum rates for these fluctuations are respectively 18% and 12% for 15- and 13-kV ignition voltage.

They are responsible for a partial reproductibility of plasma conditions.

A theory-experiment comparison is given in the same figure for sulfur and bromide ions in hydrogen plasma. Figure 3 summarizes the SSM calculations compared with experimental results. A quasilinear variation that has already been observed for heavier elements is then confirmed. Corresponding EPS values may be larger than 3, in some cases.

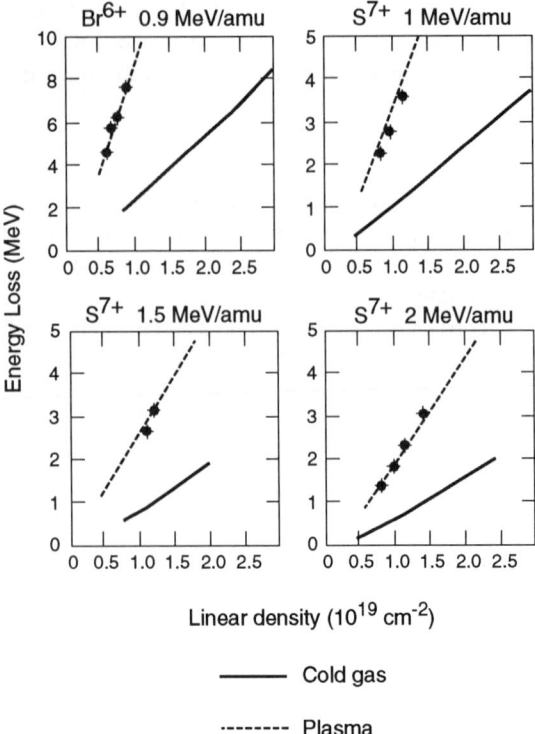

FIGURE 3. Energy losses as a function of free plus bound-electron density in a hydrogen plasma for Br[6+] with 0.93 MeV/a.m.u and also S[7+] with 1, 1.5, and 2 MeV/a.m.u, respectively.

Data on the stopping of C^{4+}, S^{7+}, and Br^{6+} ions in the initial energy range 0.7-4 MeV/a.m.u are of considerable interest for the final stopping phase, near the end of range, for particle-driven inertial fusion. Those data could provided a deeper insight into ion-beam transport in the reactor chamber containing the pellet, as well as into the beam-target compression process envisioned within the so-called classical energy deposition scenario. Within the SSM, our measuremments may be extrapolated to other ion species through the scaling relationship

$$\frac{dE'}{dx}(Z',M',E') = \left(\frac{Z'}{Z}\right)^2 \frac{dE}{dx}(Z,M,(M/M')E')$$

for two ion projectiles with charge, mass, and kinetic energy (Z,M,E) and (Z',M',E') respectively.

Results shown in Figure 3 definitively confirm the conspicuous and strongly enhanced plasma stopping in a fully ionized hydrogen target, at several inertial projectile energies. It should be appreciated that a previous attempt using intense D$^+$ beams by Young *et al.* has anticipated some of our findings about EPS in the field of light-ion-driven (ICF). Another important feature, in full agreement with the Bethe expression, is the rather weak electron temperature dependence of the ion energy loss, provided the plasma ionization is kept constant. These facts allow us to produce stopping data of interest for inertial thermonuclear fusion with a few-eV plasma target.

Similar results have been obtained for heavier ion projectiles in D.H.H. Hoffmann Group at GSI-Darmstadt (Germany) (11).

ENHANCED PROJECTILE IONIZATION IN PLASMA (EPIP)

The above SPQRII setup is essentially suited for energy loss of strongly ionized light and medium ions, provided further projectile ionization may be followed through an accurate Monte-Carlo code. For heavier elements, one has to pay a due metrological attention to the determination of projectile charge distributions at plasma exit.

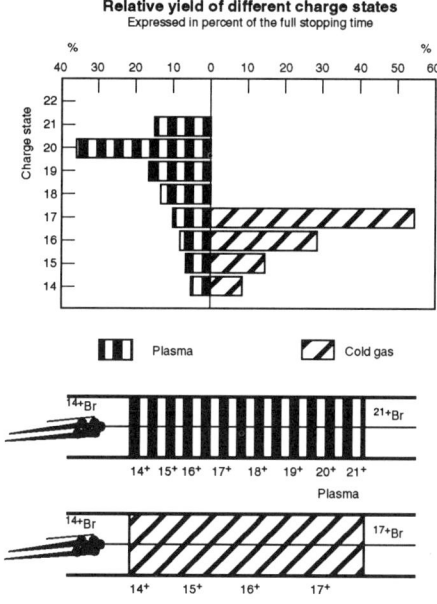

FIGURE 4. Computed projectile charge distributions in plasma ($n_e=4\times10^{17}$ cm^{-3}, T=2 eV) and cold gas for an incident ion Bi^{14+}

A similation pictured on Figure 4 gives an accurate idea of the contrasting fates of an initial Br^{14+} ion at 0.9 MeV/a.m.u, respectively stopped in a 36 cm long cold hydrogen column and a fully ionized plasma with $n_e = 4\times10^{17}$ e-cm^{-3} at T = 2 eV. In the latter case, the projectile charge gets shifted up by four units. On the other hand, the so-called SPQRI project (13) developed between CEN-Bruyères Le Chatel and Orsay is specifically dedicated to investigating the fine structure of those charge distributions by synchronizing a laser ablated plasma with a suitably collimated and bunched Cu^{9+} beam, out of a Tandem accelerator.

The ablation of nonhydrogenic targets such as C and Si, with a typical temperature in the few tens of eV range, provides heavy ion stopping in target conditions much closer to those of a real ICF working target. The charges distribution, at plasma exit, for an incoming ion projectile Cu^{9+} at 0.7 MeV/a.m.u partially stopped in a 2 mm thick and strongly ionized C plasma at 60 eV with $n_e \sim 10^{19}$ e-cm^{-3} is shown on Figure 5, in terms of the fractional charge distribution Yq for a given charge state q.

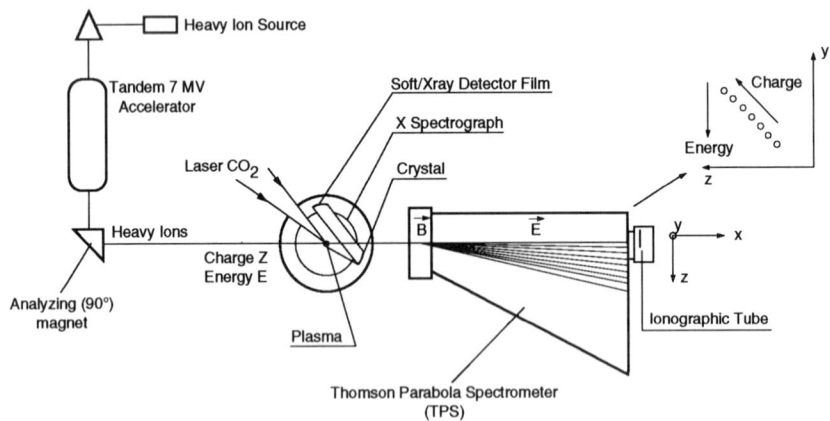

FIGURE 5. General layout of the SPQRI setup (Adapted from ref. 13).

The considered distribution is typical of a nearly equilibrium one, close to a Gaussian (see Fig. 6).

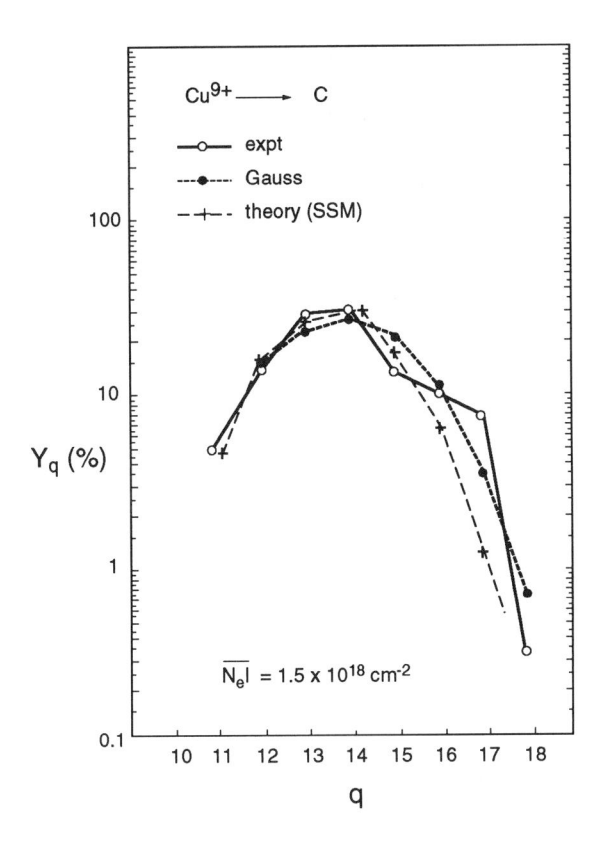

FIGURE 6. Y_q distribution in a C target with a Gaussian fit. The SSM theoretical result is obtained for a density $n_{c_{s+}} = 3 \times 10^{17}$ cm^{-3}.

FUTURE PROSPECTS

We are now currently developing a new ion-plasma setup designed for simultaneous measurements of energy losses and charge state distributions on the SPQRII beam line at IPN Orsay. We intend to make use of a sophisticated electrooptical detection which will allow to suppress the time of flight at beam line end.

We also intend to accelerate negative and positive cluster ions such as Au_n^+ and C_{60}^n in the few tens of keV/a.m.u range in order to probe the correlated stopping of ion fragments resulting from the impact with a dense plasma target.

ACKNOWLEDGMENTS

GDR-CNRS-918 includes the experimental division (Institut de Physique Nucléaire, Orsay), the theory Group (Laboratoire de Physique des Gaz et Plasmas, Orsay) and the dense discharge team (Groupement pour l'Energétique des Milieux Ionisés, Orléans).

REFERENCES

1. Bangerter, R.O., Mark, J.W.K., and Thiessen, A.R. , *Phys. Lett.* **88A**, 225 (1982); Young, F.C., Mosher, D., Stephanakis, S.J., and Goldstein, S.A. *Phys. Rev. Lett.*, **49**, 549 (1982);
2. Deutsch, C., *Ann. Phys. (Paris)*, **11**, 1 (1986)
3. Mehlhorn, T.A., *J. Appl. Phys.*, **52**, 6522 (1981); Nardi, E., Peleg, E., and Zinamon, Z., *Apply. Phys. Lett.*, **39**, 46 (1981); Maynard, G., and Deutsch, C., *Phys. Rev. A.*, **26**, 665 (1982), **27**, 574 (1983).
4. Meyer-Ter-Vehn, J., Witkowski, S., Bock, R., Hoffmann, I. Hofmann, D.H.H., Müller, R.W., Arnold, R., and Mulser, P., *Phys. Fluids B*, **2**, 1313 (1990).
5. Hewett, D.W., Kruer, W.L., and Bangerter, R.O., *Nucl. Fus.* , **31**, 431 (1991).
6. Nardi,E., and Zinamon, Z., *Phys. Rev. Lett.*, **49**, 1251 (1982); see also Bailey, D., Lee, Y.T., and More, R.M., *J. Phys. (Paris)*, Colloq. **44**, C8-149 (1983).
7. Deutsch, C., Maynard, G., Bimbot, R., Gardès, D., Della-Negra, S., Dumail, M., Kubica, B., Richard, A., Rivet, M.F., Servajean, A., Fleurier, C., Sanba, A., Hoffmann, D.H.H., Weyrich, K., and Wahl, H., *Nucl. Instrum. Methods*, **A 278**, 38 (1989).
8. Maynard, G., and Deutsch, C., *J. Phys. (Paris) Colloq.* **49**, C7-151 (1988).
9. Gardès, D., Bimbot, R., Della-Negra, S., Dumail, M., Kubica, B., Richard, A., Rivet, M.F., Servajean, A., Fleurier, C., Sanba, A., Deutsch, C., Maynard, G., Hoffmann, D.H.H., Weyrich, K., and Wahl, H., *Europhys. Lett.*, **8**, 701 (1988).
10. Hoffmann, D.H.H., Weyrich, K., Wahl, H., Peter, Th., Meyer-Ter-Vehn, J., Jacoby, J., Bimbot, R., Gardès, D., Rivet, M.F., Dumail, M.D., Fleurier, C., Sanba, A., Deutsch, C., Maynard, G., Noll, R., Hass, R., Arnold, R., and Maurmann, S., *Z. Phys. A*, **30**, 339 (1988).
11. Hoffmann, D.H.H., Weyrich, K., Wahl, H., Gardès, D., Bimbot, R., and Fleurier, C., *Phys. Rev. A*, **42**, 2313 (1990).
12. Gardès, D., Servajean, A., Kubica, B., Fleurier, C., Hong, Deutsch, C., and Maynard, G., *Phys. Rev. A*, **46**, 5101 (1992).
13. Couillaud, C., Deicas, R., Nardin, Ph., Beuve, M.A., Guihaumé, J.M., Renaud, M. Cukier, M., Deutsch, C., and Maynard, G., *Phys. Rev. E*, **49**, 1545 (1994).

Surface-Wave Sustained Plasmas: Toward a Better Understanding of RF and Microwave Discharges

*Michel Moisan, #Carlos M. Ferreira, *Joseph Hubert, *Joëlle Margot and ‡Zenon Zakrzewski

*Groupe de physique des plasmas, Université de Montréal, Montréal H3C 3J7, Québec, #Centro de Electrodinâmica da Universidade Técnica de Lisboa, Instituto Superior Técnico, Lisboa 1000, Portugal and ‡Polish Academy of Sciences, IMP-PAN, 80-952 Gdansk, Poland

Abstract. An approach is presented that unifies the description of the various existing RF and microwave discharges. It is based on two essential facts: (i) it is not the spatial distribution of the high frequency (HF) electric field intensity but its spatial average that plays in the power transfer to the plasma; (ii) the power θ required to maintain an electron in the discharge is governed by charged particle losses, which are independent of the HF E-field. This enables one to model separately the maintenance processes of HF discharges and the electrodynamic properties of HF circuits sustaining the plasma, although the discharge and the HF field are actually coupled self-consistently. The influence of the field frequency on the properties of these plasmas is also summarized.

The use of RF and microwave sustained discharges (hereafter referred to jointly as HF discharges) in industrial applications is increasing. One also notices, for example in microcircuit fabrication and lighting, that there is still much design and development work being carried out on these plasmas. One question naturally arises: to what extent are all these HF discharges really different one from another? Although their field applicator (the device providing the HF field in the discharge vessel) is different and although they operate at different frequencies and some of them are using a static magnetic field, is it possible to establish some unified picture of these various systems? Our presentation summarizes the work that has been accomplished in that direction thanks to the exceptional features of a specific type of HF discharge, the surface-wave sustained discharge (SWD).

We shall begin by recalling briefly the main characteristics of SWDs and underlining their advantages in the determination of the common features of HF plasmas. Then we will concentrate on showing that the central and common parameter in electrical discharges is the power lost per electron, θ_l, and draw conclusions from this fact. Further on, we give indications on a simplified model for HF discharges that enables one to model separately the discharge and the HF circuit used to sustain it. Finally, we examine the influence of the operating frequency on the discharge parameters.

SUMMARY OF SURFACE-WAVE DISCHARGE PROPERTIES

Figure 1 shows the typical arrangement for achieving a SWD (1). The discharge tube is generally cylindrical in shape and preferably made from a low-loss dielectric material. The field applicator in such a system excites a wave guided by the discharge tube and the plasma, the plasma being sustained by the wave electric field: the wave generates its own propagating structure. The wave power flux P(z) decreases away from the launching gap in the axial direction (z axis) as the wave transfers energy to the discharge, as shown in Fig. 2 (1); the (cross-section averaged) electron density n(z) decreases accordingly from the launcher, until it reaches a minimum density n_D below which the surface wave cannot propagate (2). The plasma column length increases with increasing $P(0) \equiv P_o$ at the launcher exit. The term surface wave indicates that the wave power flux is mostly concentrated at the surface of the propagating structure; it also means that the field intensity decreases transversely (radially here) in an exponential-like fashion away from the propagation structure (2). We underline below the properties of interest in our quest for a general model of HF discharges.

The description of SWDs is a fully self-consistent problem. Because in such discharges the surface wave generates its own propagating structure, any change

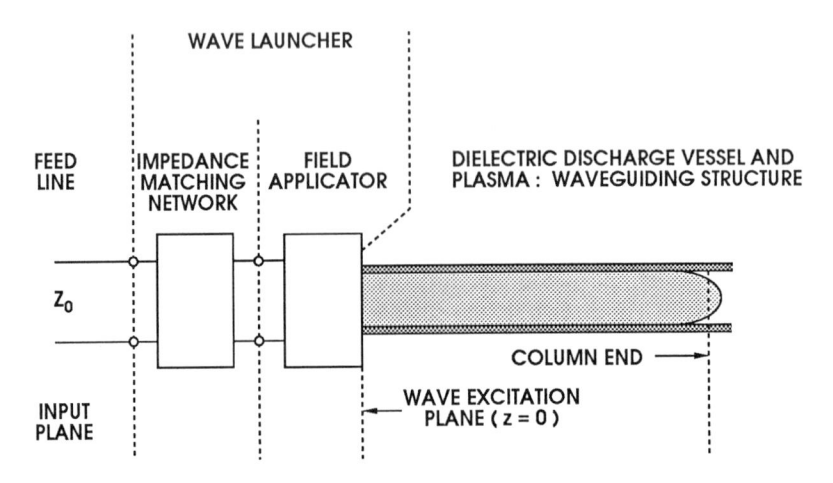

FIGURE 1. Essential elements of a surface-wave plasma source (1).

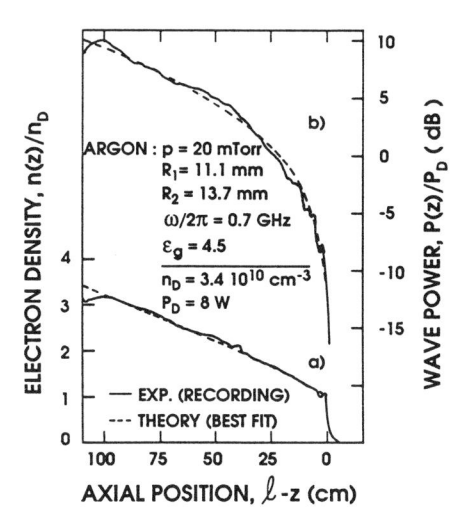

FIGURE 2. Experimental (full curve) and theoretical (borken curve) axial dependence of the wave power flux and electron density in a SWD of length ℓ. n_D (surface-wave resonance density) and P_D are normalizing parameters independent of P(0) (1).

in the **operating conditions** (nature and pressure of the gas, dimensions and material of the tube, frequency of the wave and, eventually, intensity B_o of the applied static magnetic field) and absorbed HF power will affect the plasma properties and accordingly modify the wave field distribution and power flow, and conversely. A priori, this means that the wave aspect of the problem cannot be treated separately from the discharge aspect.

The surface wave equation couples self-consistently with the discharge through the equivalent permittivity $\varepsilon_p(r,z)$ of the plasma (r is the radial position). This enables one to compute simultaneously the field intensity and electron density spatial distributions.

The discharge can be sustained by a wave propagating in a single given mode. Surface wave propagation can take place in different modes (2). In cylindrical configuration, the wave mode is identified by the integer m in the wave field phase term, which can be written as $e^{im\varphi}$ where φ is the azimuthal angle. The discharge can be sustained in the azimuthally symmetric (m=0) mode throughout the wave frequency range provided the condition f R ≤ 2 GHz cm is obeyed (f, wave frequency and R, tube inner radius) (3). With SWDs, the occurrence of a given E-field configuration does not depend on the amount of power flowing from the applicator in contrast to resonant cavity plasma sources for example, where there can be mode jumping and hysteresis.

SWDs have the broadest range of operating frequency among HF discharges. SWDs have been achieved at frequencies as low as 200 kHz and as high as 10 GHz, using different wave launchers but the same discharge tube (4); provided the aforementioned mode condition f R ≤ 2 GHz cm is obeyed, the field configuration applied to the discharge is that of the m=0 mode whatever the frequency. This has enabled us to conduct a systematic and rigorous investigation of the influence of f on HF discharges over the RF domain ($\approx$ 1 to 300 MHz) and over the most commonly used part of the microwave spectrum (300 to 2450 MHz) (5). To our knowledge, this is a unique feature of SWDs.

SWDs operate readily with an externally applied static magnetic field. The range of f in this case is as broad as with $B_o=0$, making these plasmas a unique case among HF magnetoplasmas. SWDs can be used advantageously as a substitute to microwave electron cyclotron resonance (ECR) plasmas for material etching (6) and also to compare for example the intrinsic efficiency of RF helicon discharges with that of ECR discharges (7).

The E-field is provided by a traveling wave. In this case, in contrast to the situation with an electrostatic field or a standing wave, one has access to a whole set of data because of the axial variation of the wave properties with z, favoring a closer and more complete comparison of theory with experiments.

The discharge geometry is axially symmetrical. The use of cylindrical discharge tubes and eventually cylindrical conducting enclosures allows one, under certain conditions, to write down analytical expressions for the E-field.

RESULTS FROM SWDs APPLICABLE TO HF DISCHARGES IN GENERAL

We present below a summary of the "doctrine" that came out from the modeling and experiments on SWDs. To simplify the presentation, we assume that: i) the plasma is far from local thermodynamic equilibrium, i.e. the average energy of electrons is much larger than that of ions: the excitation and ionization of atoms and molecules occur through electron impact on them; ii) the electron energy distribution function (EEDF) is locally uniform in the plasma, i.e. spatial gradients are neglected; iii) there is no transport of kinetic energy across the plasma; iv) the field frequency is high enough (typically above 10 to 20 MHz) so that only the electrons are accelerated by the HF electric field and also the EEDF is stationary; v) the charged particles diffuse ambipolarly to the wall where they recombine. A more complete treatment can be found in the literature (e.g. 5,8-10).

The Key Parameter in the Energy Balance Process: θ

The central point when analyzing the energy balance in HF and DC (f = 0) discharges is the mechanism by which the charged particles are removed from the plasma. Considering discharges operating at pressures typically lower than a few torrs, we can assume that diffusion is the predominant loss process. Its equilibrium with ionization is described by the *charged particle balance equation* (8):

$$\langle v_i \rangle = D_{se}/\Lambda^2 \qquad (1)$$

where $\langle v_i \rangle$ is the average ionization frequency by electron impact, the bracket representing averaging over the EEDF, D_{se} is some effective diffusion coefficient (e.g. ambipolar) and Λ is the diffusion length (R/2.405 in cylindrical discharge tubes long compared to R). This equation means that under steady-state conditions, ionization of the gas exactly compensates for the losses of charged particles by diffusion. For a given gas, D_{se} depends on the EEDF and on N, the density of neutral atoms: Equation (1) is an implicit function of the EEDF.

The EEDF is obtained from the Boltzmann equation. According to our assumptions, this equation reduces to its homogeneous and stationary form, which can be expressed as:

$$-\frac{2}{3}\frac{d}{du}\left[u^{3/2} v_c(u) u_c \frac{dF_o(u)}{du}\right] = S_o(F_o), \qquad (2)$$

where

$$u_c \equiv \frac{(eE_{rms})^2}{m_e\left(v_c^2(u) + \omega^2\right)} \qquad (3)$$

and S_o is the collisional term which accounts for electron-neutral collisions (both elastic and inelastic) and electron-electron collisions; $v_c(u)$ is the collision frequency for momentum transfer at energy u and m_e, the electron mass. For given operating conditions (specifically here nature of the gas, N and ω), Eq.(2) depends on E_{rms}, the root mean square intensity of the HF electric field in the discharge. When further assuming that the E-field is radially constant (homogeneous model), coupling Eq.(1) with Eq.(2) provides both the value of this field in the plasma, $\bar{E}$, and the EEDF (including $\langle u \rangle$). We summarize the above with:

Statement 1: for given operating conditions and in a steady-state system, whatever the HF power flowing to the discharge, the HF field intensity in the plasma adjusts so as to compensate exactly for the losses of charged particles. This means, for example, that one cannot utilize a resonant cavity to get a higher field intensity in the plasma: E_{rms} is an operator-independent parameter.

The radial distribution of E_{rms} can be obtained by coupling Eqs.(1) and (2) with the surface wave equation, and using the appropriate boundary conditions. In absence of a static magnetic field, the intensity of the total electric field E_{rms} (which includes E_z, E_r and E_φ components in general) increases radially toward the wall. This radial profile becomes steeper with increasing electron density, hence as a function of distance from the end of the plasma column (see Fig. 2): clearly the E-field in SWDs is strongly nonuniform. Nonetheless, Ferreira's self-consistent calculations have shown that E_{av}, the cross-section averaged value of $E_{rms}(r)$, is very close (except at the end of the column) to the value $\bar{E}$ obtained with the homogeneous model (11). Figure 3 shows that E_{av} and $\bar{E}$ are almost equal as a

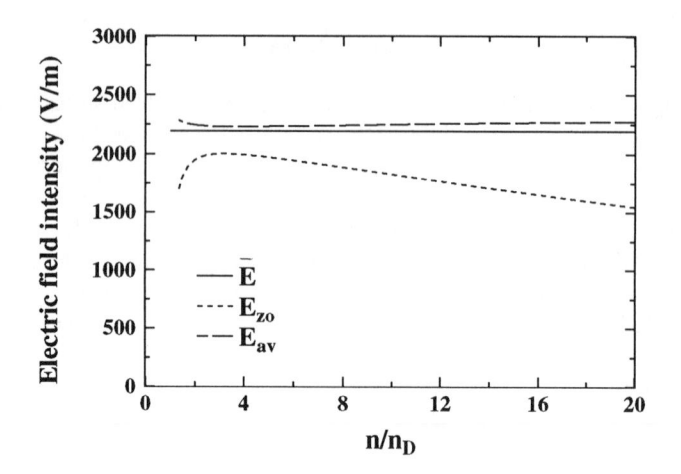

FIGURE 3. Calculated root mean square value of the total electric field intensity in the plasma as a function of electron density: $\bar{E}$, uniform field value (homogeneous model) and E_{av}, cross-section averaged value for a surface wave field; E_{zo} is the surface wave axial component at r=0 (f = 600MHz, R = 13 mm).

function of n, although E_z(r=0) varies substantially. Zakrzewski had reached a similar conclusion before by arguing that (ambipolar) diffusion is a global process of the plasma bulk (12).

Statement 2: in HF discharges, the intensity of E_{rms}(r) adjusts so that E_{av} is independent of electron density (hence of absorbed power) and almost equal to $\bar{E}$ of the homogeneous model.

The Power Balance per Electron

Multiplying Eq.(2) by u and integrating over energy does not provide an independent equation but it enables one to bring out the power balance per electron (9). Following statement 2, we assume the E-field intensity to be radially uniform and equal to E_{av}, and we then get:

$$\sigma_r E_{av}^2 = n\theta_l, \qquad (4)$$

where σ_r is the real part of the electric conductivity (which depends linearly on n). The parameter θ_l comes from the integration of the product of $S_o(F_o)$ and u, and the left-hand term represents the absorbed power density provided by the wave. When assuming excitation and ionization to occur from direct electron impact on the atom in the ground state, θ_l can be expressed as (8):

$$\theta_l = \frac{2m_e}{M}<v_c(u)u> + \sum_j <v_j(u)>eV_j + <v_i(u)>eV_i, \qquad (5)$$

where M is the mass of heavy particles, V_j and V_i are the threshold potentials for excitation to level j and for ionization by electron impact. Recalling that the HF energy is transferred to the discharge through acceleration of the electrons, θ_l represents the power loss (on the average) per electron to the plasma, which occurs through collisions with heavy particles. For a given gas (i.e. given collision cross-sections), the value of θ_l in Eq. (5) is totally dependent on the EEDF, and consequently on the charged particle losses. Under steady-state conditions, the power loss θ_l is compensated by the power absorbed (on the average) per electron from the HF electric field:

$$\theta_a \equiv \sigma_r E_{av}^2 / n, \tag{6}$$

and we can write the *power balance per electron* as:

$$\theta_a = \theta_l, \tag{7}$$

the common value being the parameter θ. It can be considered as the power required to maintain an electron in the discharge.

Properties of the Parameter θ

Similarity Law. An interesting feature of θ is that, under certain conditions, the plot of θ/p as a function of pR yields a similarity law, i.e. a relation between a combination of variables. In the present case, it means that one only needs to determine (theoretically or experimentally) a few different values of θ that are well distributed within the pR range of interest in order to obtain later on θ for any set of p and R values. Figure 4 is an example of such an experimental plot obtained from measuring θ_a with SWDs, at three different values of R: the similarity law is well verified. Recalling that the $E_{rms}(r)$ profile varies strongly

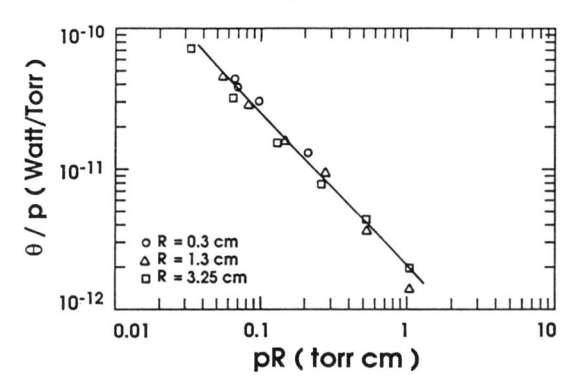

FIGURE 4. Measured θ/p values as a function of the pR product, at a constant frequency (200 MHz), for three different values of the discharge tube radius, showing the existence of a similarity law (5).

with R, Fig. 4 is an experimental confirmation of our assumption that θ_1 is related to E_{av} and not to the profile of $E_{rms}(r)$. Nonetheless, there are cases where this similarity law is not obeyed in the simple present form: i) when excitation and ionization of the gas do not result from direct electron impact on the atom in the ground state but proceeds through intermediate atomic energy levels; ii) when charged particles are lost predominantly by recombination in the plasma volume: θ_1 then depends on n; iii) when the EEDF is sensitive to the field frequency (see below).

Exact Power Balance per Electron. It was stated above that $\theta_a=\theta_1$. This is not rigorously true since in general additional terms contribute to energy losses. One of these terms deals with ion acceleration in the space charge electric field (including the plasma sheath), thus inducing gas and wall heating. Another source of energy losses is related to kinetic energy transport in the plasma (13). This term is particularly important at low gas pressure (typically below 50 mtorr). To simplify the presentation, we neglect these contributions in the following.

The Diffusion Regime in HF Magnetoplasmas. The θ parameter keeps on playing the central role in describing energy transfer in HF magnetoplasmas as well (7). It also enables one to distinguish between classical ambipolar diffusion and anomalous (Bohm) diffusion which occurs specifically in magnetoplasmas. The transition betwen these two regimes occurs at large enough values of B_o (7). We have plotted the boundary line from one regime to the other over a $B_o p$ vs. pR diagram, using magnetized SWD results; using this novel diagram, we can predict which diffusion regime is actually taking place in other RF and microwave magnetized plasmas submitted to an axial static magnetic field (7).

The comparison of SWDs with helicon discharges strongly suggests that the value of θ is the same in all HF magnetoplasmas under similar operating conditions, in contrast to what is usually believed. The influence of frequency on the EEDF could affect this result but this is unlikely to happen since these plasmas have a high ionization degree favoring electron-electron collisions, and thus the existence of a Maxwellian EEDF (see below).

θ under ECR Conditions. We have investigated magnetized SWDs further to examine the power balance per electron under ECR conditions, which imply a frequency match, $\omega_{ce} \approx \omega$ (ω_{ce} is the electron cyclotron angular frequency) and low collisions, $v_c/\omega \ll 1$, in order for the resonant absorption to be dominant (see below). Figure 5 shows that θ_a decreases steadily with increasing ω_{ce}, essentially because diffusion losses are reduced by the magnetic confinement of charged particles; there is no sign of extremum for θ_a at the ECR frequency match condition over the gas pressure range examined. This behavior corresponds to the fact that the flow of energy inside the discharge is still controlled by the diffusion of charged particles. Recent theoretical results obtained from a hydrodynamic model based on three-moment equations confirm this result (13).

Statement 3: in a diffusion controlled HF discharge, with or without a static magnetic field, the transfer of energy from the high frequency field to the plasma is well described by the parameter θ. It is independent of the field applicator

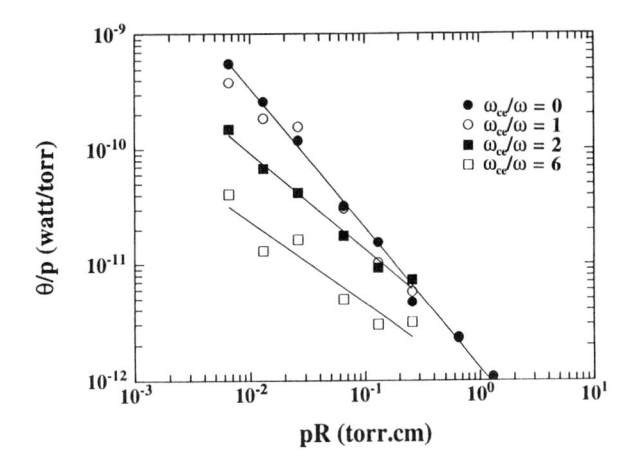

FIGURE 5. Measured absorbed HF power per electron normalized to gas pressure as a function of pressure times tube radius in a surface-wave magnetoplasma, at f=600 MHz in argon gas for R=13 mm and for different values of B_o.

configuration (which determines the E-field distribution), of the level of power flow incident on the discharge and of the power transfer mechanism which can be either collisional or resonant.

Dependence of the E-field Intensity on θ_l

We compare below the influence of collisional absorption and resonant absorption on the E-field intensity in the discharge.

When an electron is accelerated in a periodically varying field, the net amount of work done over a complete period is zero: the electron only gains energy when its periodic motion is interrupted by a collision. This is expressed quantitatively by using in Eq. (6) the real part of the electric conductivity for collisional power transfer:

$$\sigma_{rc} = -\frac{2}{3}\frac{e^2 n}{m_e} \int_0^\infty \frac{u^{3/2} \nu_c(u)}{\nu_c^2(u) + \omega^2} \frac{dF_o}{du} du \qquad (8)$$

which can be used to calculate θ_a (Eq. (6)).

When collisional absorption is too low, resonant absorption can be used to sustain a discharge. It requires a static magnetic field and a propagating wave. The electrons undergo cyclotronic motion around the B_o-field lines; the wave must have an E-field component perpendicular to B_o, and this component needs to rotate in the same direction as electrons. Then provided $\omega_{ce} = \omega$, the electrons "see" a constant E-field in their reference frame, and thus continuously gain energy; in this case, the rate of collisions can be very low but the power transfer nonetheless very high. This contrasts with collisional absorption under B_0 values close to but outside the ECR frequency match condition; in this case, for $v_c/\omega \ll 1$, complete periods of oscillations occur between two successive collisions: collisional transfer is then comparatively less efficient than resonant absorption. As a result, E_{av} can be lower under resonant absorption since it is required to compensate for the same power loss per electron, θ. To show this quantitatively, we use the concept of effective electric conductivity which is introduced (14) to express resonant absorption of HF power in the same form as collisional absorption (Eq. (4)):

$$Re(\sigma_{eff})E_{av}^2 = \theta n. \qquad (9)$$

The detailed expression for $Re(\sigma_{eff})$ can be found in (14). Figure 6 shows how this effective conductivity varies with ω_{ce}/ω: there is a sharp maximum at the ECR frequency match, and this maximum decreases with decreasing operating frequency. This explains why ECR is more easily achieved with microwave power than with RF power. For a given power density ρ_a, θn is constant (by definition $\rho_a = \theta n$), and thus from Eq. (9), E_{av} must go through a minimum as $Re(\sigma_{eff})$ passes through a maximum. This is the reason why it is easier to sustain low pressure plasmas at ECR and why the setting of ω_{ce} is quite sharp.

When people started to model the positive column of DC discharges, they used the parameter E/p; this parametrization was later on extended to microwave discharges. It should be clear that this description is related in a simple way to that of θ only under collisional absorption.

Statement 4: the E-field intensity is a passive parameter: it adjusts to compensate for the particle loss per electron through diffusion. It is more practical to use the parameter θ to model energy transfer in electrical discharges.

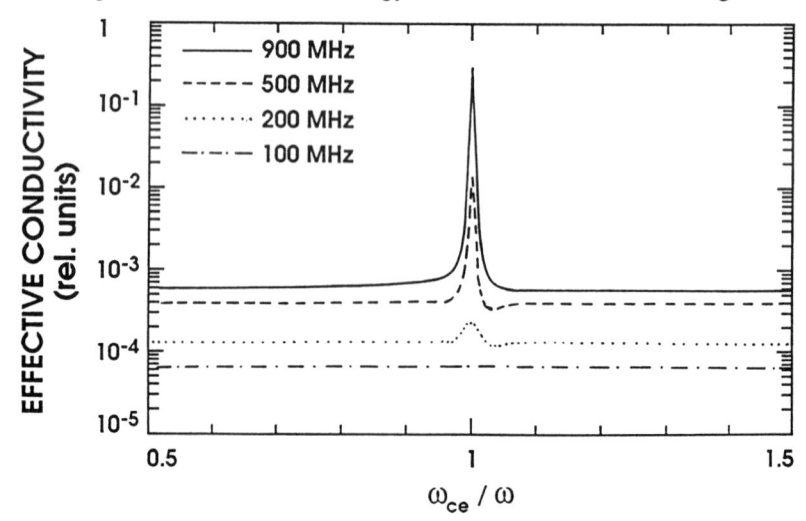

FIGURE 6. Real part of the effective electric conductivity as a function of ω_{ce}/ω for the HE_{01} fundamental mode at four values of f, for $\omega/\omega_{pe}=0.3$ (ω_{pe}, angular electron frequency) (14).

Simplified Model for HF Discharges

We have seen above that the HF field and the discharge are strongly self-consistently coupled in SWDs. Nonetheless, the fact that θ and E_{av} do not depend on the actual radial distribution of E_{rms} open the way to breaking self-consistency when describing the energy processes in electrical discharges. It means that the discharge maintenance processes can be modeled separately from the electromagnetic properties of the HF applicator containing the plasma, and the results merged at the end. This is the essence of the so-called simplified model for HF discharges.

This model (15) covers differently the two existing categories of HF discharges: i) HF discharges with a localized active zone (no E-field phase difference between any two points in the discharge) as, for example, in resonant cavity plasma sources; ii) traveling-wave HF discharges as is the case with SWDs. The simplified model is detailed in (10) for the first category and in (4) for the second category. Because of the lack of space, we limit ourselves to a brief description of the first category of these HF plasma sources. We describe the procedure to determine the properties of the discharge and then we indicate a method to obtain the electrodynamic properties of the HF plasma source.

Analysis of the Discharge Plasma

This analysis rests on the two balance equations presented above.

The equation for charged particle balance in the case of a diffusion controlled plasma (Eq. (1)) provides: i) the relative spatial distribution, or profile, of the electron density ñ(r,z)/n̄ (ñ is the local density value and n̄ is the average density in a volume V of plasma). This is a first approximation result since it is obtained assuming the E-field to be spatially uniform; consideration of the actual $E_{rms}(r,z)$ profile yields a slightly different electron density profile (11); ii) the value of E_{av} and the EEDF, which are obtained from coupling the charged particle balance with the homogeneous Boltzmann equation.

The power balance per electron (Eq. (7)) yields θ either by calculating $θ_a$ (Eq. (6)) from E_{av} and the EEDF, or by calculating $θ_1$ (Eq.(5)) with the corresponding EEDF. Then, since by definition, the total power absorbed in the discharge is:

$$P_A = \bar{n} θ V, \tag{10}$$

one obtains the average density n̄; the spatial distribution ñ(r,z) is then known from its radial profile ñ(r,z)/n̄.

The most complicated step in this procedure is generally the calculation of the EEDF. For this reason, in cases where electron-electron collisions are important, one may want to assume the EEDF to be Maxwellian. In this case, the balance of charged particles determines T_e instead of E_{av}. The latter is then obtained from the expression for $θ_a$ after having calculated $θ_1$ since $θ_a = θ_1$.

Analysis of the HF circuit containing the plasma

We wish to determine the electrodynamic properties of the HF plasma source, i.e. the impedance seen at the source input as functions of the operating conditions of the discharge and power flow incident on the circuit, given the field applicator and matching network. We do this by the equivalent circuit method instead of solving a full set of electromagnetic equations. In this method, the various circuit elements or impedances correspond to electromagnetic energy storage or dissipation processes occurring in various parts of the system. The plasma is one such impedance, which is expressed using the plasma complex conductivity representation; this conductivity depends on ñ and on the average or effective electron-neutral collision frequency for momentum transfer. The analysis yields the fraction $\eta = P_A/P_I$ of the electromagnetic energy incident on the discharge that is absorbed in the plasma for given operating conditions. We have used the equivalent circuit method extensively (1); examples of this procedure can be found in (10) for open and closed cylindrical tube discharges placed in a microwave circuit element.

Influence of the Applied Frequency ($B_o=0$)

There have been in the past many papers, mostly experimental in nature, dealing with the factors responsible for the dependence of the plasma properties on ω in HF discharges. The main problem with these investigations was the impossibility of carrying out a true frequency experiment, i.e. one in which the operating conditions are held constant except for ω. It seems that only SWDs offer the right conditions for such an experiment, as mentioned in the introduction. We begin this section by summarizing experimental results concerning this effect on the plasma deposition and etching of polymers in SWDs. Then we present the theoretical basis of the frequency effect.

Experimental Results

Deposition Experiments. The deposition of polymers by plasma CVD has been investigated in an HF field whose frequency f was varied between 10 and 400 MHz; the upper frequency limit was set by the value of R such that a m=0 mode SWD could be used. Two monomers, C_4H_8 and C_4F_8, have been employed at 23 % concentration in an argon based mixture. The results show that the deposition rate R_d (normalized to P_A) increases with f until a plateau is reached. In the case of isobutylene, two sets of data have been purposely plotted, one at constant P_A, the other at P_A values increasing with f (5). The agreement between these two sets of values validates the normalization of the deposition rate: the underlying question is the necessity of holding either n or P_A constant to perform a true frequency experiment. However, as a rule, these experiments were actually conducted in conditions such that the highest possible value of R_d was obtained in each case. Since the normalization used makes the experiment akin to constant P_A operation, recalling that n increases linearly with P_A at constant θ since $\theta_a \equiv n\theta$, it suggests that the deposition rate increases linearly with n at a given frequency. These experiments were performed at different gas pressures and extended to krypton and helium as carrier gas: an increase of R_d/P_A typically by a factor 2 to 5 has been observed over the frequency range investigated.

Etching Experiments. The etching of polyimide in a CF_4 discharge with added O_2 has been conducted for frequencies between 13.56 and 2450 MHz. The cylindrical discharge tube was tapered: the substrate was located in the large radius section while the surface wave was excited over the small radius section such that even at 2450 MHz, the highest frequency used, the wave was excited in a m=0 mode. It is found that the etch rate R_e (normalized to P_A) goes through a maximum around 50 MHz. As in the deposition experiment, two different sets of P_A values have been recorded; in the present case, the R_e/P_A normalization is not as good as with the deposition experiment, but nonetheless, the existence of a maximum etching efficiency close to f = 50 MHz is well demonstrated.

Theoretical Analysis

Understanding in detail the above experiments would require an accurate knowledge of all the elementary processes occurring in the plasma volume and at the substrate surface. Our goal is not to provide such an individualized analysis but to describe the basic mechanisms responsible for the frequency effect. The starting point of the model is the Boltzmann equation for the EEDF, using the same set of five assumptions listed earlier.

Influence of f on the EEDF. Equation (2) shows that the transfer of power, $\nu_c(u) u_c$, from the HF field to the electrons is maximum at given ω for an energy u_m such that $\nu_c(u_m) = \omega$ because of the term $\nu_c(u) / (\nu_c^2(u) + \omega^2)$ in Eq. (2). Then, when ω is varied, u_m also varies, and the EEDF $F_o(u)$ changes with ω. To bring out the essential results of this mechanism, we consider two limiting cases, $\nu_c(<u>)/\omega \gg 1$ ("DC" case) and $\nu_c(<u>)/\omega \ll 1$ ("microwave" case). Figure 7 shows typical results of such a calculation in argon (6). The "DC" EEDF (curve A) has less electrons of low energy than the "microwave" EEDF (curve H) but has more of them in the tail (this part of the EEDF corresponding to the energy range above the lowest energy threshold for excitation, eV_1); in particular, there are much more electrons in the energy interval defined by eV_1 and eV_i, the energy threshold for ionization. As a result, for a given electron density, the density $\dot{n}_j$ of atoms excited to level j per second is higher in the DC case since:

$$\dot{n}_j \equiv n(C_j N), \qquad (11)$$

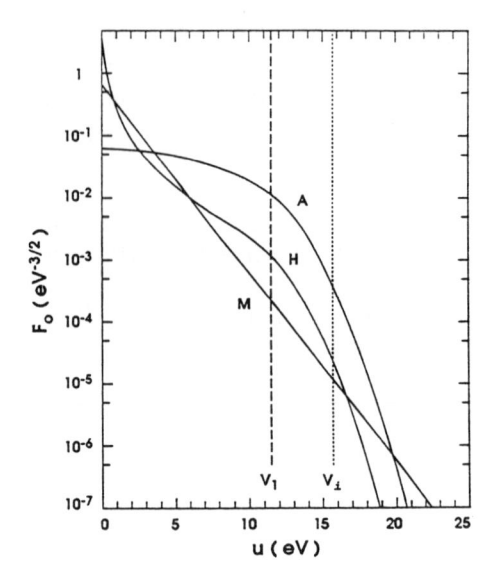

FIGURE 7. Electron energy u distribution function calculated from a self-consistent model for an argon discharge in a long cylindrical tube under ambipolar diffusion conditions for pR=1 torr.cm, in three limiting cases. Curve A: no electron-electron collisions and $\omega \to 0$ (DC case); curve H: no electron-electron collisions and $\omega \to \infty$ (microwave case); curve M: dominating electron-electron collisions (maxwellian distribution) (5).

where the excitation rate C_j is obtained from:

$$C_j = (2e/m_e)^{1/2} \int_{V_j}^{\infty} \hat{\sigma}_j(u) F_o(u) u \, du, \tag{12}$$

$\hat{\sigma}_j(u)$ denoting the corresponding excitation cross-section by direct electron impact. The above two EEDFs have been calculated assuming no electron-electron collisions. When such collisions are included and n/N increased continuously, curves A and H start to merge at n/N $\geq 10^{-4}$, tending (low energy electrons first) toward a Maxwellian EEDF as n/N goes on increasing. In argon, the EEDF is Maxwellian at n/N $\geq 10^{-2}$. Clearly then ω no longer acts on the EEDF.

Influence of f on θ. Figure 8 shows the calculated value θ/p as a function of pR. We see that θ decreases with increasing frequency when electron-electron collisions are neglected. The smallest value of θ is achieved in the microwave case provided pR is not too large, otherwise θ is the lowest with the Maxwellian EEDF. The influence of electron-electron collisions is discussed in detail in (16).

Statement 5: the most efficient argon ion source that can be realized with an electrical discharge is achieved at microwave frequency when pR is low or with dominating electron-electron collisions when pR is high.

Maximum of $\dot{n}_j$ *at constant n and constant* P_A. Clearly from Eq. (11) and Fig.

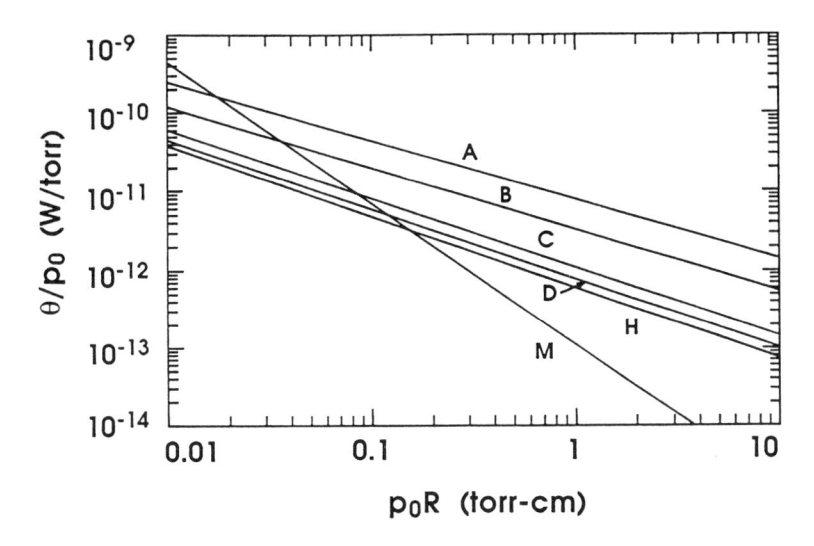

FIGURE 8. Plot of θ/p as a function of pressure times radius, at different values of the effective electron-neutral collision frequency normalized to ω: A (∞, "DC" case), B (6.67), C (1.25), D (0.67), H (0, "Microwave" case). Curve M is for a Maxwellian EEDF (9).

7, the maximum of $\dot{n}_j$ at constant n is obtained in the "DC" case. In the case where θ_a is kept constant, C_j and n vary in opposite direction. To deal with this matter, a perturbative approach has been used: a trace gas in introduced in argon, which by definition does not affect the carrier gas. Two different types of cross-sections were used for the trace gas, and their threshold energy varied continuously from 0 to above eV_i. The results are that at low pR, the highest value of $\dot{n}_j$ is obtained in the microwave case (low electron-electron collisions) while at large pR, it comes from a Maxwellian EEDF (high electron-electron collisions).

Statement 6: operating argon discharges with DC or RF excitation is inefficient unless the electron-electron collision frequency is high enough and pR is not too small.

CONCLUSION

Considering the case of SWDs where the E-field can be strongly non-uniform, we have shown that θ weakly depends on the E-field configuration in the discharge. This means that this parameter is the same in all HF discharges under given operating conditions and, in that sense, no HF discharge is intrinsically more efficient than others. We have seen that the frequency of the HF field affects the EEDF when electron-electron collisions are not too important as to yield a Maxwellian EEDF. Based on the analysis of an argon diffusion controlled discharge, it appears that DC and RF discharges with low plasma density (non

Maxwellian EEDF) are less efficient than microwave sustained discharges or discharges at any frequency with a high enough degree of ionization. Nonetheless, there remain differences between HF discharges, essentially in the spatial distributions of the various species owing to different spatial distributions of their HF field (17).

ACKNOWLEDGMENTS

The authors would like to thank Irène Pérès for numerical calculations and Jean-Eudes Samuel for preparing most of the figures. Thanks are due to the Fonds pour la Formation de Chercheurs et l'Aide à la recherche (FCAR) and the Conseil National de Recherches en Sciences Naturelles et en Génie for funding this work.

REFERENCES

1. Moisan, M., and Zakrzewski, J. Phys. D: Appl. Phys. **24**, 1025 (1991).
2. Moisan, M., Shivarova, A., and Trivelpiece, A. W., Plasma Phys. **24**, 1331 (1982).
3. Margot-Chaker, J., Moisan, M., Chaker, M., Glaude, V.M.M., Lauque, P., Paraszczak, J., and Sauvé G., J. Appl. Phys. **66**, 4134 (1989).
4. Moisan, M., and Zakrzewski, Z., M. Moisan and J. Pelletier eds., in *Microwave Excited Plasmas*, Amsterdam, Elsevier 1992, chap. 5.
5. Moisan, M., Barbeau, C., Claude, R., Ferreira, C. M., Margot, J., Paraszczak, J., Sá, A. B., Sauvé, G., and Wertheimer, M. R., J. Vac. Sci. Technol. **B 9**, 8 (1991).
6. Bounasri, F., Moisan, M., St-Onge, L., Margot, J., Chaker, M., Pelletier, J., El Khakani, M.A., and Gat E., J. Appl. Phys. **77**, 4030 (1995).
7. Margot, J., and Moisan, M., J. Vac. Sci. Technol. A (to appear in December issue) (1995).
8. Ferreira, M. C., and Moisan, M., Physica Scr. **38**, 382 (1988).
9. Ferreira, M., and Moisan, M., M. Moisan and J. Pelletier eds., in *Microwave Excited Plasmas*, Amsterdam, Elsevier 1992, chap. 3.
10. Zakrzewski, Z., Moisan, M., and Sauvé, G., M. Moisan and J. Pelletier eds., in *Microwave Excited Plasmas*, Amsterdam, Elsevier, 1992, chap. 4.
11. Ferreira, C. M., C. M. Ferreira and M. Moisan eds., in *Microwave Discharges: Fundamental and Applications* (NATO ASI Series), New York, 1993, pp. 313-337.
12. Zakrzewski, Z., J. Phys. D:Appl. Phys. **16**, 171 (1983).
13. Pérès, I., Fortin, M., and Margot, J. to be submitted to Phys. Rev. Lett.
14. Margot, J., and Moisan, M., J. Phys. D: Appl. Phys. **24**, 1765 (1991).
15. Moisan, M., and Zakrzewski, Z., J. M. Proud and L. H. Luessen eds., in *Radiation Processes in Discharge Plasmas*, (NATO ASI Series), New York, Plenum, 1986, pp. 381-430.
16. Sá, P.A., Loureiro, J., and Ferreira, C.M., J. Phys. D: Appl. Phys. **25**, 960 (1992).
17. Ricard, A., Barbeau, C., Besner, A., Hubert, H., Margot-Chaker, J., Moisan, M., and Sauvé, G., Can. J. Phys. **66**, 740 (1988).

The present state and development trends of discharges. Physics and chemistry of plasmas.

Andrei A. Ivanov

Russian Research Center "Kurchatov Institute", Moscow, 123182 Russia

Abstract. Various discharges are considered with objective to apply to plasma chemistry and technology. The discharges are divided in two classes: equilibrium discharges, such as arc, radio frequency, microwave discharges; and discharges with parameters far from equilibrium, such as glow, electron cyclotron, beam-plasma discharges and discharge in crossed electric and magnetic fields. Conditions for optimal parameters are proposed.

INTRODUCTION

Though various classifications of discharges are used, for the case under consideration the most appropriate one to our opinion will be as follows:

I. The discharges where state of a plasma is close to equilibrium, such as arc discharge, radio frequency and microwave discharges at high pressure.

II. The discharges, where state of a plasma is far away from equilibrium, such as glowing discharge, microwave discharge, electron cyclotron discharge, magnetron discharge, beam plasma discharge, discharge in crossed electric and magnetic fields.

One should bear in mind that the two classes introduced have no sharp boundary, e. g., radio frequency and microwave discharges at low pressure can be far enough from equilibrium. The systems of the class I can be described, neglecting to the first approximation slight deviation from equilibrium. All the laws of chemistry, all the concepts and definitions could have been applied immediately, the plasma realizing heat transfer to chemical reagents.

Systems of the class II. need more detailed information for the description, since the equations of chemical thermodynamics are no longer valid, distributions of particles are far from the equilibrium ones. Thus, one should use self-consistent kinetic approach for all the plasma components, such as electrons and ions, atoms and molecules, interaction of the particles with fields. Sometimes, computer simulation of such a system is inevitable.

EQUILIBRIUM SYSTEMS

Plasma permits to achieve the temperatures three - five times higher than those of traditional chemistry, and since rate coefficient k obeys the Arrenius law $k = k_0 \exp(-E/T)$, one gets essential increase of reaction rate. Here E is an activation energy, temperature being measured in energetic units. Concentrations of reagents do not depend on the temperature, so reaction rate w can be written in the next form

$$w = w_0 \exp(-E/T); \quad w_0 = k_0 \cdot n_{A_1}^{\eta_1} \cdot n_{A_2}^{\eta_2} \cdot \ldots \tag{1}$$

Here $n_{A_i}^{\eta_i}$ is the concentration of the reagent of kind A_i, η_i is the stechiometric coefficient of the corresponding reagent of stechiometric equation of chemical reaction

$$\sum_{i=1}^{k} \eta_i A_i \longleftrightarrow \sum_{j=1}^{l} \gamma_j B_j. \tag{2}$$

Let's note, that reaction rate we have introduced corresponds to forward reaction, η_i, A_i and γ_j, B_j corresponding to initial substances and final products of reaction respectively. In equilibrium state the rates of forward and reverse reactions are equal (see (2)), so one can introduce the concept of activation energy for direct reaction E_f and for reverse reaction E_r. Thus for equilibrium we have

$$k_{0f} \exp(-E_f/T) \cdot n_{A_1}^{\eta_1} \cdot n_{A_2}^{\eta_2} \cdot \ldots = k_{0r} \exp(-E_r/T) \cdot n_{B_1}^{\gamma_1} \cdot n_{B_2}^{\gamma_2} \cdot \ldots$$

The rate coefficients k_{0f} and k_{0r} depend on temperature, but the dependence appears to be weak compared to the exponential one. It follows from the last equation, that

$$\frac{n_{B_1}^{\gamma_1} \cdot n_{B_2}^{\gamma_2} \cdot \ldots}{n_{A_1}^{\eta_1} \cdot n_{A_2}^{\eta_2} \cdot \ldots} = \frac{k_{0f}}{k_{0r}} \exp(-(E_f - E_r)/T) = K_c(T), \tag{3}$$

i. e., we have obtained the well known law of mass action, the constant $K_c(T)$ being called chemical equilibrium constant. Making use of the principle of detailed balance and quantum mechanics ab initio calculations, one could have found rate coefficients of forward and reverse reactions both. But as a rule the rate coefficients are determined experimentally at equilibrium conditions.

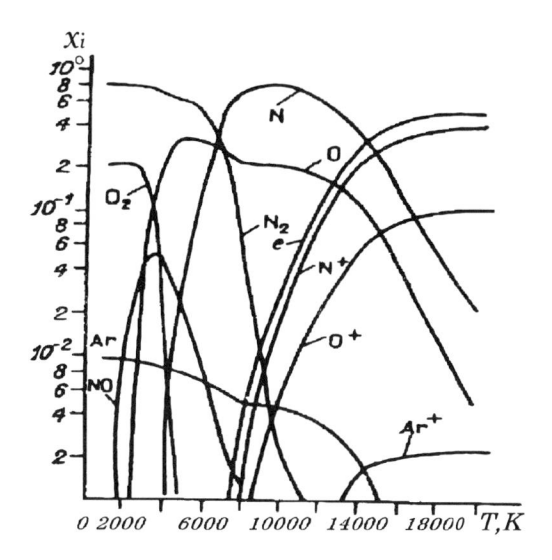

Figure 1: Dependence of the air chemical composition on temperature, $x_i = n_i / \sum_k n_k$ being the corresponding specific density.

Let's demonstrate typical features of plasma chemistry production for case of NO synthesis. One can use equations of high temperature chemistry for the following set of chemical reactions:

$$N_2 + O_2 \leftrightarrow 2NO; \quad O_2 + M \leftrightarrow 2O + M; \quad NO + M \leftrightarrow N + O + M;$$

$$N_2 + M \leftrightarrow 2N + M; \quad N + O_2 \leftrightarrow NO + O; \quad O + N_2 \leftrightarrow NO + M.$$

Here M is a background gas of inert molecules. The rate coefficients, that appear to fit existing experimental data are most accurately presented by Wray [1]. Corresponding differential equation, e. g., for the first reaction has the following form

$$\frac{dn_{NO}}{dt} = k_f \cdot n_{N_2} \cdot n_{O_2} - k_r \cdot n_{NO}^2.$$

Solution of this set of equations is represented in Figure 1. Reversibility of reactions considered will cause the transition to initial substances, provided that cooling is slow enough. Fast quenching could essentially improve situation, e. g., rate of quenching of $10^7 K/s$ gives up to 15% of equilibrium concentration of NO for optimal mixtures and pressures. Actually, output of NO is less than 7% due to the difficulty of quenching.

The Arc

Arc discharge appears to be low voltage self sustained discharge, cathode voltage drop being of the order of ionization potential. Direct ion bombardment provides essential heating of cathode, thus causing thermoelectric emission. The arc currents are high enough ($1 - 10^5 A$), current densities being $10^2 - 10^7 A/cm^2$. Since arc plasmas are at equilibrium in broad range of densities and temperatures [2], physics of the discharge can be described in frames of model of Steenbeck [3], where electric field E is supposed to be constant, so that radial distribution of conductivity σ is determined by temperature distribution, which one can find from heat conductivity equation:

$$-\frac{1}{r}\frac{d}{dr}(r \cdot \kappa \frac{dT}{dr}) = \sigma(T)E^2. \qquad (4)$$

Since the temperature of plasma is much higher than that of wall $T_w = T(R)$, one can put $T(R) = 0$, where R is radius of chamber, $\frac{dT}{dr} = 0$ at $r = 0$ due to symmetry of system. Because of the strong dependence of conductivity on temperature

$$\sigma(T) \sim \exp(-I/2T) \qquad (5)$$

we can make use of the difference in radial distributions of σ and T (Fig. 2).

We can approximate σ by a rectangular function (Fig. 2), $T(r)$ at $r \leq r_0$ being approximated by the following parabolic function

$$T(r) = -(T_c - T_0)\frac{r^2}{r_0^2} + T_0 = -\Delta T \frac{r^2}{r_0^2} + T_0. \qquad (6)$$

Here $T_0 = T|_{r=0}$, $T_c = T|_{r=r_0}$. The right hand side of (4) is a specific power supply, so the total power per unit length of the system W will be obtained by integrating both sides of (4) over the whole volume and then dividing by the length of the system

$$W = -2\pi \cdot r_0 \cdot \kappa_0 \cdot \frac{dT}{dr}|_{r=r_0} \stackrel{see\ eq.\ (6)}{=} 4\pi \cdot \kappa_0 \cdot \Delta T. \qquad (7)$$

It can be seen, that in the vicinity of $r = r_0$ one can write

$$\frac{d\sigma}{dT}|_{T=T_c} \approx \frac{\sigma(T_c)}{\Delta T}. \qquad (8)$$

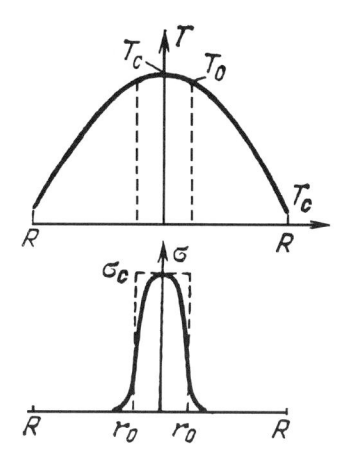

Figure 2: Dependence of temperature and conductivity on radius.

Excluding ΔT from (7) and (8), we obtain

$$4\pi \cdot \kappa_0 \cdot \sigma_c = W \frac{d\sigma}{dT}|_{T=T_c}. \tag{9}$$

Making use of (5), (9) we get finally

$$T_c = \sqrt{(I/8\pi\kappa_0)W}. \tag{10}$$

Thus the power per unit length, potential of ionization or effective potential of ionization and heat conductivity permit us to estimate the temperature and hence all the parameters of arc discharge. Making use of (5), (10), Ohm's law $i = \sigma_c \cdot E \cdot \pi r_0^2$ and Joule's law $W = i^2/\pi r_0^2 \sigma_c$ one can obtain $i - E$ plot of arc. Figure 3 shows the comparison of experimental data with calculated ones for the case of nitrogen arc at atmospheric pressure with chamber radius 1.5 cm [4]. The results obtained are valid when plasma is in equilibrium with respect to temperature, viz., when electron temperature T_e is not much larger than that of heavy particles T.

Estimate can be made by taking into account that electron heating is σE^2, electron cooling being of the order of $(T_e - T) \cdot n_e \cdot \nu \cdot m/M$.

The contemporary processings are organized as a rule in such a way, that plasma flows and those of raw materials are injected separately in reaction chamber, so turbulence is necessary for their effective mixing. It means that transport processes and chemical kinetics should be considered in self consistent approach, which could be hardly realized [5]. Another problem is fast quenching, providing fixation of final products, which can be solved by cooling in heat exchanger, by gas streams, in supersonic nozzle. Process of nitrogen fixation mentioned above has a very long history. The process

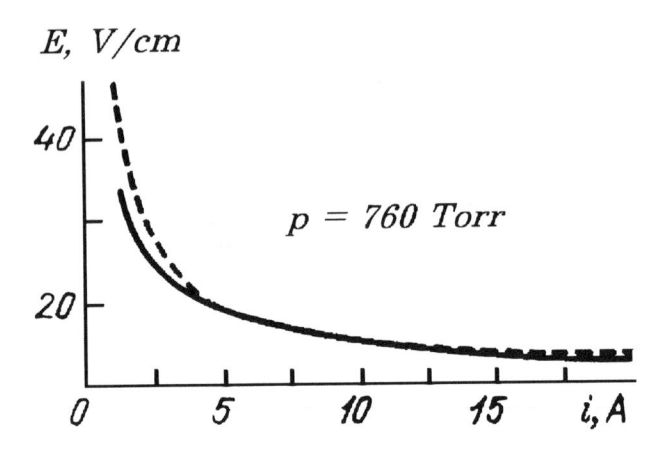

Figure 3: $i - E$ plot of arc discharge in nitrogen for radius 1.5 cm. Solid curve corresponds to experiment, dashed curve is the result of calculation.

was realized in Norway in 1900 by making use of arc discharge. It was the first plasma chemistry application, the large scale devices having been used up to 1930. Since that time the ammonia process appears to be the main industrial one, fixation of nitrogen from air being rather a problem of academical study. However, it is shown, that plasma chemistry process could have been competitive when the device power is within the range $10 - 100\ MW$ [6]. There is another large scale process, viz., synthesis of acetylene from natural gas and from wastes of fuel industry. This process, called as well pyrolysis of methane, is described by Kassel's scheme

$$CH_4 \to CH_2 + H_2;\quad CH_2 + CH_4 \to C_2H_6;\quad C_2H_6 \to C_2H_4 + H_2;$$

$$C_2H_4 \to C_2H_2 + H_2;\quad C_2H_2 \to 2C + H_2\,,$$

where intermediate compound is ethylene. Theory predicts maximum output of acetylene 40-50%, experiment gives 16%. The first and the biggest industrial application is a Huels mill, which was on in 1940, nowadays productivity being $10^5 t/year$ of acetylene from methane, provided by 17 reactors, each having power 8.2 MW. Other byproducts are ethylene ($10^4 t/year$) and HCN ($1.6 \cdot 10^3 t/year$). Another large scale production of acetylene is based on device of power 11 MW, productivity being $2.2 \cdot 10^4 t/year$ (Du Pont de Nemur).

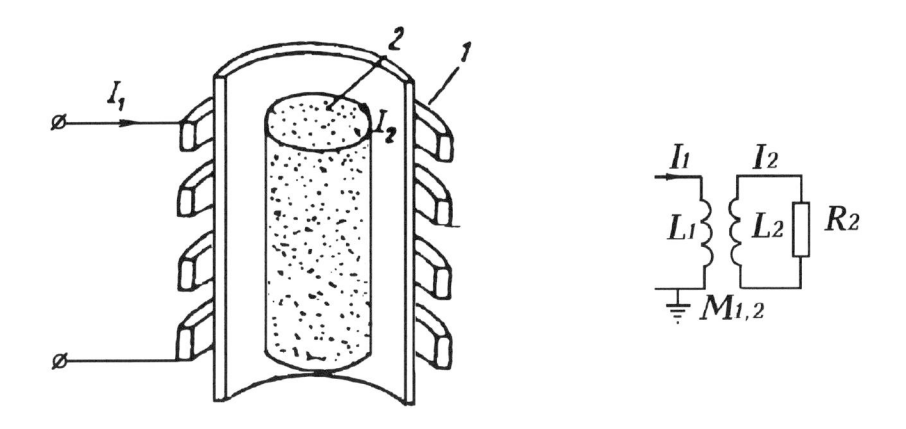

Figure 4: Scheme of induction RFI discharge. 1: magnetic coils, 2: plasma column. The electrical parameters having subscript $_1$ refer to outlet circuit, the parameters with subscript $_2$ refer to plasma.

Radio frequency discharges

There are two types of high frequency devices, used in plasma chemistry, differing by the frequencies: devices, having radio frequency range from 100 kHz up to 100 MHz, power source making use of valve generators; ultra high frequency UHF or microwave devices, having frequencies higher, $1 - 100$ GHz, the power source being microwave generators. Absence of electrodes permits to decrease level of pollutions in plasma, to increase variety of media, creating plasmas, and to achieve practically unlimited life time of devices. The foundation of theory of radio frequency devices was created in the first half of XX century [7], [8]. The success achieved in early forties provided appearance of technology of radio frequency devices, viz., the first of them was Reed's "radio frequency plasma torch" [9]. Gaseous stabilization of plasma discharge and its ignition were proposed ten years later. From the very beginning it was established, that two types of discharge can exist in valve, immersed in radio frequency fields: the discharge, brought about by electric capacity coupling (RFE), or induced electric fields (RFI). Electrical circuit of RFI device is shown in Figure 4.

Energy transfer to plasma in RFI devices is realized by oscillating magnetic field. Electric current in plasma ranges from several hundred up to one thousand amperes. If heat flux to wall does not exceed 5 W/cm^2, one will use quartz chamber without special cooling; when heat flux is higher (up to 20 W/cm^2), water cooling chambers can be used. To avoid plasma contact with the wall of the chamber, one makes use of gaseous stabilization,

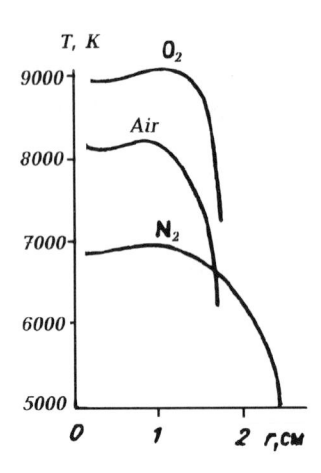

Figure 5: The typical radial temperature distribution for various gases in the RFI discharge.

when the flux of cold gas injected ousts the plasma from the wall. A special apparatus is usually used for formation of axial and vortex stabilization both. Sometimes an interesting method of stabilization is used, when chamber walls are made porous, so penetration of cold gas brings the ousting of plasma from the wall about. Maximum power of known systems is 1 MW. The general efficiency is about 55–65%.

Since power of device is limited by the heat flow on the wall, conditions of stable volume discharge are satisfied at a relatively narrow range of parameters, the maximum values of temperature and plasma flow rate being almost the same for RFI devices of different types. Figure 5 shows characteristic distributions of temperatures for various plasma forming gases. It should be noted, that maximum of temperature is located at the zone of maximum current density, displaced towards the wall due to skin-effect. Velocity of plasma flow is about $25-50$ m/s. Nozzle is used sometimes to increase the velocity.

Steenbeck model was used during a long period as a logical standard for assessment of not only the arc devices but also RFI discharge devices. Generally speaking, it is obvious that the first RFI discharge devices were designed by the channel model. But the main disadvantage of this model was poor temperature prediction. In the 60th a simple but very effective model of RFI discharge was proposed [10]. The boundary conditions were as follows: $T_w \approx 0$ at $r = R$ as the device was cooled and $H(0) \approx 0$ because of the skin-effect. Introducing the F-function strongly affected by the temperature: $F(T) = \int \sigma(T) k(T) dT$, one gets

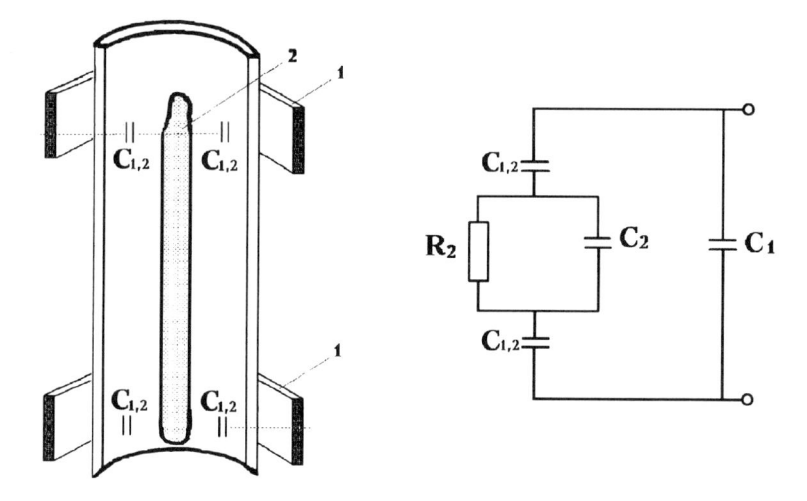

Figure 6: Scheme of electrical RFE discharge. 1: electrodes, 2: plasma column. The electrical parameters having subscript $_1$ refer to outlet circuit, the parameters with subscript $_2$ refer to plasma.

$$\frac{c^2}{64\pi^2}H^2(R) = \left(\frac{I_0 N}{2}\right)^2 = F(T). \tag{11}$$

Thus it is possible to connect the current of inductor I_0 and the number of winds per unit length N directly to the plasma temperature T. This method is broadly used for primary estimation of the RFI discharge devices.

The power source of capacity discharges connects to plasma by electric capacity of coaxial system, created by outer electrodes and plasma column. Equivalent circuit diagram is shown in Figure 6. Distance between electrodes is of the same order of magnitude as diameter. Discharge current ranges from several amperes up to several tens of amperes, the gradient of voltage along the axis being $20-200$ V/cm. It stipulates the appearance of non equilibrium plasma in the device.

The devices, suitable for application, appeared in early 70s. There are two forms of RFE discharge, viz., the volume one and the channel one. The volume discharge is similar to glow discharge, the homogeneous distributions of parameters occurring. At higher power bright spoke appears at the axis, diameter of the latter being roughly ten times less than that of chamber. Usually the regime corresponds to the nominal one. Efficiency is about 30–60 %.

Measurements made in argon plasma at atmospheric pressure [11] showed that quasi equilibrium took place. Study of nitrogen plasma at pressure

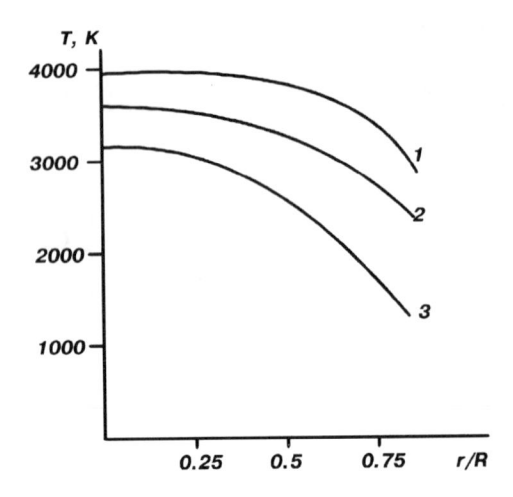

Figure 7: Radial distribution of gas temperature in the RFE discharge. 1: $P = 15.7\ kW$, $I = 8.0\ A$; 2: $P = 11.2\ kW$, $I = 6.0\ A$; 3: $P = 7,1\ kW$, $I = 5.3\ A$, P being power.

$100 - 600\ Torr$ [12] permits one to say the electron temperature ($7000\ K$) to be twice as much as that of heavy particles ($4000\ K$). The increase of pressure made the temperatures almost the same, the difference taking place for degree of ionization and population of metastable states only. The profile of gas temperature is shown in Figure 7.

Now we are in a position to introduce total description of equilibrium systems, such as arc discharge, RFI discharge, RFE discharge to some extent, the latter being sometimes very far from equilibrium. Traditionally one has to write the set of hydrodynamics equations, describing the gas and plasma flows, heat transfer equation, equations of chemical kinetics of the type we had discussed, to make the estimates of validity of Saha equation for each special case, bearing in mind, that equilibrium in radiation is not valid as a rule, and on solving the complete system of nonlinear equations to compare solution with experiment.

NON EQUILIBRIUM SYSTEMS

Systems can become more non equilibrium, when total pressure is decreased, so the problem of quenching can be solved. However, one needs more detailed information for the system to be described, the equations of chemical thermodynamics being valid no longer. Different approaches should be used in calculations for equilibrium and non equilibrium, where step-wise ionization is more important. Ionization is called direct, if it occurs for electron collision with an atom, being in ground state; otherwise, if many collisions are necessary for ionization, one deals with step-wise ionization. The ratio of the rate coefficient of direct ionization k_f to that of step-wise ionization k_{st} is as follows [13]:

$$\frac{k_f}{k_{st}} \sim \left(\frac{T_e}{I}\right)^{7/2}.$$

Life time of excited atom is small, so if photon mean free path is larger than characteristic length of a plasma, emission of radiation will provide essential losses of energy. However, capture of radiation by non excited atoms is resonant process with its cross section being of the order of λ^2, where λ is photon wave length. Diffusion of radiation is described by theory [14], which shows the losses to occur only in narrow outer layers of plasma.

Non equilibrium discharges seem to be rather attractive from view-point of chemistry. Temperature of electrons is much higher than that of heavy particles, vibration temperature of molecules T_v can be close to that of electrons. In principle it can bring about the decrease of activation energy E_a, so rate coefficients in terms of equilibrium chemistry could have been written as

$$k \sim \exp\left(-\frac{E_a - \alpha \epsilon_v}{T}\right), \quad E_a > \epsilon_v.$$

Here α is semi empirical coefficient, which permits to describe processes without detailed information on dynamics of collisions [15] [16].

Thus flow of energy is directed from power source to electrons, then from electrons to vibration levels of molecules, and finally to chemistry. Power is to be spent as well to sustain steady state discharge, i. e., to produce electrons in amount sufficient for compensation for their depletion due to diffusion and attachment.

The Glow Discharge

A breakdown of a gas initiates Townsend discharge of a few micro-amperes, and in between this discharge and arc discharge there is a discharge in which the strength of current passing is measured in milli-amperes. This is a glow discharge (Fig. 8).

The glow discharge is made up of different parts, as can be seen of the luminous properties (Fig. 9): the Crook dark space (CM), the negative glow (MN), the Faraday dark space (NO), the positive column (OA).

The conditions of existence of self-sustained stable discharge are well known [17]. Two coefficients α and γ were introduced. First one α describes a number of pairs of charged particles produced by electrons on the unit of length, so the number of pairs on path d, equal to length of cathode

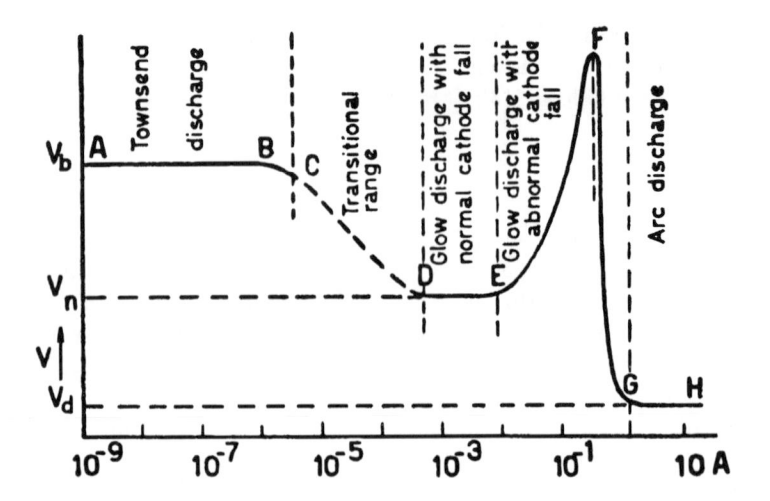

Figure 8: The characteristic of a self-sustaining gas discharge, the voltage V being on a linear and the current i on a logarithmic scale. V_b: breakdown voltage. V_n: normal cathode fall of potential. V_d: arc voltage.

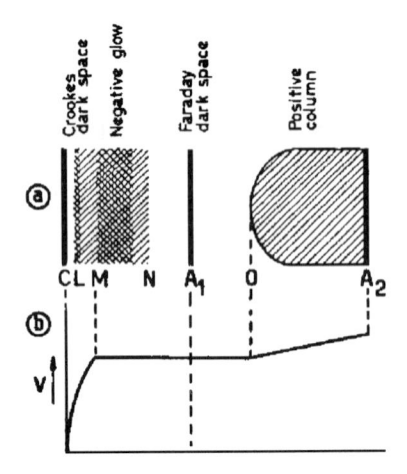

Figure 9: Glow discharge between cathode C and anode A for different positions of the latter, A1 and A2. Diagram (a) shows the optical phenomena, luminous intensity being indicated by the density of the hatching, diagram (b) is the potential distribution between cathode and anode.

layer, will be $(\exp(\alpha d) - 1)$. The ions reaching the metal surface of cathode can strip the electron with bound energy of the order of ionization potential of the atom. The production of electrons near cathode is described by the second Townsend coefficient γ, which is the ratio of number of free electrons to number of ions. As a rule, the second coefficient shows a weak dependence on ion energy, i. e., on electric field strength near the cathode. The introduced coefficients provide us a condition for existence of self-sustained discharge $\gamma(\exp(\alpha d) - 1) = 1$, or $\alpha d = \ln(1 + 1/\gamma)$. The relation obtained permits to find condition of gas breakdown, when constant electric field is applied at a distance d.

The most interesting is positive column of glow discharge, where parameters of plasma do not depend upon near electrode layers. Positive column plasma density varies from $10^8 cm^{-3}$ up to $10^{11} cm^{-3}$, so degree of ionization Z_i is $10^{-8} - 10^{-7}$ at pressure range $1 - 10\ Torr$. Electron temperature T_e is of the order of several eV. Relative simplicity of measurements in glow discharge permits to check different self-consistent models for plasma chemistry.

It is necessary to investigate electron distribution function for averaged description of glow discharge. In positive column electric field E can be put constant, so the electron distribution function is usually represented in the following form $f(t, v, \theta) = f_0(t, v) + \cos\theta f_1(t, v)$, where θ is the angle between electron velocity and electric field, v being absolute value of velocity. Standard procedure gives

$$\frac{\partial f_0}{\partial t} = \frac{1}{v^2}\frac{\partial}{\partial v}\left(\frac{e^2 E^2}{3m^2 \nu_t} v^2 \frac{\partial f_0}{\partial v}\right) + St(f_0), \qquad (12)$$

Since drift velocity of electrons is $u = -eE/m\nu_t$, one can estimate diffusion coefficient in velocity space as $D_v = \frac{1}{3}u^2 \nu_t$, the difference being taken into account between transport frequency of electron-neutral collisions ν_t and electron-neutral collision frequency ν. The last term of (12) is called collisional term and describes both elastic and inelastic collisions of electrons. For the case of low degree of ionization and low temperature of electrons one can neglect electron-electron and electron-ion collisions, not take into account excitation of energetic levels of heavy particles and use Focker-Planck collisional term

$$St_{el}(f_0) = -\frac{m}{M} \cdot \frac{T}{v^2} \cdot \left[v^3 \nu_t \left(\frac{f_0}{T} + \frac{1}{mv}\frac{\partial f_0}{\partial v}\right)\right]. \qquad (13)$$

Solution of (12) for steady state will be

$$f_0 = const \cdot \left[-\int_0^v \frac{mv\, dv}{T + \frac{1}{3}(\frac{eE}{m\nu_t})^2 M}\right]. \qquad (14)$$

If frequency of elastic collisions does not depend upon velocity, one gets for large electric fields

$$f_0 \sim \exp\left(-\frac{mv^2}{2T_{eff}}\right), \qquad T_{eff} \sim M \cdot u^2. \tag{15}$$

It means that mean electron energy is of the order of $T_{eff} \gg T$, so they can heat atoms up to the temperature T', $n_a(T - T') \sim n_e \cdot T_{eff}$. Thus result obtained is valid if $\dfrac{T_{eff}}{T} \ll \dfrac{n_a}{n_e}$. Since atomic density in glow discharge is much larger than that of electrons, effective electron temperature is high. If $\nu_t(\epsilon) = const$, one will get Maxwell distribution with $T_e = T_{eff}$; if mean free path $\lambda = v/\nu_t(\epsilon)$ is a constant, one will get Druyvestein electron distribution function

$$f_0 \sim \exp\left[-\frac{\epsilon^2}{\frac{1}{3}\frac{M}{m}(eE\lambda)^2}\right].$$

When electron energy exceeds corresponding threshold, inelastic collisions can occur. It brings about decrease of electron temperature as well. Let's discuss some of inelastic processes.

a) Excitation of vibration levels ($e - V$) process.

A simple example can be considered, when electrons lose their energy due to excitation of the first vibration level [19], [18]. The corresponding collisional term will have a differential form, if mean electron energy is larger than the energy of excitation $\hbar\omega_0$

$$St_{\hbar\omega_0} = \frac{\hbar\omega_0}{m}\frac{1}{v^2}\frac{\partial(v\nu_{exc}f_0)}{\partial v}. \tag{16}$$

Equation (12) in a stationary case with additional inelastic collisional term takes the following form

$$\frac{1}{v^2}\frac{\partial}{\partial v}\left(\frac{e^2E^2}{3m^2\nu_t}v^2\frac{\partial f_0}{\partial v}\right) + St_{el}(f_0) + St_{\hbar\omega_0} = 0.$$

Thus instead of (13) one gets, neglecting low gas temperature T

$$f_0 = const \cdot \exp\left[-\int_0^v \frac{mvdv}{\frac{1}{3}(\frac{eE}{m\nu_t})^2 M}\left(1 + \frac{M}{m}\frac{\nu_{exc}}{\nu_t}\frac{\hbar\omega_0}{mv^2}\right)\right].$$

One can see, that everything now depends on the ratio $\dfrac{\nu_{exc}}{\nu_t} \sim \dfrac{\sigma_{exc}}{\sigma_t}$. If excitation occurs without formation of auto ionization state, $\dfrac{\sigma_{exc}}{\sigma_t} \sim \left(\dfrac{m}{M}\right)^{1/2}$, T_{eff} will be of the same order of magnitude as before. Situation will change, when formation of auto ionization state takes place, viz., $\sigma_{exc} \leq \sigma_t$. Taking into account that electron energy in glow discharge is several eV, i. e., of the order of atomic unit energy E_a, and $\hbar\omega_0 \sim E_a\left(\dfrac{m}{M}\right)^{1/2}$, one gets that T_{eff} will decrease $\left(\dfrac{m}{M}\right)^{1/2}$ times. It is very difficult to describe in theory cross section of excitation of vibration levels even by the order of magnitude, since the existence and width of auto ionization level in many cases are unknown [5], [26]. Thus, one uses experimental results and semi empirical formulae if available. It should be pointed out, that precise measurements were made for excitation only from ground vibration level, where existence of auto ionization level for some molecules was established [20].

b) Excitation of electronic states of atoms and molecules.

Energy of excitation is of the order of ten eV, so total amount of fast electrons, which could excite or ionize atom or molecule, is not so large. Hence, losses of energy of fast electrons do not change drastically distribution function in the range of low electron energies and collisional term can be written as

$$St_u = -\nu_u f_0 = -n_a v \sigma_u(v) f_0. \qquad (17)$$

Here σ_u is cross section of inelastic scattering, $\sigma_u \sim (\epsilon - E_{exc})^{1/2}$, or $\sigma_u \sim (\epsilon - I)$, where E_{exc} and I are the excitation energy and the potential of ionization respectively. If electron density is small and atoms are in ground states, (12) for atomic gases takes the form

$$\frac{1}{v^2}\frac{\partial}{\partial v}\left(\frac{e^2 E^2}{3m^2\nu_t}v^2\frac{\partial f_0}{\partial v}\right) = \nu_u f_0.$$

The last equation is valid for $v \geq v_0$, where v_0 is the threshold velocity for excitation or ionization of the atom. Putting $f_0 = A \cdot \exp(s(v))$, and neglecting of the second derivative of $s(v)$, one gets

$$f_0 = A \cdot \exp\left[-\int_{v_0}^{v} \sqrt{3}\,\frac{m\sqrt{\nu_t\nu_u}}{eE}dv\right].$$

c) Dissociative recombination and attachment of electrons.

Dissociative recombination ($e + AB^+ \to A + B$) has the highest rate coefficient $k_r^{ei} \sim (10^{-7} - 10^{-6}) cm^3/s$ at room temperature, demonstrates a weak dependence on temperature and strong dependence on vibration excitation [21]. Dissociative attachment ($e + AB \to A^- + B$) brings about the formation of negative ions and is provided mainly by fast electrons. Corresponding cross sections were measured and can be found in literature [21].

d) Diffusion.

In general case electron distribution function depends upon coordinates, so corresponding losses of charged particles should be taken into account. Life time of charged particles τ_L can be expressed in terms of ambipolar diffusion coefficient D_a and so called diffusion length Λ_D, determined by the geometry of system, viz., $\tau_L = \Lambda_D^2/D_a$. Thus, for self-consistent solution it is necessary to add $-f_0/\tau_L$ to the collisional term in (12) [18].

Electron distribution function depends on distribution function of the working gas over quantum states $F(k)$. Thus, kinetic equation (12) is to be solved along with the system of equations, describing population of quantum states. Special attention should be drawn to the vibration kinetics of the molecules. Equations, taking into account one quantum $V - V$ exchange, $V - T$ relaxation and multi quantum transition for $e - V$ processes were proposed in [22]. The most important for low degree of ionization Z_i are $V - V$ processes, providing population of upper vibration levels, for $Z_i \geq 10^{-4}$ $e - V$ process dominates [23]. $V - T$ process is of less importance since its rate coefficient is of the order of $10^{-17} cm^3/s$, meanwhile corresponding coefficients for $V - V$ and $e - V$ processes are $10^{-12} cm^3/s$ and $10^{-8} cm^3/s$ respectively.

e) $V - V$ processes.

In glow discharge $e - V$ processes bring about mainly excitation of low vibration levels, the population of high levels, i. e., chemically active states, occurs due to $V - V$ exchange. Taking into account anharmonism x_e of oscillation of high levels, one obtains Treanor distribution [24]

$$F(v) \sim \exp\left(-\frac{\hbar\omega_0}{T_v} + x_e \frac{\hbar\omega_0}{T}\right). \qquad (18)$$

Fraction of highly excited molecules could be much more than that of Boltzmann.

f) $V - T$ processes.

As a rule, this process is slow due to the smallness of probability of transition, which is proportional to $\exp(-2\pi\omega_0 a/v_M)$ according to Landau-Teller formula, where v_M is velocity of molecule, a is a radius of action of molecular forces [25]. The value $\xi = \omega_0 a/v_M$ is called Massey parameter, which is large for broad range of gas temperature.

Self-consistent description of glow discharge appears to be rather complicated problem, so several parameters are sometimes considered as given, e. g., gas temperature T, electron temperature T_e, degree of ionization Z_i [5]. Study the kinetics of population of vibration levels, taking into account $V - V$ and $V - T$ processes, shows for nitrogen, that at high degree of ionization ($Z_i \geq 10^{-4}$) effective excitation of upper vibration levels takes place, the vibration temperature being equal to that of electrons. Unfortunately, degree of ionization in glow discharge is much less, so it is necessary to seek other

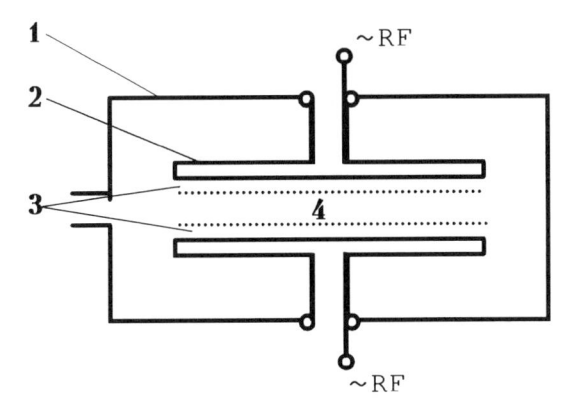

Figure 10: Scheme of RF discharge with planar geometry. 1: chamber; 2: RF electrodes; 3: space charged plasma layers; 4: plasma.

discharges for chemistry. Nevertheless, more recent results show a progress in self-consistent description of glow discharge, which permits to find the rate coefficients for various elementary processes along with experiment [26].

Radio frequency and microwave discharges at low pressure.

Detailed discussion of kinetics permits us to use formulae obtained for description of low pressure discharge with time oscillating electric fields $E = E_0 \cos \omega t$. Standard procedure for spherically symmetric part of time averaged electron distribution function gives us for steady state (compare with (12))

$$\frac{1}{v^2}\frac{\partial}{\partial v}\left(\frac{e^2 E^2}{3m^2 \nu_t} \cdot \frac{\nu_t^2}{2(\nu_t^2 + \omega^2)} \cdot v^2 \frac{\partial \overline{f_0}}{\partial v}\right) + St(\overline{f_0}) = 0. \quad (19)$$

We see that coefficient of diffusion in velocity space D_v is replaced now by

$$D_\omega = D_v \cdot \frac{\nu_t^2}{2(\nu_t^2 + \omega^2)}. \quad (20)$$

If $\omega \gg \nu_t$, one gets $D_\omega = v_\sim^2 \cdot \nu_t / 6$, $v_\sim = eE_0/(m\omega)$.

Now it is plane that one can use the main equations (13)-(18), substituting $\overline{f_0}, D_\omega$ and E_0 for f_0, D_v and E, so radio frequency discharge at low pressure can be considered as a special type of glow discharge.

This discharge is broadly used for etching and deposition processes, the planar geometry being common (Fig. 10). Physics of radio frequency dis-

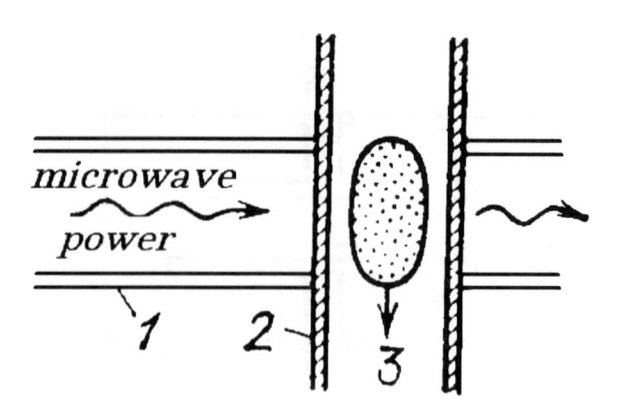

Figure 11: Schematic representation of microwave discharge. 1: waveguide; 2: dielectric tube; 3: plasma column.

charge was studied in survey [27], the typical parameters of discharge being: $T_e/T > 10$, wide working pressure range from $10^{-3} Torr$ up to $10^2 Torr$, almost plane plasma parameters space distribution over plane electrode wafer surface, existence of a flux of energetic ions towards the surface under treatment [28]. However, plasma density $\sim 10^{11} cm^{-3}$ is not high enough, so excessive bias of the order of several hundred volts leads to the limitation of chemical processes due to surface damage. Recent researches have shown the formation of great amount of particles with sizes of the order of 1 μm, which appeared due to bombardment the plates treated by ions, the quality of wafer being deteriorated. A great deal of papers concerning the phenomenon, were submitted to a special conference [29]. Moreover, skin effect brings about decreasing of heating zone. Attempts have been made to introduce magnetic field, to use whistlers, to create multi electrode RF system, for the parameters of discharge to improve [30]. For microwave discharge amplitude of electron oscillations $v_\sim/\omega$ is much less compared to characteristic length of device, so charged particles do not reach walls, i. e., for wavelength $\sim 10\ cm$, pressure $\sim 1\ Torr$, $E_0 \sim 500\ V/cm$ amplitude $v_\sim/\omega \sim 3 \cdot 10^{-3} cm$, so in fact it is much less than characteristic length of microwave device, which is of the order of wavelength. Thus, one can get rid of the bias, increase plasma density by order of magnitude up to $\omega_{pe} = \left(\dfrac{4\pi n_e e^2}{m}\right)^{1/2} = \omega$, so for frequency 2.4 GHz critical electron density is $6.7 \cdot 10^{10} cm^{-3}$.

Typical scheme of microwave discharge is shown in Figure 11. Here the waveguide is crossed by dielectric tube, transparent for microwave radiation. For the parameters mentioned above plasma column diameter is of the order of 1 cm. The experiments on synthesis of NO from air showed the output less than 7%, the gas flow rate and the energetic cost of NO molecule having been $5.6 \cdot 10^{21}$ particles of air per second and 24 eV respectively [31].

The Electron Cyclotron Discharge.

Electron cyclotron discharge (ECR) is the microwave discharge in a magnetic field. Microwave power falling on the plasma along magnetic lines is absorbed in the vicinity of the ECR surface, where condition $\omega = \omega_{He} = \frac{eH}{mc}$ is satisfied. ECR discharge exists at an extremely wide range of operating pressure (from 10^{-5} up to 10^2 $Torr$) and input power $10 - 10^5 W$.

ECR discharge at low pressure $10^{-6} - 10^{-4} Torr$, when electron free path length is much larger than chamber dimensions, is characterized by extremely high electron energies (up to relativistic energies) and nonmaxwellian electron distribution function. Mirror trap configuration of the magnetic field is commonly used for improving of plasma confinement [32], [33].

In the case of the collisional regime, corresponding to the chamber pressure $p > 10^{-4}$ $Torr$, electron-neutral collisions prevent the electrons from acceleration up to high energies, because this mechanism needs many regular oscillations of the electrons with the conservation of energy and without random phase shifts. Electrons are accelerated by the electric field of the electromagnetic wave and their heating is a result of the electron-neutral collisions, which come to the random phase shifts of electron with respect to the electromagnetic wave. In this regime the electron mean free path is less than system dimensions, but $\omega \gg \nu_t$. The power of the right-hand polarized electromagnetic wave, absorbed by unit plasma volume, is given by equation:

$$P = \frac{n_e e^2 E_0^2}{2m\nu_t^2} \cdot \frac{\nu_t^2}{(\omega - \omega_{He})^2 + \nu_t^2}. \tag{21}$$

In the vicinity of the resonant point, where $|\omega - \omega_{He}| < \nu_t \ll \omega$ power absorption is much larger than that of collisional microwave power absorption without magnetic field $((\nu_t/\omega)^2 \ll 1)$ and is equal to that in the case of static electric field. The microwave frequency of 2.45 GHz is commonly used; corresponding resonant magnetic field is 875 Gs. ECR discharge is characterized by low ion temperature and high plasma density for relatively low operating pressure; plasma density is much higher than that of low-frequency RF or DC discharges; plasma processing can be carried out without electrode or chamber wall interactions as potential sources of impurities; absence of large RF sheath potentials which takes place in the case of low frequency or DC discharges and brings about the surface destruction by the energetic ions [37].

This is the reason for using of this discharge for different surface treat-

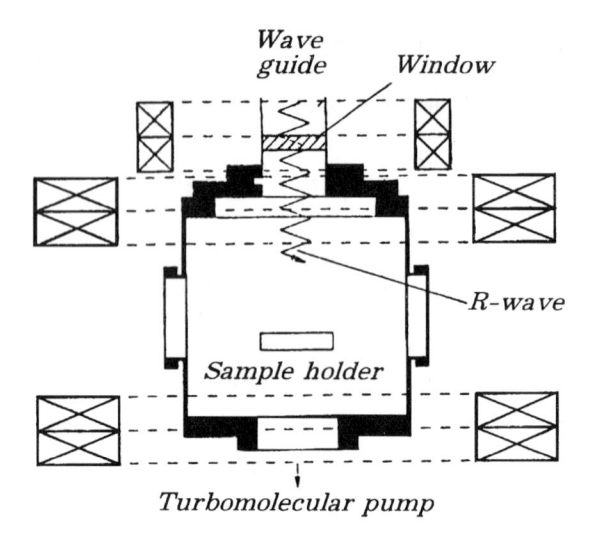

Figure 12: Technical set-up of a standard ECR reactor.

ments or micro structuring by reactive ion etching; various ion sources. The ion energies of ECR plasma at the boundary of solid are about $4-5\ T_e$ (few tens of eV) due to sheath ambipolar potential acceleration and can be increased by plasma acceleration in the divergent magnetic field [37]. Obtained high flux of low energy ions can be used for treatment of sensitive materials ($Si:H$ layers, doped Si, $GaAs$ etc.), SiO, $SiON$ coating, deposition of diamond layers [37]. High energy ion fluxes can be achieved by DC or RF-biasing of the substrate.

Standard ECR reactor consists of a cylindrical vessel inside the water-cooled magnetic field coils, as shown in Figure 12.

Usually the right hand circularly polarized electromagnetic wave is irradiated along magnetic field lines through a quartz window. In order to reduce the amount of reflected power one would have to enlarge the scale length of the magnetic field variation, especially in the resonant region [37]. The divergence of the magnetic field behind ECR zone leads to the plasma acceleration onto substrate. In this configuration ECR discharges in Ar, H_2, O_2, CH_4 have been investigated to obtain hydrogenated carbon layers as a model system for sheath deposition. Typical parameters of CH_4 discharge (at substrate position) are: microwave power is $0.2-0.5\ kW$, pressure is $5 \cdot 10^{-4} - 3 \cdot 10^{-3} Torr$, $n_e \sim (0.2-0.5) \cdot 10^{13} cm^{-3}$; ion energy is $10-50\ eV$ (according to **grad** B) and $10-200\ eV$ (with DC bias); ion flow is $(2-5) \cdot 10^{16} cm^{-2} s^{-1}$. The density can be increased by increase of power [37]. Experiments with $10-100\ kW$ power showed that $n_e = 2 \cdot 10^{13} cm^{-3}$ at $20\ kW$ and $9 \cdot 10^{13} cm^{-3}$ at $90\ kW$ with $T_e = 2\ eV$ and 80% ionization [38].

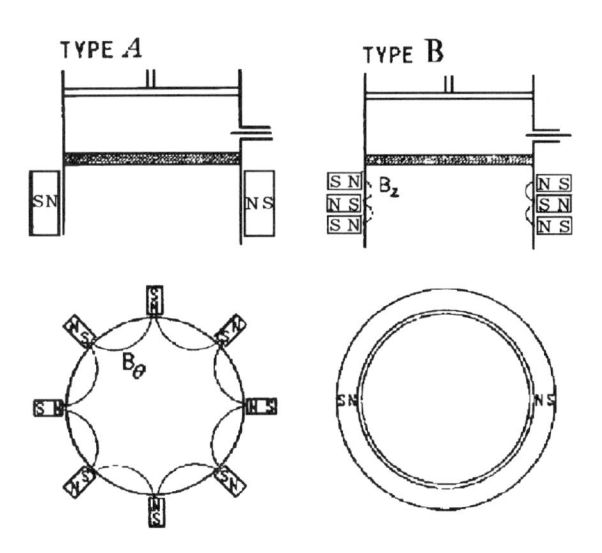

Figure 13: Schematic representation of ECR devices with permanent magnets.

Microwave ECR discharge is characterized by relatively high ion flow densities along the magnetic field gradient and large variability of the ion energy spectrum. Plasmas with the small energies $\sim 10\ eV$ obtained in the configuration with a weakly divergent magnetic field are used for deposition and etching. Medium and high ion energy plasmas can be obtained by acceleration in the divergent magnetic field and/or DC(or RF) biasing of the substrate. **grad** B acceleration occurs only for low gas pressure, because electron-neutral collisions prevent this mechanism from realization [37].

In large-scale plasma processing it is necessary to obtain unmagnetized large-diameter plasmas providing a sufficiently uniform flow in front of substrates. Usually a high magnetic field is generated by solenoidal coils. In large scale uniform processing coil diameter should be large. This leads to the increase in weight and volume of the devices and electric power supplied to the coil. Also, a strong magnetic field exists in front of the substrates even if they are placed far from the coils.

The problems mentioned above can be solved by using permanent magnets, which provide a sufficiently strong magnetic field in the narrow areas in the vicinity of their location. There are various kinds of such ECR plasma sources depending on the magnet arrangement and microwave coupling. Two of them are shown in Figure 13.

A classification of these devices and plasma parameters obtained is given in [39]. Plasma confinement near ECR region is due to the crossed static magnetic and electric fields of the sheath. Charged particles drift in the

$\vec{E} \times \vec{B}$ direction. So, drift direction depends on the magnets location. The azimuthal drift is more preferential because closed drift orbits could be easily established in this case. Typical plasma parameters, obtained in such devices, are [39]: electron density n_e is $(1.5 - 2) \cdot 10^{11} cm^{-3}$; electron temperature T_e is $(2-5)$ eV; gas pressure is $(0.5-2)$ $Torr$; electromagnetic power is $100-1500$ W, plasma uniformity is $1-12\%$.

High pressure ECR discharge is used for plasma-chemistry applications. Experiments for ECR discharge ignition for the pressures up to atmospheric have been performed for $N_2 - O_2$ mixture [40]. Plasma densities $n_e \sim 3 \cdot 10^{12} cm^{-3}$ in continuous regime (microwave power was $30W$) and $8 \cdot 10^{13} cm^{-3}$ in pulsed one (power was 30 kW) with electron temperature $(1-5)$ eV have been obtained in the discharge at the gas pressure 50 $Torr$. Absorption coefficient of the input power was $70-90\%$ ($3-5\%$ without resonance) and discharge has been observed at the pressures up to atmospheric ones.

The Beam Plasma Discharge

Beam plasma discharge (BPD) was first predicted in papers [41], where instability of collisionless plasma penetrated by electron beam was described. Since that time thorough investigation of the instability was made both theoretically and experimentally [42]. It was shown, that growth rate of instability γ was given by

$$\gamma = \frac{\pi}{2} \cdot \omega_{pe} \cdot \frac{n_b}{n_e} \cdot \frac{\omega_{pe}^2}{|k| \cdot k} \cdot \left(\frac{1}{n_b} \frac{\partial f_b}{\partial v} \right)_{v = \omega_{pe}/k}.$$

Here f_b is distribution function for beam electrons, $n_b \ll n_e$ is electron beam density, the instable wave having frequency ω equal to plasma frequency ω_{pe}, the wave vector being parallel to the velocity of electron beam, wave number $k = \omega/v_b$, where v_b is the velocity of beam electrons. One can see the strong dependence of growth rate on derivative of distribution function of beam electrons over velocity, viz., when the derivative is zero, stabilization could occur. This circumstance brought about the appearance of quasi linear theory, which took into account influence of Langmuir plasma waves of finite amplitude on distribution function of beam electrons and described slow evolution of Langmuir waves and distribution function self-consistently [43], [44]. For the simplest one dimension case equations are as follows:

$$\frac{\partial f_b}{\partial t} = \frac{\partial}{\partial v} \left(\frac{4\pi e^2}{m^2} \pi \frac{W_k}{|v|} \frac{\partial f_b}{\partial v} \right); \qquad (22)$$

$$\frac{\partial W_k}{\partial t} = W_k \pi \omega_{pe} v^2 \frac{1}{n_e} \frac{\partial f_b}{\partial v}. \qquad (23)$$

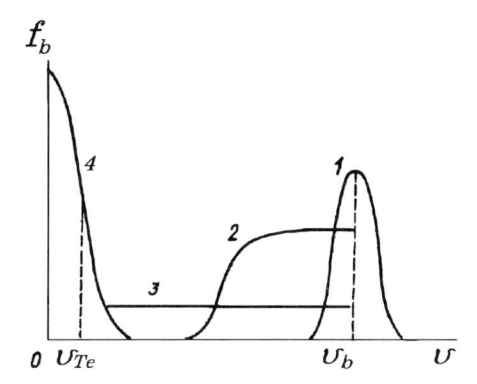

Figure 14: Scheme of electron beam relaxation in plasma. 1 is initial distribution function of electron beam; 2 and 3 represent the stages of beam evolution; 4 is plasma electron distribution function.

The first equation describes evolution of distribution function, the second corresponds to the growth rate in kinetics with f_b dependent on time. One can see that self stabilization occurs when $\dfrac{\partial f_b}{\partial v} = 0$, so final state corresponds to plateau for distribution function, energy of Langmuir waves W_k can be determined from the laws of conservation (Fig. 14).

From the view-point of collisionless plasma Langmuir waves can not transfer their energy to main body plasma electrons, the situation changing when collisions of electrons are taken into account. It is necessary, of course, for instability to take place, viz., $\gamma \geq \nu_t$, that mean free path of beam electrons be larger, than the length of system, i. e., beam energy should be high enough [45]. Special attention was drawn to condition of ignition of BPD discharge, which occurs when secondary electron density due to ionization of neutral gas by electrons of the beam exceeds the beam density, the diffusion losses of plasma with magnetized electrons being taken into account. Magnetic field was introduced for transportation of electron beam into plasma volume from electron gun, as well as for decrease of diffusion across magnetic field, because for magnetized electrons characteristic length of ambipolar diffusion Λ_D is of the order of length of the device along magnetic field.

Now we can consider heating of main body plasma electrons in the same way as for the case of microwave discharge (see (19), (20)), where $\omega = \omega_{pe} \gg \nu_t$, and for Langmuir waves

$$W = \frac{E_0^2}{4\pi} = \int W_k dk = \int \frac{mv^2}{2}\left(f_b(v, t=0) - f_b(v, t=\infty)\right) dv.$$

Experiments show, that more than half of beam energy comes into en-

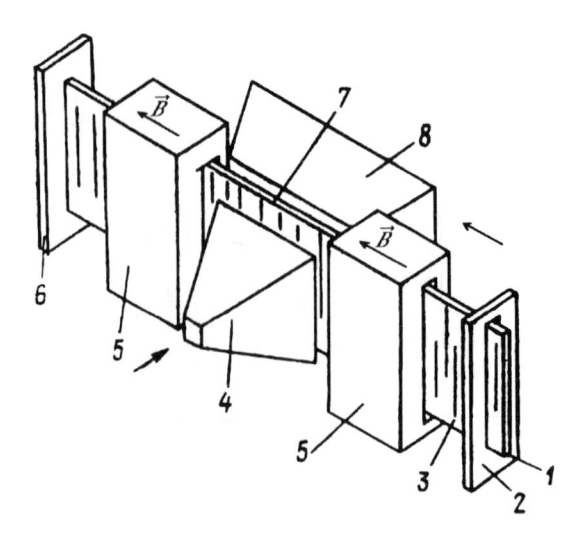

Figure 15: Set-up of experimental device with ribbon beam configuration. 1: cathode; 2: anode; 3: electron beam; 4: gas supply; 5: magnetic coils; 6: electron beam collector; 7: plasma discharge area; 8: product collector.

ergy of the waves, relaxation length being of the order of several tens of centimeters [46], [47], [48]. Since degree of ionization could be higher than that in other types of discharges, one should add electron-electron collisional term [45]. Typical experimental device is shown in Figure 15. Ribbon beam configuration was used to increase cross section of discharge.

Beam penetrates into chamber by orifices and a system of differential pumping provides low pressure in electron gun, magnetic field has a mirror configuration with the mirror ratio 2 and intensity 1 kGs in the center of device. The key objective was to check BPD for various chemical reactions. For NO synthesis from air at relatively low gas pressure $\sim 10^{-1} Torr$, beam current ~ 1 A, beam energy $1-5$ keV, the output is shown in Figure 16 for various gas flow rates.

Another class of reactions, dealing with etching, deposition and micro mechanics was thoroughly studied [48]. Various configurations of the beam were used for technology. BPD plasma of sheet-like configuration has been studied and used for wafer processing [49]. In this system BPD with RF source of small power was used for etching and deposition processes. High effectiveness of the discharge for surface treatment have been proved. High etching rates were obtained for both isotropic (up to 10 $\mu m/$ min) and anisotropic (0.6 $\mu m/$ min) etching of silicon. Other materials such as SiO_2, Si_3N_4, $GaAs$, SiC, refractory metals were etched, $\alpha - Si : H$, SiO_2, Si_3N_4, diamond-like films have been deposited successfully as well. All these quite different pro-

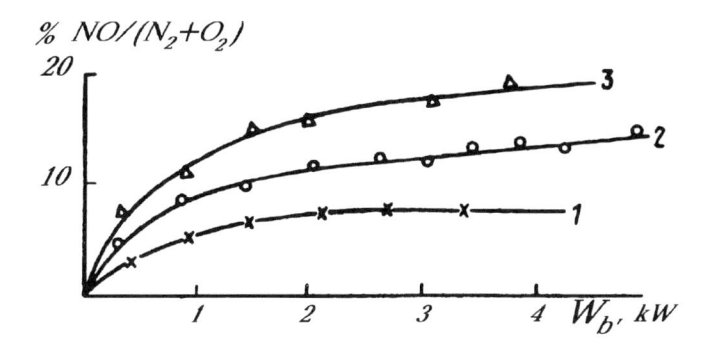

Figure 16: Yield of NO vs beam power. 1: gas pressure $p = 2\cdot 10^{-1}$ $Torr$, gas flow rate G=1500 l/s; 2: $p = 1\cdot 10^{-1} Torr$, G=1500 l/s; 3: $p = 5\cdot 10^{-2}$ $Torr$, G=120 l/s.

cesses could be realized in the system due to high values of BPD parameters ($n_e = 10^{13} cm^{-3}$, $T_e \sim 3-5$ eV) being easily controlled.

The main disadvantages of BPD are expensive gas transport at low pressure and problems of creating of electron gun of high power, working in magnetic field. The first problem is of less importance for expensive processes, e. g., for reduction of rare earth oxides and fluorides by hydrogen when high purity is of interest, etching and deposition for microelectronics. The second problem was recently solved, electron gun with power 100 kW was built and applied for beam plasma discharge used as the microwave amplifier [50].

It should be noted that BPD is studied now for glow and vacuum arc discharges, where an electron beam is formed in near cathode region, making an appreciable contribution in ionization of working gas. Study of greed key elements for thermoionic transformer showed the electron beam with energy of several tens eV to be necessary to explain high plasma density, observed in experiments [51].

Discharges in Crossed Electric and Magnetic Fields

There is a variety of discharges in crossed constant electric and magnetic fields, now used for various purposes, such as magnetron discharge, plasma thrusters, rotating plasma discharge or plasma centrifuge. Though there is a difference in configurations and obtained plasma parameters, we shall try to show their likeness so description can be done by unique approach.

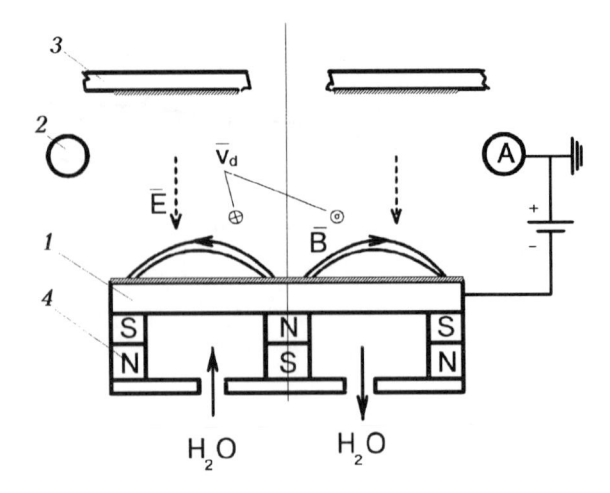

Figure 17: Scheme of magnetron ion sputtering system. 1: cathode with a sputtered target; 2: ring anode; 3: substrate; 4: magnets. v_d is the direction of electron drift.

The Magnetron Discharge

To begin with let's start from so called magnetron discharge (Fig. 17). The cylindrical geometry with radius of the order of 100 cm and height of the order of several centimeters with the magnetic field directed along radius, i. e., parallel to the surfaces treated, voltage being applied to the surfaces. Thus one has $\vec{E} \times \vec{H}$ azimuthal drift of magnetized electrons with velocity $v_d = \frac{c}{H^2} \left(\vec{E} \times \vec{H} \right)$. For typical parameters $E \sim 10^3 \, V \cdot cm^{-1}$, $H \sim 100 \, Gs$, drift velocity being $v_d \sim 6 \cdot 10^8 cm \cdot s^{-1}$. This velocity corresponds to 100 eV electron energy, exceeding several times energy of ionization of working gas Ar, ion directed energy being of the order of 1 keV. Gas pressure is about $10^{-3} Torr$, degree of ionization Z_i is rather high and can be varied from 10^{-2} up to 10^{-1}. Ions are non magnetized, so the current is provided mainly by the ions, ion current density being of the order of $0.1 - 0.2 \, A \cdot cm^{-2}$. It permits to provide rate of deposition several times larger than that of cathode sputtering [52]. Though many technological and experimental devices are broadly used now, self-consistent description of the discharge has not yet been presented.

The Plasma Thruster

Another type of discharge is a well known plasma thruster, which in principle is similar to that of described above [53].

The difference is in geometry of device (Fig. 18) and parameters of plasma

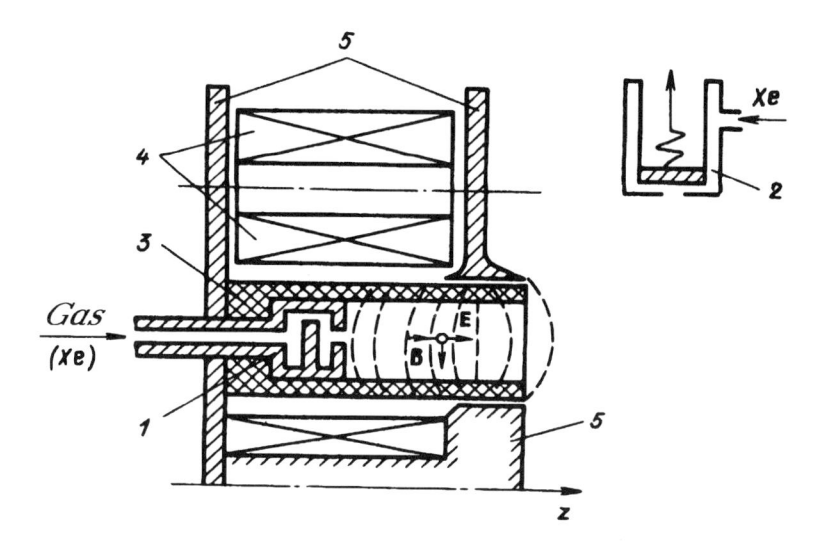

Figure 18: Technical set-up of plasma thruster device. 1: anode; 2: cathode; 3: discharge chamber; 4: magnetic system; 5: magnetic circuit.

obtained. Width of channel along magnetic field is now less than that of along electric field, plasma density is ten times less ($10^{10} - 10^{11}$ cm^{-3}), electric field $E \sim 10^2\,V \cdot cm^{-1}$. Gas pressure is of the same order and due to the smallness of the channel width compared to mean free path of electrons, transport of electrons across magnetic field cannot be provided by electron-neutral collisions. A very interesting mechanism of surface conductivity was proposed [54], in accordance with a theory presented a part of electrons, colliding with the surface of channel, provides electric current. The current appears to be distributed across the channel with characteristic length equal to Larmor radii of these electrons, total current being $\sim 2\pi \cdot R \cdot \rho_e \cdot v_d \cdot n_e \cdot e$. Here R is radius of the channel (width of the channel is much less than its radius), ρ_e is Larmor radius of electron at a temperature $T_e \sim 10\ eV$, determined by the energy acquired by electrons passing by near wall sheath. Attempts have been made to use discharge described for surface treatment and deposition.

The Plasma Centrifuge

Behavior of plasma in crossed electric and magnetic fields in the case when ions are magnetized, was studied during a long period of time from view-point of creation of optimal configuration for thermonuclear synthesis. In 1971 Lehnert proposed to use rotating plasma for isotope separation (Fig. 19) [55]. Later the rotating plasma was proposed for separation of final

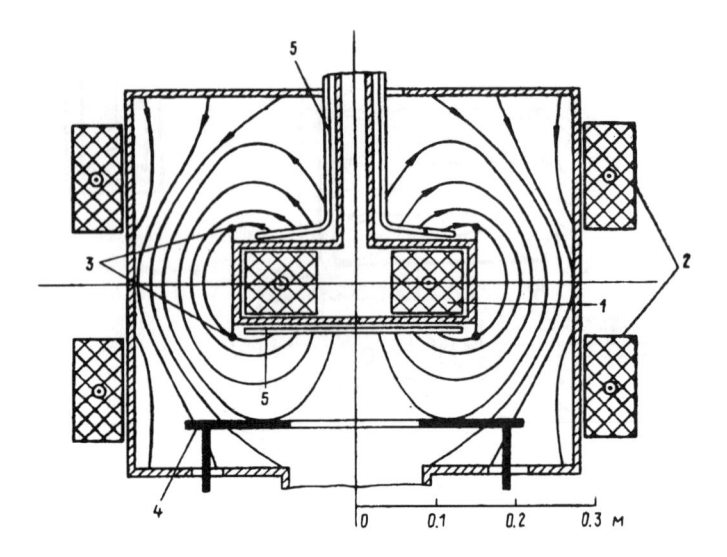

Figure 19: Discharge chamber of FI installation. 1: main coil; 2: auxiliary coils; 3: ring anode; 4: cathode; 5: insulators.

products of plasma chemical reactions with objective to alleviate the problem of quenching [45].

Mechanism of isotope separation in fully ionized plasma of high density up to $10^{15} cm^{-3}$, due to ion-ion collisions was investigated [56], [45], [48] and verified in pulsed devices [57], [58]. Maximum magnetic field was 10 kGs, initial gas pressure density was $10^{15} cm^{-3}$, since only at this density it was possible to ignite discharge, which was sustained during $10^{-3} s$. Making use of natural mixture of ^{20}Ne and ^{22}Ne isotopes the following results were obtained: $v_d = 1.3 \cdot 10^6 cm/s$, T_e was several eV, maximum enrichment coefficient was 1.15 during $10^{-2} s$, the enrichment coefficient having been introduced in usual way as ratio $\left(\dfrac{n_h}{n_l}\right)|_R \div \left(\dfrac{n_h}{n_l}\right)|_{r_0}$. Here R and r_0 are the maximum and minimal radii of the system respectively, h and l correspond to heavy and light ions. Since the interest for technology is a stationary system, attempts were made to create stable arc discharge with high degree of ionization [55], [59]. To improve situation it was proposed to use BPD in crossed fields configuration, the electron beam having been directed along axis of the system (Fig. 20) [60].

The discharge appeared to be stable at pressure $10^{-4} - 10^{-3} Torr$. Magnetic field in the chamber was 6 kGs, beam electron energy and current were 10 keV and 1.5 A respectively, $v_d = 5 \cdot 10^5 cm/s$, ion temperature was about 5 eV, radial current was about 6 A, enrichment was 1.3, plasma density $n_e \simeq 10^{13} cm^{-3}$ at periphery of the chamber and several times less

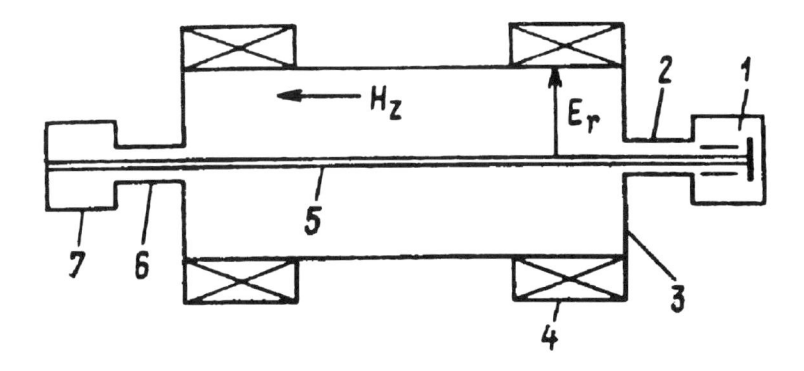

Figure 20: Set-up of beam plasma discharge device with crossed fields configuration. 1: electron beam gun; 2: pumping; 3: discharge chamber; 4: magnetic coils; 5: electron beam; 7: electron beam collector.

in the center. High ion temperature and relatively low density of plasma were incentives to look for another mechanism of isotope separation, viz., the mechanism of surface ion currents [61]. Neutrals coming into plasma from ends of the system were ionized on the ionization length

$$l_{ion} \sim \frac{u}{\tau_{ion}} \sim \frac{1}{n_e \sigma_{ion}} \cdot \frac{u}{v_{Te}}$$

Here u is the velocity of neutral, σ_{ion} is the cross section of ionization. Plasma ions having velocity $C_s = \sqrt{T_e/M}$ reach the butts of device, where recombination takes place, the neutrals returning to plasma being ionized during the time of ionization τ_{ion}. Later on the ion displaces in radial direction on a distance $\sim v_{d/\omega_{Hi}}$ during the time $\sim \omega_{Hi}^{-1}$. Hence the ion passes the distance $v_{d/\omega_{Hi}}$ during the total time $\tau_{ion} + \omega_{Hi}^{-1} \approx \tau_{ion}$. It brings about polarization drift of ions in near end regions with radial velocity $v_r \simeq v_d \cdot (\omega_{Hi}\tau_{ion})^{-1}$, so that the velocity is proportional to ion mass. More details can be found in [48], [62].

CONCLUSION

At present time results and experience obtained in study of controlled thermonuclear fusion are to be analyzed with objective to use them for plasma chemistry and plasma technology. Development of plasma chemistry brought about the necessity of understanding the problems in systems far from equilibrium, since using equilibrium or quasi equilibrium systems appears to be not effective for many processes. Quite a number of processes are bound to be realized by means of equilibrium and quasi equilibrium discharges such as arc, radio frequency and microwave ones, especially in cases, when plasma is used as an efficient heater, the problem being considered by methods of high temperature chemistry. Electrons with density found from Saha formula provide the heating up to the temperatures lower than 1 eV due to Joule dissipation. Concepts of kinetics of chemical reactions permit to predict the existence of desired products at a given temperature.

However, when high population of vibration levels is necessary for chemical reaction, effectiveness of systems mentioned above is poor. Glow discharge with electron and vibration temperatures of several eV turned out to be a system far from equilibrium and was used for etching and deposition processes, direct reduction of metals from oxides and chlorides. An essential disadvantage of glow discharge is low pressure and small surface of active zone and consequently small gas flow rate, preventing from large productivity. Moreover, low electron density demands an increase of residence time of molecule in active zone in order to be either excited or dissociated. Meanwhile the gas flow rate in arc devices reach 10^{23} particles per second, in glow discharge it is less by three-four orders of magnitude. To estimate optimum electron density one can make use of a simple following consideration. Characteristic time of inelastic collision τ_u, causing excitation of vibration levels is $\tau_u \sim (n_e \sigma_u v_{Te})^{-1}$, where σ_u is the cross section of excitation. This time determines the time of residence in active zone of discharge, the characteristic size of active zone, i.e, the thickness of plasma slab, being estimated as $d \sim u\tau \sim 1/n_e \sigma_u \cdot u/v_{Te}$, where u is the translation velocity of molecule. For usual parameters of glow discharge $n_e \leq 10^{11} cm^{-3}$, $u \sim 10^5 cm/s$, $v_{Te} \sim 10^8 cm/s$, $\sigma_u \leq 10^{-16} cm^2$ value d is about $10^2 cm$, while radius of discharge is several centimeters. It means that molecule collides with a wall of the chamber sooner than it is excited high enough. The conditions for discharge to be optimum could apparently be formulated as follows:

a) electron density n_e and degree of ionization Z_i should be high enough, viz., d should be less than characteristic length of chamber;

b) electron temperature should be high enough to provide vibration excitation. The temperature, however, should not exceed some value to prevent from full ionization;

c) density of molecules n_M cannot be too high to prevent from relaxation of vibration energy into rotation and translation ones;

d) gas flow rate Q should be of the order of 10^{24} particles per second, at any rate, for large scale chemistry at relatively low pressure, so cross section of discharge should be rather large, since $Q = n_M \cdot u \cdot S$;

e) though quenching problem is easier for low pressure systems, the problem demands a special consideration.

Analyzing from this point of view the trends in discharge study, one can conclude that more and more elements well known in the physics of high temperature plasma appear in technological devices, viz., using of magnetic field, implementation of electron cyclotron resonance, using of plasma instabilities to heat electrons, etc. Technology calls for increasing of plasma density and electron temperature, so discharges sometimes appear first in application without profound theoretical study. Investigation on thermonuclear fusion is of great interest with this respect because thorough study of highly noneqilibrium plasma both theoretically and experimentally was done. Apparently self-consistent description, including approaches of high temperature plasma along with kinetics of internal degrees of freedom of atoms and molecules is of current interest.

ACKNOWLEDGMENTS

The authors would like to thank A. V. Baitin, L. I. Elizarov, V. Yu. Fedotov, M. D. Karetnikov, K. M. Kuzanyan, Yu. F. Nasedkin, and A. A. Serov for helpful discussions.

REFERENCES

1. Wray, K. L., *Air. Research Report* **104**, 1-15 (1961).

2. Pfender, E., *Mater. Res. Soc. Proc.* **30**, 13-36, (1983).

3. von Engel, A. and Steenbeck, M., *Electrischegasenentladungen. Ihre Physik und Technik*, 2 Bande, Berlin, 1932/1934, Ann Arbor, 1944.

4. Raizer,Yu.P., *Physics of gas discharge* [in Russian], Moscow: Nauka, 1987, ch. 10, pp. 443-446.

5. Polak, L.S. et al., *Theoretical and applied plasma chemistry* [in Russian], Moscow: Nauka, 1975, ch. 4, pp. 199-205.

6. Gauvin, W. H., and Choi, H. K., *Mater. Res. Soc. Proc.* **30**, 101-116, (1983).

7. Townsend, J. S., and Donaldson, R. H., *Philos. Mag.* **2**, 674-682 (1926).

8. Babat, G. I., *Vestnik Electrop.* [in Russian] **3**, 47-51 (1942).

9. Reed, T. B., *J. Appl. Phys.* **32**, 821- 833, (1961).

10. Rovinskyi, R.E.,Grouzdev, V.A., Sobolev, A.P., *Zh. Prikl. Mat. Tekh. Fiz.* **1**, 143-150 (1967).

11. Egorova, C. A., *Zh. Prikl. Spectr.* **6**, 22-26, (1967).

12. Berdichevskyi, M. G., and Marousin, V. V. *Izv. SOAN SSSR, ser. Techn. Fiz.* **8**, 72-79 (1979).

13. Smirnov, B. M., *Physics of Weakly Ionized Gas* [in Russian], Moscow: Nauka, 1972, ch. 9, pp. 345-352.

14. Biberman, L. M., *Zh. Eksp. Teor. Fiz.* **17**, 416-425 (1947); **19**, 584-595 (1949).

15. Levitskyi, A. A. et al., *Khim. Vys. En.* **17**, 625-632 (1983).

16. Kondratiev, V. N., *Rate Constants of the Reactions in Gaseous Phase* [in Russian], Moscow: Nauka, 1971, ch. 5, pp. 251-255.

17. von Engel, A., *Ionized Gases*, Oxford: 1955.

18. Ivanov, A. A., and Soboleva, T. K., *Noneqilibrium Plasma Chemistry* [in Russian], Moscow: Atomizdat, 1978, ch. 2, pp. 119-124.

19. Smirnov, B. M., *Physics of Weakly Ionized Gas* [in Russian], Moscow: Nauka, 1972, ch. 6, pp. 261-263.

20. Schulz, G. J., *Phys. Rev.* **A135**, 988-995 (1964).

21. Rusanov, V. D., and Fridman, A. A., *Physics of Chemically Active Plasma* [in Russian], Moscow: Nauka, 1984, ch. 3, pp. 89-99.

22. Rich, J.W., *J. Appl. Phys.* **42**, 2719-2726 (1971).

23. Polak, L.S. et al., *Theoretical and applied plasma chemistry* [in Russian], Moscow: Nauka, 1975, ch. 2, pp. 59-66.

24. Treanor, C. E., Rich, I. W., Rehm, R. G., *J. Chem. Phys.* **48**, 1798-1805 (1968).

25. Landau, L. D., Teller, E., *Phys. Z. Sowjetunion* **10**, 34-45 (1936).

26. Popov, A. M., Rakhimov, A. T., Rakhimova, T. V., *Fiz. Plasmy* **19**, 1241-1267 (1993).

27. Godiak, V. A., Kouzovnikov, A. A., *Fiz. Plasmy* **1**, 496-505 (1975).

28. van Voorst Vader, P. J. Q., "Magnetic Enhanced RIE of Si and SiO_2 in CF_4," presented at the *ISPC-10 Symposium on Plasma Chemistry*, Bochum, Germany, August 4-9, 1991.

29. Garscadden, A. et al., *Plasma Sources Sci. Technol.* **3**, 239-245 (1994).

30. Mieno,T., Shoji, T., Kadota, K., "Generation of High Density Plasma and Film Deposition by Using an RF Whistler Wave Discharge," presented at the *ISPC-10 Symposium on Plasma Chemistry*, Bochum, Germany, August 4-9, 1991.

31. Czernichowski, A. et al., "On Experimental Study of Synthesis of Nitric Oxides in Microwave Plasma," presented at the *ISPC-3 Symposium on Plasma Chemistry*, Limoges, July 13-19, 1977.

32. Ivanov, A. A., Spektor, M. D., Frank-Kamenetskyi, D. A.,*Pisma Zh. Eksp. Teor. Fiz.* **11**, 136-138 (1970).

33. Timofeev, A. V., *Voprosy Teor. Plasmy*, vol. **14** [in Russian], Moscow: Atomizdat, 1986, pp. 56-66.

34. Zhiltsov, V. A. et al., *Fiz. Plasmy* **17**, 771-784 (1991).

35. Geller, R., Jackquot, B., Pauthenet, R., *Rev. Phys. Appl.* **15**, 995-1001 (1980).

36. Golovanivsky, K. S., *Plasma Sources Sci. Technol.* **2**, 240-250 (1993).

37. Wilhelm, R., in *Microwave Discharges: Fundamentals and Applications*, New York: Plenum Press, 1993, pp. 161-179.

38. Petty, C. C., Smith, D. K., *Rev. Sci. Instrum.* **57**, 2409-2414 (1986).

39. Iizuca, S., Sato, M., *Jpn. J. Appl. Phys.* **33** , 4221-4225 (1994).

40. Rusanov, V. D., and Fridman, A. A., *Physics of Chemically Active Plasma* [in Russian], Moscow: Nauka, 1984, ch. **6**, pp. 224-227.

41. Akhiezer, A. I., Fainberg, Ya. B., *Zh. Eksp. Teor. Fiz.* **21**, 1262-1269 (1951).

42. Vedenov, A. A., Riutov, D. D., *Voprosy Teor. Plasmy* **6** [in Russian], Moscow: Atomizdat, 1972, pp. 3-75.

43. Vedenov, A. A., Velikhov, E. P., Sagdeev, R. Z., *Nuclear Fusion. Supplement*, p. 2, 464-476 (1962).

44. Drummond, W. E., Pines, D., *Nuclear Fusion. Supplement*, p. 3, 1049-1057 (1962).

45. Ivanov, A. A., *Fiz. Plasmy* **1**, 147-159 (1975).

46. Zakatov, L. P., Plachov, A. G., *Zh. Eksp. Teor. Fiz.* **60**, 588-593 (1971).

47. Levitsky, S. M., Shashurin, I. P., *Zh. Eksp. Teor. Fiz.* **52**, 350-356 (1967).

48. Ivanov, A. A., *Itoghi Nauki Tekhn.* **3**, Moscow: VINITI, 1982, pp. 176-238.

49. Nasedkin, Yu. F.,Serov, A.A. et al., "Study of Efficiency of the New Beam Plasma Discharge for Plasma Chemistry Applications," presented at the *ESCAMPIG-8*, Greifswald, GDR, August 26-29, 1986.

50. Fainberg, Ya. B., *Fiz. Plasmy* **20**, 613-619 (1994).

51. Baksht, F. G., Kostin, A. A., *Fiz. Plasmy* **9**, 628-636 (1983).

52. Marakhtanov, M. K., *Magnetron System of Ion Sputtering* [in Russian], Moscow: MGTU Press, 1990, ch. 2, pp. 30-35.

53. Morozov, A. I., and Shubin, A. P., *Itoghi Nauki Tekhn.* **5**, Moscow: VINITI, 1984, pp. 178-260.

54. Morozov, A. I., *Zh. Prikl. Mat. Tekh. Fiz.* **3**, 19-22 (1968).

55. Lehnert, B., *Nuclear Fusion* **11**, 485-533 (1971).

56. Bonnevier, B., *Plasma Physics* **13**, 763-774 (1971).

57. James, B. W., Simpson, S. W., *Phys. Lett.* **46A**, 347-348 (1974).

58. Belousov, A. V., Karchevsky, A. I. et al., *Fiz. Plasmy* **5**, 1239-1259 (1979).

59. Nathrath, N., "The plasma centrifuge III: measurements of rotating uranium plasmas," presented at the *XIII-ICPIG Conference*, Berlin, GDR, 1977.

60. Babaritsky, A. I., Zhuzhunashvili, A. I., Ivanov, A. A. et al., *Fiz. Plasmy* **4**, 840-850 (1978).

61. Ivanov, A. A., Leiman, V. G., *Fiz. Plasmy* **4**, 668-673 (1978).

62. Ivanov, A. A., Timchenko, N. N., *Fiz. Plasmy* **16**, 1491-1497 (1990).

Fundamental Electron Collision Processes Relevant to Low-Temperature Plasmas

Kurt H. Becker

Physics Department, City College of C.U.N.Y.
Convent Avenue and 138th Street, New York, NY 10031-9198 U.S.A.

Abstract. This paper attempts to elucidate the mutually beneficial interaction between electron collision physics and the physics and chemistry that govern the properties of low temperature plasmas, in particular technologically relevant low-temperature processing plasmas. We hope to demonstrate how recent developments in collision physics contributed to an improved understanding of the fundamental collision processes in low-temperature plasmas, how they made possible more sophisticated modelling efforts of such plasmas and how they advanced the development of more sensitive plasma diagnostics techniques. At the same time, we hope to show how some of the many unanswered questions and challenges faced by the plasma processing community have stimulated new developments and novel approaches in atomic and electron collision physics.

I. INTRODUCTION AND BACKGROUND

Low-temperature processing plasmas have gained prominence through their widespread use in many key technologies such as the plasma-assisted etching of microstructures, the deposition of thin films and in a variety of surface modification applications. The development of low-temperature plasma technology has been achieved largely empirically through a trial-and-error approach, while the understanding of the relevant plasma processes in terms of the fundamental interactions in the plasma have been lagging behind. Ever more complex processes and plasma reactors, ever higher demands on the processed materials in terms of shrinking feature size, increased selectivity and uniformity, and the need for faster throughput and lower rejection rates render it highly desirable to improve our understanding of the basic plasma processes as well as our plasma diagnostics capabilities. These goals can only be achieved, if we drastically improve the

© 1996 American Institute of Physics

existing data base on electronic and atomic collision processes relevant to the physics and chemistry that govern the properties of these plasmas.

Glow discharge plasma processing utilizes plasmas which are characterized by mean electron energies ranging from 0.5 eV to 5 eV and charge densities of 10^8 cm^{-3} to 10^{12} cm^{-3}. Collision processes of interest in such plasmas include a multitude of processes involving electrons, ions, and photons as projectiles and atoms, molecules, free radicals and excited species (metastables, electronically and vibrationally excited species) as targets. The list of specific targets of interest is comprehensive and ranges from atomic hydrogen, oxygen and nitrogen to very polyatomic molecules such as Si-organic and metal-organic compounds. Many species in between these two extreme categories are relevant in a variety of diverse plasma-assisted processes: atoms such as Al, Si and the halogens, simple diatomics such as H_2, Cl_2, HBr, polyatomics such as all partially and fully halogenated methane compounds, BCl_3, SF_6 and SiH_4, free radicals such as CF_x, SiH_x (x=1-3) to name just a few. The rest of this paper will concentrate mainly on electron collision processes and their relevance to low-temperature processing plasmas.

The relevance of electron collision processes in low-temperature plasmas is largely determined by the overlap of the electron energy distribution function (EEDF) with the respective electron-impact cross section. This overlap, in turn, is determined by two factors, by the threshold and shape of the respective electron-impact cross section and by the shape of the EEDF. Electron-impact dissociation and ionization processes have threshold energies of typically 5 eV to 15 eV, which means that the effectiveness of these processes for the plasma properties is crucially dependent on the shape of the high-energy tail of the EEDF. The high-energy tail of the EEDF, in turn, is determined to a large extent by low-energy electron collisions such as vibrational excitation of the feedgas molecules and dissociative attachment which are among the most effective energy-loss mechanisms for the plasma electrons. In cases where the cross sections for vibrational excitation and/or dissociative attachment are small, these energy loss channels are weak and, as a consequence, there is a pronounced high-energy tail of the EEDF. This usually results in a situation where single electron dissociation and ionization processes are important. Such a situation is schematically shown in the upper diagram of fig. 1. Molecules with large vibrational excitation cross sections and/or large dissociative attachment cross sections, on the other hand, provide efficient energy-loss mechanisms for the plasma electrons and, as a consequence, the high energy tail of the EEDF is significantly reduced in such cases. Consequently, single electron dissociation and ionization are less significant fundamental collision processes in the plasma under these circumstances. This situation is schematically shown in the lower diagram of figure 1. We will restrict the discussion in the rest of this paper to electron-impact ionization and dissociation processes as the two most important "high-energy" electron collision processes relevant to low-temperature plasmas and to collisions with targets in excited states which is another important, albeit often neglected, "low-energy" loss mechanism for the plasma electrons.

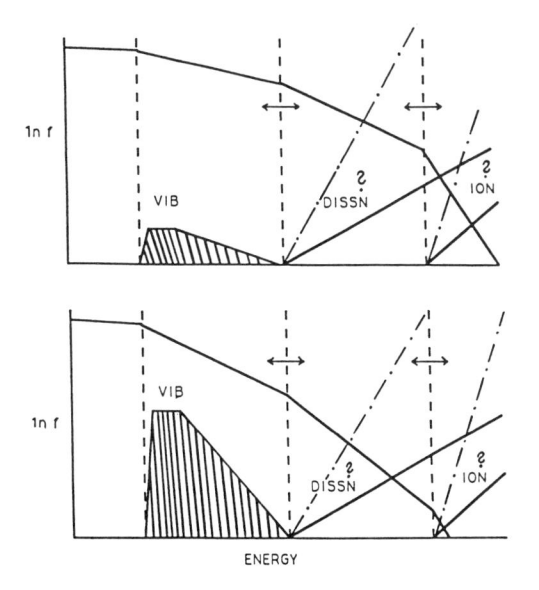

Fig. 1: A schematic diagram illustrating the overlap between the EEDF f (plotted on a log-scale vs. the electron energy) and various inelastic electron-impact cross sections in the low-energy regime.

II. THE STATUS OF ELECTRON-IMPACT IONIZATION PROCESSES

A recent topical review on the subject of electron-impact ionization of atoms, molecules, ions, and transient species discussed several aspects of the experimental and theoretical status of electron-impact ionization processes from the viewpoint of their application to low-temperature plasmas [1]. We refer the reader to this review and to references therein to earlier general reviews and to specialized research papers for further detailed information on this subject. We will only address a few selected aspects of electron-impact ionization processes in this paper which are of special importance to selected low-temperature processing plasmas.

Recently, in the context of plasam-assisted remediation of SO_2, modelling efforts and diagnostics of plasma processes and reactors for environmental clean-up demonstrated an urgent need for reliable electron-impact ionization data for the SO_2 molecule as well as for the SO radical which is an abundant constituent in SO_2-containing discharges. A database search found no collision data on the SO free radical. The two most recent measurements of absolute partial ionization cross sections for the SO_2 molecule revealed differences in the reported cross sections of as much as a factor of 4. As part of an ongoing research collaboration between our group at City College and a group in Greifswald, Germany, comprehensive studies of the electron-impact ionization of SO_2 were carried out independently by the two groups using two different experimental techniques [2], a fast-beam apparatus at

CCNY and a high-resolution mass spectrometer at Greifswald. Excellent agreement at the 10-15% level between the two independent measurement was obtained for the parent SO_2^+ ionization cross section as well as for all dissociative

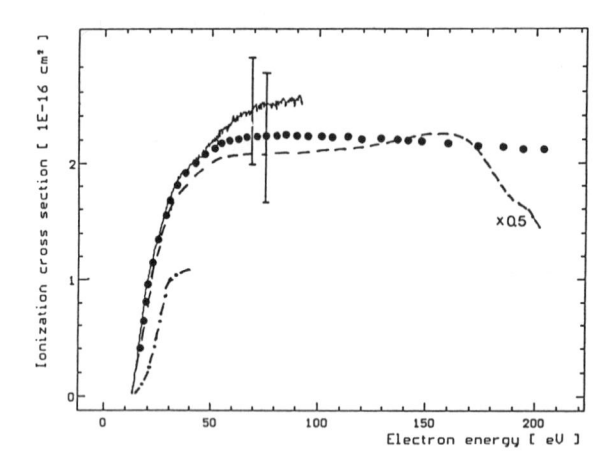

Fig. 2: Absolute SO_2^+ parent ionization cross section as a function of electron energy. (See text for further details.)

ionization cross sections. Our results for SO_2^+ are shown in fig. 1 along with the earlier measurements by Orient and Srivastava [3] and Smith and Stefenson [4]. While there is excellent agreement between the fast-beam data obtained by our group (full circles) and the mass spectrometer data measured at Greifswald (solid trace), both measured cross sections are more than a factor of 2 larger than the earlier data of Smith and Stefenson (dash-dot line) in the energy range from threshold to 40 eV (which was the highest electron energy used by these authors). At the same time, our data are about a factor of 2 smaller than the cross section reported by Orient and Srivastava (dashed line - note that the data of Orient and Srivastava in fig. 2 were multiplied by 0.5 for clarity of presentation) and there are serious differences in the energy dependence. While there was much better agreement for the partial ionization cross section SO^+, S^+ and O^+ between our measurements and the data of Orient and Srivastava, the large discrepancy in the absolute value and in the shape of SO_2^+ cross section leads to a marked difference in the total (single) SO_2 ionization cross section as shown in fig. 3. The total (single ionization cross section, in turn, is a crucial quantity that influences the ionization balance in SO_2-containing discharges. Support for the cross sections measured independently by our group and by the Greifswald group comes from a calculated cross section based on a recently modified additivity rule [5] (see figure 3).

There are two important reasons why ionization processes involving free radicals are as important as the electron-impact ionization of the stable parent molecules. Free radicals are abundant constituents in technological plasmas and are, in fact, the reactive species which make plasma-assisted processing a viable and widely used technology. In addition, threshold ionization mass spectroscopy

(TIMS) has evolved as a powerful plasma diagnostics technique [6,7] which enables the absolute determination of radical species concentrations in plasmas.

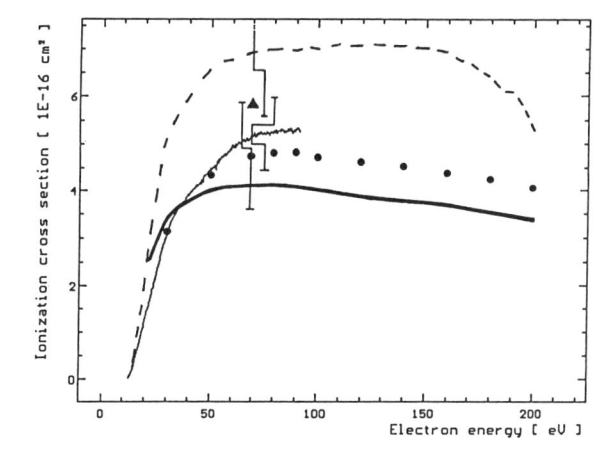

Fig. 3: Same as figure 1 for the total single SO_2 ionization cross section. The triangle represents the total SO_2 ionization cross section measured by Cadez et al. (27). Also shown is a calculated cross section (solid line) based on a modified addtivity rule.

TIMS requires a quantitative knowledge of radical ionization cross sections. The fast-beam apparatus used by our group at City College provides a unique tool to measure absolute cross sections for the electron-impact ionization and dissociative ionization of free radicals and other transient species. Detailed descriptions of our fast-beam apparatus and of its performance characteristics have been given in several publications to which we refer the reader for further details [8,9]. Suffices to say, that the apparatus is capable of generating well-characterized, mass-selected beams of a large number of free radicals which serve as a target for subsequent electron-impact ionization studies. Fig. 4 shows the absolute cross sections for the

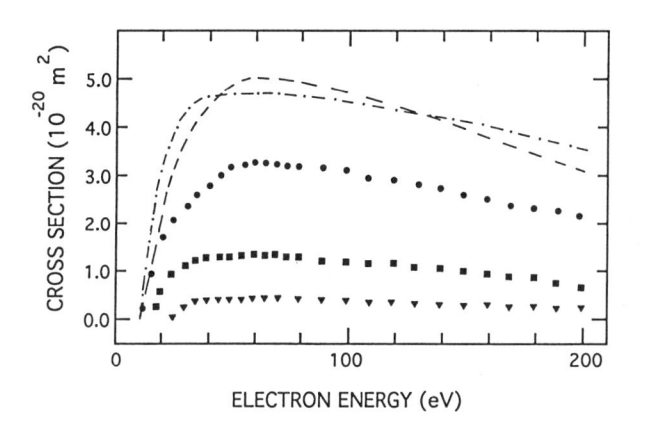

Fig. 4: Absolute cross sections for the formation of the SO^+ parent ions (●) and the S^+ (■) and O^+ (▼) fragment ions from SO. Also shown are the experimetally determined total single ionization cross section (dashed line) and a calculated total single SO ionization cross section (dash-dot line)

electron-impact ionization and dissociative ionization of the SO free radical [10] which was carried out in conjunction with our studies of the SO_2 ionization. There are no other measurements with which we can compare our experimental results for SO. However, we found excellent agreement between the total single SO ionization cross section derived from the measured partial ionization cross sections and the result of a calculation using the same modified additivity rule that was applied before in the case of the SO_2 molecule (see fig. 3).

Other free radicals relevant to low-temperature processing plasmas which have been studied in our fast-beam apparatus include the fluorine-containing species SiF_x, CF_x, and NF_x (x = 1-3). (See [11] and references therein to earlier work for details regarding available ionization cross section data for these free radicals.) Measurements for the hydrogen-containing radicals SiH_x, CH_x, and NH_x (x = 1-3) are currently underway.

III. THE STATUS OF ELECTRON-IMPACT DISSOCIATION PROCESSES

We would like to divide electron-impact dissociation processes into two categories, dissociative excitation and dissociation into neutral ground-state fragments. Dissociative excitation refers to those dissociation processes which leave the fragment(s) to be detected in an excited state which subsequently decays radiatively. Dissociative excitation processes can be studied with relatively little experimental effort by monitoring and analyzing the optical emissions. There is a reasonably broad data base on the dissociative excitation of many molecules relevant to low-temperature plasma processing, particularly for halogen-containing molecules as discussed in detail in a recent review [12]. Even some of the more complex polyatomic molecules such as the Si-organic molecules HMDSO, TMS, and TEOS, which are widely used in deposition plasmas, have been studied recently [13]. Figure 5 shows the optical emission spectrum in 250 - 500 nm region recorded with a FWHM resolution of 2 nm produced by 100 eV controlled single electron impact on tetramethylsilane (TMS). The CH A $\rightarrow$ X and B $\rightarrow$ X band systems, the H Balmer lines, and the Si lines at 251 nm and 288 nm are readily observed. The HMDSO emission spectrum looks very similar to the TMS spectrum, while we did not find any indication of the Si lines in the TEOS spectrum. This can be understood in terms of the molecular structure of the three target molecules. A detailed analysis of the emission spectra in terms of identification of the emitting species of the various features, absolute emission cross sections of the spectral features and appearance energies provide insight into the collision-induced break-up mechanism of these complex species. This provides invaluable information for the interpretation of data from plasma induced emission spectroscopic studies and when combined with mass spectrometric data can eludicate the dominant plasma processes that govern the deposition process.

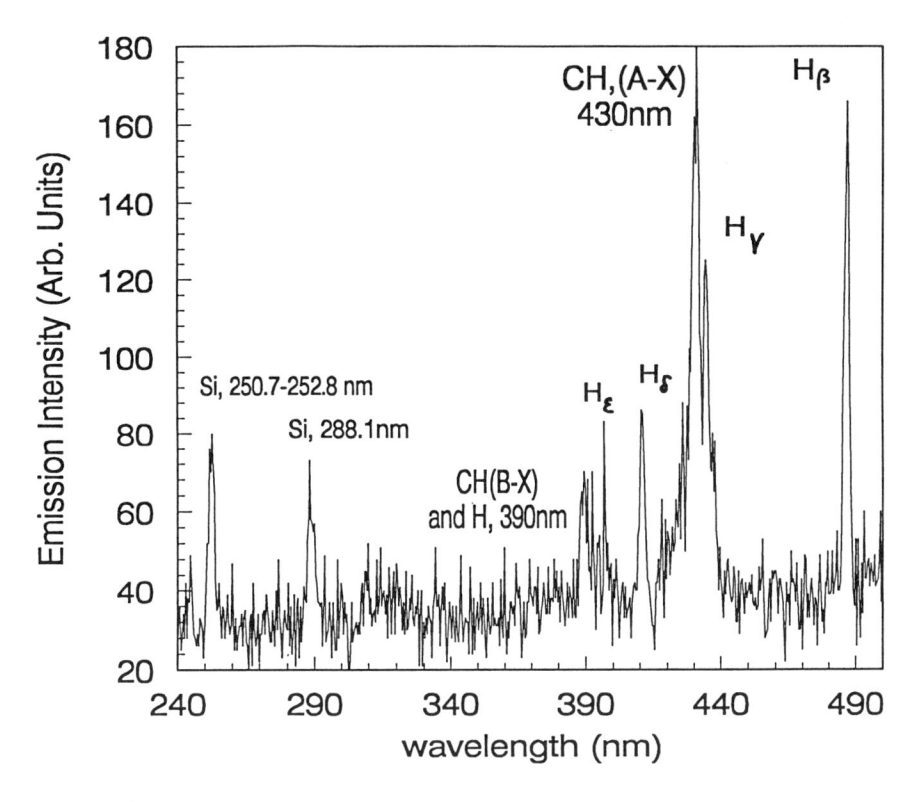

Fig. 5: Optical emission spectrum produced by 100 eV electron impact on TMS. The main emission features have been labelled (see text for more details).

The break-up of a molecule into two or more neutral ground-state fragments caused by the impact of a single electron is one of the most fundamental colllisional interactions between electrons and molecules. Dissociation into neutral ground-state species is also an important process from an application-oriented view-point, e.g. in low-temperature processing plasmas, where these processes are responsible for the formation of a multitude of reactive species from the often rather inert parent molecules. It is, therefore, astonishing that these processes have not received much attention by the collision physics community until recently. The reason for the lack of research activities relating to the study of electron-impact dissociation processes by theorists and experimentalists alike is due to the serious difficulties associated with the investigation of these processes compared to other electron collision processes. Neutral dissociation fragments in the ground state carry neither a charge nor any excess energy that could easily be exploited for the quantitative detection of these species in experimental studies. Complex target molecules and processes such as dissociation and ionization with a multitude of final states are inherently difficult to handle in ab initio, fully quantum mechnical theoretical calculations. The need to understand electron-impact dissociation processes into neutral ground-state fragments in low-temperature plasmas stimulated innovative and novel approaches by the collision physics community to study these processes.

A brief review of the experimental techniques used to study electron-impact dissociation into neutral ground-state fragments reveals the following:
- The chemical getter technique pioneered by Winters and collaborators [14] uses a chemical getter to trap the dissociation products. The drawbacks of this approach are a very limited range of target molecules which can be studied and a lack of selectivity as far as the dissociation products are concerned (i.e. only total dissociation cross sections can be obtained)
- The two-electron-beam technique in which the first electron beam dissociates the target molecule and a second electron beam which is spatially separated from the first one probes the dissociation fragments. The main drawback of this technique is the problem of absolute calibration
Both techniques were developed about 25 years ago and have rarely been used recently because of the many problems associated with their application.

Recently, Cosby and collaborators developed a new approach using a fast neutral beam in conjunction with a correlated product detection scheme [15]. This techniques enables the measurement of absolute electron-imapct dissociation cross sections for diatomic molecules or polyatomic molecules whose dissociation is dominated by a two-fragment break-up process with relative ease and has been successfully applied to molecules such as CO and Cl_2. The method is obviously inadequate for the investigation of dissociation processes where more than two fragments are formed. In a different approach, McConkey and collaborators modified the two-electron beam technique mentioned above by replacing the second (probing) electron beam with a tunable dye-laser beam and used laser-induced fluorescence (LIF) to detect the electron-imapct produced neutral ground-state dissociation products [16]. In this approach, the laser beam probes the dissociation fragments directly in the interaction region of the electron beam and the gas beam. Processes studied so far include OH formation from H_2O and CN formation from HCN. The main problem with this techniques lies in obtaining a reliable absolute calibration. The limited wavelength range covered by currently available tunable dye-lasers is another limitation (no tunable vacuum ultraviolet lasers are routinely available as yet !). However, one can argue that this approach is perhaps the most versatile and most promising technique for the future in view of the rapid developments in laser technology.

Sugai and collaborators [17] developed a new variant of the previously described two-electron-beam technique. They were able to overcome many of the problems associated with earlier approaches by using the technique of threshold ionization mass spectroscopy to probe the dissociation fragments produced by the interaction of the first electron beam with the target gas. The absolute calibration of the dissociation cross sections obtained by this technique requires a knowledge of the dissociative ionization cross sections of the parent molecules (which are available in most cases) and of the fragment (radical) ionization cross sections

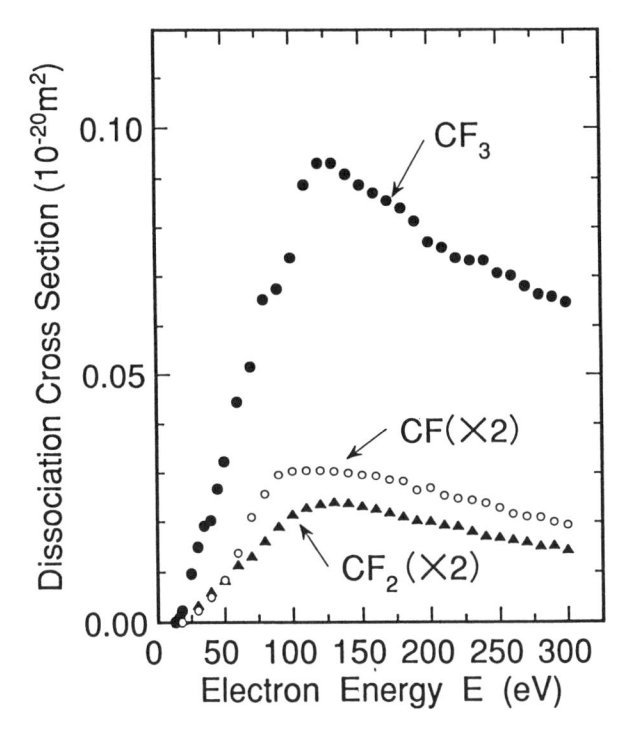

Fig. 6: Absolute cross sections for the formation of CF3, CF2, and CF free radicals by electron-impact dissociation of CF4.

(which are available only in a few cases). Molecules relevant to low-temperature processing plasmas for which absolute dissociation cross sections have been obtained by this technique include SiF4, CH4, CHF3, SiH4 and CF4. The most recent CF4 data [18] are shown in figure 6. In some cases such as SF6 and SiH4 only relative neutral dissociation cross section could be reported due to a lack of the necessary radical ionization cross sections.

On the theoretical side, first attempts are underway to treat electron-imapct dissociation processes in the framework of quantum mechanical theories. The appilcation of variational methods such as the complex Kohn method or the Schwinger method (in conjunction with the use of massively parallel computers) has produced first results for the dissociation of Cl2, BCl3 and NF3. In the case of Cl2 and NF3, the results of the calculations seem to agree rather well with experimental results [19].

IV. THE STATUS OF ELECTRON COLLISIONS WITH EXCITED-STATE SPECIES

Electron collisions with target species in excited states have been recognized as very important collision processes in low-temperature plasmas. Due to complications related to the preparation and characterization of excited state targets experimental studies of electron collisions with excited targets are scarce as are theoretical calculations. A survey of the existing experimental data base on electron collisions with excited targets reveals the following findings:

1. Ionization: data are available for the excitation of metastable rare gases, metastable H and O, laser-excited Na and Ba as well as some qualitative results on the ionization of vibrationally excited molecules - very recently, the first account of quantitative experimental studies was reported (see below)

2. Excitation: data are available for metastable He (see below), laser-excited Na, Rb, Cs, and Ba as well as some qualitative work on vibrationally excited molecules

3. Dissociation: there are essentially no data available at this point in time

4. Dissociative Attachment: there have been numerous qualitative and quantitative investigations [20]

On the theoretical side, there are first attempts to treat the excitation of excited targets [21] and some work regarding the effects of vibrational excitation of molecular targets on ionization cross sections has been carried out (see below).

Very recently, Bergmann and collaborators reported the first quantitative experimental study of the effect of vibrational excitation on measured molecular ionization cross sections [22]. These authors measured the ionization rate of Na_2 dimers which were excited by laser radiation to form distributions of vibrationally excited species corresponding to average vibrational quantum numbers of $v = 8.5$, 14.5, and 20. Bergmann and collaborators found a shift of the ionization threshold towards lower impact energies and an enhanced ionization rate for the vibrationally excited target molecules compared to target molecules in the vibrational ground state. The enhancement in the ionization rate reached a factor of 3.5 for the distribution characterized by $v = 20$, but the effect was limited to a narrow range of impact energies up to about 3 eV above the ionization threshold. The maximum in the Na_2 ionization cross section was essentially unaffected by the vibrational excitation of the target molecules. It should be noted that the obseved effect on the ionization cross section in the near-threshold region, if confirmed for other molecules, can drastically enhance the efficiency of electron-impact ionization processes in low-temperature plasmas. A shift of the ionization threshold threshold to lower energies and an enhanced ionization cross section in the near-threshold region can result in a significant increase of the overlap of the EEDF of the plasma electrons with the ionization cross section and thus lead to a substantial increase in the efficiency of electron-impact ionization processes which, in turn, will influence the ionization balance in the plasma (see figure 1).

Qualitative support for the observation reported by Bergmann and collaborators in Na$_2$ comes from our recent measurements [23] of the ionization and dissociative ionization of deuterated methane, CD$_4$. The charge neutralization of primary CD$_4^+$ ions in our fast-beam apparatus produces vibrationally excited neutral CD$_4$ target species with an internal energy of up to 2 eV. However, our measured ionization cross sections above about 30 eV agreed very well with data obtained by other authors using techniques where only methane molecules in the vibrational ground state (or in a vibrational distribution corresponding to room temperature) are present. The vibrational excitation of the CD$_4$ target molecules in our experiment resulted in a shift of the ionization threshold to lower energies and left ist mark on the measured cross section shapes only for impact energies below about 30 eV. In particular, the maximum in the cross section curves did not show any dependence on the vibrational excitation of the target molecules. The above observations are also supported by theoretical calculations by Capitelli and collaborators [24,25] for molecules such as N$_2$ and H$_2$. Only the calculations for H$_2$, which has the largest vibrational spacing of all molecules, revealed a change in the calculated maximum of the ionization cross of ±20% when going from $v = 0$ to $v = 10$.

Electronic excitation of excited target species such as metastable rare gas atoms is an important processes in low-temperature plasmas and it is a collision processes of fundamental importance as well. Metastable rare gas atoms, e.g. He (2^{3}S) atoms represent targets which are different from the spherically symmetric,

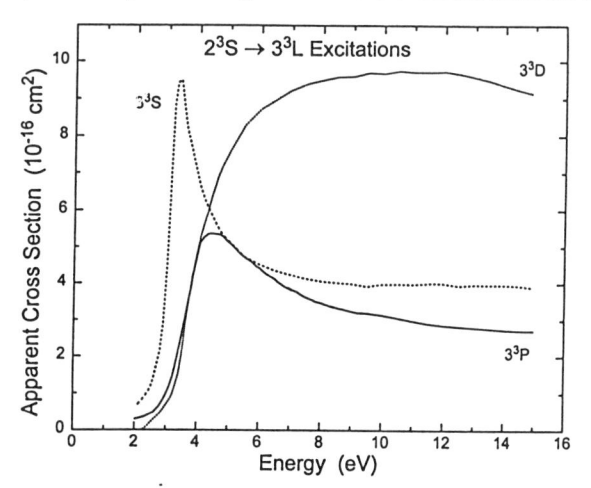

Fig. 7: Absolute cross sections for excitation of various He 3^{3}L states from the metastable 2^{3}S state by electron impact at low electron energies.

tightly bound He (1^{1}S$_0$) ground state. Lin and collaborators at the University of Wisconsin have carried out extensive experimental investigations of the electron-impact excitation out of the He metastable states [26]. In a first generation of experiments [26], these authors used a hollow-cathode discharge to produce a mixture of metastable and ground state He atoms. Collisions with electrons whose energy was kept below the onset energy of the excitation of ground state He atoms present in the beam resulted in the collisionally induced population of higher-lying levels in He which subsequently decay radiatively. The detection of the emitted radiation allows the determination of the cross sections for exciting the higher levels out of the metastable level. Once the energy of the electron beam exceeds the energy necessary to excite the same levels from the ground state, this experimental technique breaks down due to the preponderance of ground state atoms in the target beam. A set of low-energy cross sections measured for excitation of the higher-lying 3^3S, 3^3P, and 3^3D states is shown in figure 7. Two observations are noteworthy: (i) the measured cross sections are larger than the cross sections for exciting the same levels from the He ground state by 1-2 orders of magnitude and (ii) the convential rules of thumb regarding the cross section shapes for electron-impact excitation of ground state atoms appear to be violated. Specifically, the following observations should be noted:

(i) for the n = 3,4,5 levels, the following pattern emerged:
- sharply peaked excitation cross sections for the n^3S levels
- less sharply peaked cross sections for exciting the n^3P levels
- broad excitation cross section cross sections for the n^3D levels

(ii) the 3^3D cross section is 4 times larger than the 3^3P cross section, even though the excitation of the 3^3P state is optically allowed whereas the excitation of the 3^3D is not.

Very recently, the same authors have succeeded in obtaining cross section data for the excitation of metastable He atoms up to impact energies of 1 000 eV in a second generation apparatus which is based on a fast-beam concept - metastable He atoms are selectively produced by appropriate charge transfer thus minimizing the

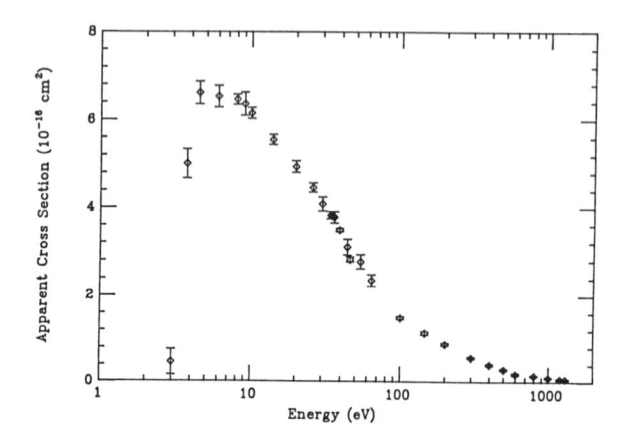

Fig. 8: Absolute cross section for excitation of the He 3^3D state from the metastable 2^3S state by electron impact for impact energies up to 1000 eV.

number of ground state He atoms in the target beam. Figure 8 shows their preliminary data for the cross section for excitation of the He 3^3D state from the 2^3S state from threshold to 1000 eV obtained by this technique.

V. CONCLUDING REMARKS

We hope to have demonstrated in this paper how collision physics, a discipline which has its home in basic science and low-temperature plasma processing, a very application-oriented technology-driven field, can benefit from each other. Fundamental research focusing on electron collisions, ion collisions and photon interactions with technologically relevant targets provides invaluable collision and spectroscopic data for plasma modelling and for CAD of plasmas processes and plasma reactors, helps identify key plasma processes and bottlenecks in reaction mechanisms, enables the development of new and more sophisticated plasma diagnostics techniques, and provides the necessary tools for a more quantitative interpretation of optical and mass spectrometric plasma diagnostics studies. At the same time, challenges, unresolved questions and problems facing the plasma processing community have stimulated novel research in collision physics. The need to understand the complex plasma processes at a microscopic level have initiated new experimental work involving complex target molecules and led to the development of new experimental techniques to generate, study and detect unstable, transient, reactive and/or corrosive species. Theory has been encouraged to treat complex target species and processes with many final states such as ionization and dissociation. The development of new and improved scaling laws and semi-empirical methods has been advanced by the need to compare data for a large number of target species and for classes of targets of similar structure. Computational methods have benefitted from ever more powerful and faster computers and the implementation of highly parallel computers.

ACKNOWLEDGMENTS

I am grateful to my collaborators over the past few years who have made significant contributions to the work presented here, in particular Dr. K.A. Blanks, Ms. J. Dike, Mr. M. Ducrepin, Dr. Z.J. Jabbour, Mr. P. Kurunczi, Mr. A. Levin, Mr. M.B. Roque, Dr. R.B. Siegel, Dr. K.E. Martus, Dr. V. Tarnovsky, and Dr. S.H. Zheng. Thanks are due to Prof. L. W. Anderson, Dr. R. Basner, Dr. P.C. Cosby, Dr. R.S. Freund, Dr. A. Garscadden, Dr. R.A. Gottscho, Prof. C.C. Lin, Dr. J. Perrin, Dr. T.N. Rescigno, Dr. M.Schmidt, Dr. S.K. Srivastava, Prof. R.A. Bonham, Prof. H. Deutsch, Prof. T.D. Märk, Prof. V.B. McKoy, and Prof. H. Sugai for many helpful and stimulating discussions. I am particularly grateful to Prof. H. Sugai and Prof. C.C. Lin for making some of their recent data available to us prior to publication. The financial support of this work by the U.S. National Foundation (NSF), the U.S. Department of Energy (DOE), the U.S. National Aeronautics and Space Administration (NASA), the American Chemical Society - Petroleum Research Fund, AT&T Bell Laboratories, a NATO Collaborative Reserarch Grant, and the City University of New York through its PSC-CUNY grant program is gratefully acknowledged.

REFERENCES

1. K. Becker and V. Tarnovsky, Plasma Sources Sci. Technol. **4**, 307 (1995)
2. R. Basner, M. Schmidt, H. Deutsch, V. Tarnovsky, A. Levin, and K. Becker. J. Chem. Phys. **103**, 211 (1995)
3. O.J. Orient and S.K. Srivastava, J. Chem. Phys. Phys. **78**, 2949 (1983)
4. O.I. Smith and J. Stefenson, J. Chem. Phys. **74**, 6677 (1981)
5. H. Deutsch, T.D. Märk, V. Tarnovsky, K. Becker, C. Cornelissen, L. Cespiva, and V. Bonacic-Koutecky, Int. J. Mass Spectrom. Ion Proc. **137**, 77 (1994)
6. R. Robertson, D. Hils, H. Chatham, and A. Gallagher, Appl. Phys. Lett. **43**, 544 (1983)
7. H. Kojima, H. Toyoda, and H. Sugai, Appl. Phys. Lett. **55**, 1292 (1989)
8. R.S. Freund, R.C. Wetzel, R.J. Shul, and T.R. Hayes, Phys. Rev. A **41**, 3575 (1990)
9. V. Tarnovsky and K. Becker, Z. Phys. D **22**, 603 (1992)
10. V. Tarnovsky,, A. Levin, and K. Becker, J. Chem. Phys. **102**, 770 (1995)
11. K. Becker, in "Electron Collisions with Molecules, Clusters, and Surfaces", H. Ehrhardt and L.A. Morgan (editors), Plenum Press, New York (1994), p. 127-140
12. K. Becker, Comm. At. Mol. Phys. **30**, 261 (1994)
13. P. Kurunczi, J.P. Michel, N. Abramzon, K.E. Martus, and K. Becker, Contributed Papers of the XXII ICPIG, K. Becker, W.E. Carr, and E.E. Kunhardt (editors), Stevens Institute of Technology Press, Hoboken, NJ (1995), pp. 2-183
14. H.F. Winters, J. Chem. Phys. **63**, 3462 (1975)
15. P.C. Cosby, J. Chem. Phys. **98**, 6813 (1993)
16. M. Darrach and J.W. McConkey, J. Chem. Phys. **95**, 754 (1991)
17. T. Nakano, H. Toyoda, and H. Sugai, Jap. J. Appl. Phys. **30**, 2908 & 2912 (1991)
18. H. Sugai, private communication (1995) and to be published in Contr. Plasma Phys. (1995)
19. T.N. Rescigno, Bull. Am. Phys. Soc. **40**, 1329 (1995)
20. P.G. Datskos and L.G. Christophorou, J. Chem. Phys. **86**, 1986 (1987)
21. E.J. Mansky, Bull. Am. Phys. Soc. **40**, 1305 (1995)
22. M. Külz, A. Kortyna, M. Keil, B. Schnellhaaß, and K. Bergmann, Z. Phys. D **33**, 109 (1995)

23. V. Tarnovsky, A. Levin, H. Deutsch, and K. Becker, J. Phys. B (1996), in press
24. M. Cacciatore, M. Capitelli, and C. Gorse, Chem. Phys. **66**, 141 (1982)
25. R. Celiberto and T.N. Rescigno, Phys. Rev. A **47**, 1939 (1993)
26. R.B. Lockwood, L.W. Anderson, and C.C. Lin, Z. Phys. D **24**, 155 (1992); C.C. Lin and L.W. Anderson, private communication (1995)
27. I.M. Cadez and M.V. Kurepa, Appl. Phys. **16**, 305 (1983)

Low Frequency Oscillations and Chaos in Plasmas

A. Piel and T. Klinger

Institut für Experimentalphysik, Christian-Albrechts-Universität zu Kiel
D-24098 Kiel, Germany

Abstract. Regular and chaotic low frequency oscillations are studied in thermionic discharges. A model for relaxation oscillations is presented that is based on particle-in-cell simulations and probe measurements. The synchronization of the oscillations by an external driver and the period-doubling route to chaos can be understood in terms of a periodically forced van der Pol oscillator. The stabilization of unstable periodic orbits by the Ott-Grebogy-Yorke technique is demonstrated. Spatio-temporal chaos is found at the onset of drift-wave turbulence. Here the transition to chaos involves the Ruelle-Takens-Newhouse scenario with a sequence of Hopf bifurcations. The spatio-temporal evolution of interacting drift-waves is studied with a novel 64-probe array.

INTRODUCTION

Plasma, with its large number of degrees of freedom and its inherent nonlinearity, has during the last decade become a convenient object to study low-dimensional dynamical behavior of a physical system. Compared to hydrodynamics or solid state physics, plasmas allow for easy and detailed local diagnostics. Some simple systems can even be successfully modelled by particle-in-cell simulations.

The study of chaos in plasmas goes back to 1987 with two seminal observations of period-doubling routes to chaos, namely in ionization waves [1] and in filament cathode discharges [2]. Both phenomena initiated a number of subsequent detailed investigations in gas discharges [3–15] as well as in filament cathode discharges [16–32]. Low dimensional chaos was also found in tokamak experiments on Mirnov oscillations [33, 34] and density fluctuations [35], but was contrasted by high dimensionality for broadband edge fluctuations [36, 37]. Low-dimensional attractors were also observed in the coupling of drift-waves with potential relaxation oscillations [38, 39]. Other systems which were studied for chaotic behavior comprise ion-sheath oscillations in dp-machines [40–46] and ionization type drift-waves in high-frequency magnetron discharges [47–50]. Earlier reviews were given in this series of conferences in [6, 7, 51].

This paper describes 3 groups of chaos phenomena: (1) relaxation oscillations, period doubling, and synchronization, (2) controlling chaos, and (3) the onset of drift-wave turbulence. The examples are mostly chosen from filament cathode discharges, where quite a good understanding of the basic mechanisms is now available from models based on PIC-simulations. These topics were selected to demonstrate the emerging detailed understanding of oscillator-like behavior, and to demonstrate two upcoming techniques: controlling chaos and diagnostics with large numbers of probes.

Routes to Chaos

Although we have now a wealth of experimental observations of chaotic phenomena in plasmas, only a few systems are beginning to be understood in their microscopic mechanisms. *Temporal chaos* is observed as fluctuations of discharge voltage, current, or probe signal. From the experimental point of view we can distinguish *undriven chaos*, which occurs when an accessible system parameter (pressure, current, emission) is gradually changed [1, 15, 18, 19, 20, 52], and *driven chaos*, which is forced by an externally applied periodic voltage of varying amplitude or frequency [4, 6, 8, 16, 25, 29, 30, 40, 44]. In both cases, standard routes to chaos, like period-doubling, quasiperiodicity, intermittency, or a series of Hopf bifurcations are found.

Spatio temporal chaos is observed in systems where the fluctuations are related to wave propagation, typically at the onset of turbulence. Ionization waves [4], drift-waves [39, 53] or ionization-drift waves [47], belong to this class. Ionization waves mostly show *temporal chaos*, which may be attributed to the locking of a certain number of wavelengths in the system [13]. This may also explain the successful modelling by the van der Pol oscillator [14].

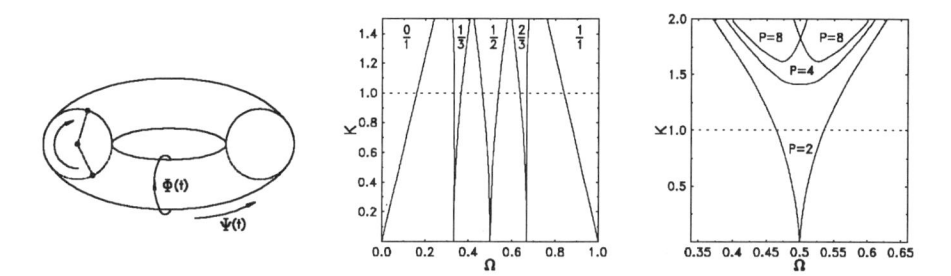

FIGURE 1: *left:* A system with two incommensurate frequencies forms a 2-torus in phase space. *middle:* The locking regimes of the sine-circle map in the (Ω, K)-plane form Arnol'd tongues. *right:* Beyond the critical line $K = 1$, period doubling and quadrupling is found.

The period-doubling route to chaos is often discussed in terms of a quadratic mapping of the unit interval [54]

$$x_{n+1} = 4r x_n (1 - x_n) \qquad (1)$$

which was originally introduced as a model for the annual evolution of a population of size x_n. The population size of the next generation x_{n+1} depends on the present size and the remaining ressources $\propto (1 - x_n)$. Hence, at large growth rates $\propto r$, period doubling (and further bifurcations) may occur because too large populations will die of starvation. We will show below that this kind of reasoning can also be applied to relaxation oscillations in plasmas.

The simplest model for a nonlinear oscillator interacting with an external periodic driver at frequency f_d is the sine-circle map

$$\theta_{n+1} = \theta_n + \Omega - \frac{K}{2\pi} \sin(2\pi \theta_n) \qquad (2)$$

where θ_n is the phase of the oscillator in the n-th period, $\Omega = f/f_d$ and K describes the driver amplitude. This model (see Fig. 1) shows quasiperiodic, frequency locked states, and a transition to chaos [55]. Beyond the critical line $K = 1$, even period doubled and quadrupled states were found [56]. We will show that synchronization behavior can, to the lowest order, be described by the sine-circle map, and that refinements can be made by other oscillator models, e.g. the van der Pol model, to describe finer details.

Intermittency is observed, if a limit cycle loses stability, and results in an almost regular (laminar) motion, which is interrupted by erratic bursts. Type-I intermittency [57] was observed in a driven filament cathode discharge [16] and characterized by an f^{-2} noise spectrum. Similar observations were reported for the undriven case [18, 19] and for discharges with negative ions [32]. Among the chaotic phenomena, intermittent behavior is least understood in terms of the microphysics behind it.

From the general theory of nonlinear dynamics the onset of turbulence in plasmas and fluids is expected to follow the Ruelle-Takens-Newhouse scenario [58], which employs a sequence of two Hopf-type bifurcations to a one and two frequency system, and a subsequent Hopf-bifurcation to a low-dimensional chaotic state. This aspect will be addressed in connection with drift-wave turbulence.

RELAXATION OSCILLATIONS IN THERMIONIC DISCHARGES

This class comprises filament cathode discharges, low pressure arcs, Q-machines, and thermionic converters, where low-frequency oscillatory states are well known [59–63]. For Q-machines and Knudsen diodes the potential relaxation instability is observed, which invokes the formation of a transient

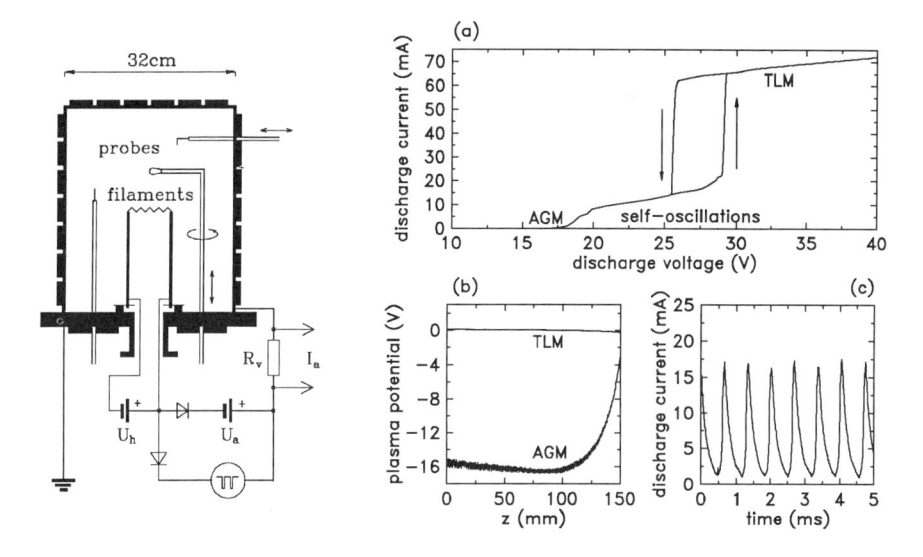

FIGURE 2: *left:* Filament cathode discharge with multi-dipole confinement. *right:* (a) The static $I_d(U_d)$ characteristics shows hysteresis. (b) The anode-glow-mode has plasma potential close to cathode potential. (c) Strong relaxation oscillations are found close to the bifurcation point.

double layer [64, 65]. The static $I_d(U_d)$ characteristic of filament cathode discharges shows different kinds of hysteretic effects [19,66–69]. The negative differential resistance behind the sudden jumps was discussed in terms of catastrophe theory [70, 71], and a relation to chaotic behavior was conjectured.

Figure 2 shows a simple experimental arrangement with a single filament cathode and multidipole confinement. There are two discharge states, the anode-glow-mode (AGM) with space charge limited current, and the temperature-limited-mode (TLM) [72, 73]. Close to the upward hysteresis point the AGM becomes unstable and performs low-frequency current oscillations. Oscillations were also found in the TLM close to the downward hysteresis point [27]. Multiple hysteresis loops are often found in discharges with many filament cathodes [27, 69, 71]. Carrère [74] has demonstrated, that for each filament an additional subloop is created.

Period Doubling

When the filament cathode discharge is operated in the AGM and additional modulation is applied, the discharge current and the plasma potential (see Fig. 3, *left*) show a period-doubling sequence to chaos [2, 24, 25] when either the pulse height or width is increased. The relationship to iterated quadratic

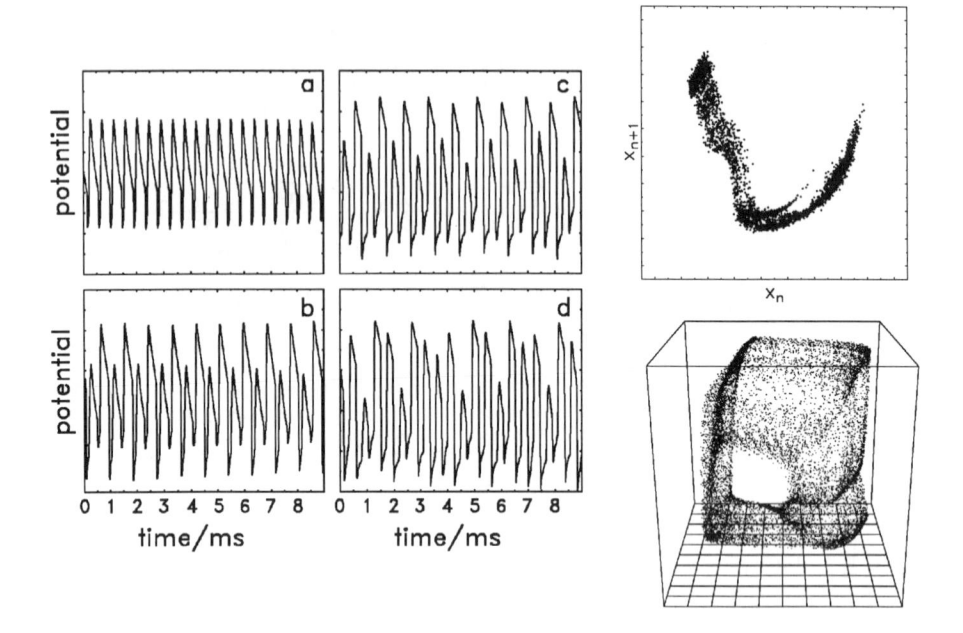

FIGURE 3: *left:* Period doubling route to chaos by increasing the width of the applied pulses: (a) period 1, (b) period 2, (c) period 4, (d) chaos. *right:* First return map shows backfolding. The chaotic phase space attractor has a dimensionality of $D_2 = 2.0 \pm 0.2$.

maps can be seen in the first return map in the chaotic state (see Fig. 3, *right*), which shows backfolding that, together with sensitivity on initial conditions, generates the chaos. The phase space attractor in a 3D reconstruction exhibits the fractal appearance expected for chaotic states. An estimate of the attractor's dimensionality results in a correlation dimension $D_2 = 2.0 \pm 0.2$ and the greatest Lyapunov exponent is found positiv $\lambda_g = 1.07$ bit/period.

Particle-in-Cell Simulation

The microphysics of the AGM oscillations can be understood by comparing with 1D particle-in-cell simulations [29]. High electron emission rates create a virtual cathode that causes space charge limitation of the electron current. Ions are created by impact ionization close to the anode and, because of the low pressure, perform a nearly free fall towards the cathode. Only a small fraction is stopped by charge exchange collisions with neutrals. These thermal ions, however, are trapped in the potential well of the virtual cathode and form a cathodic plasma (see Fig. 4).

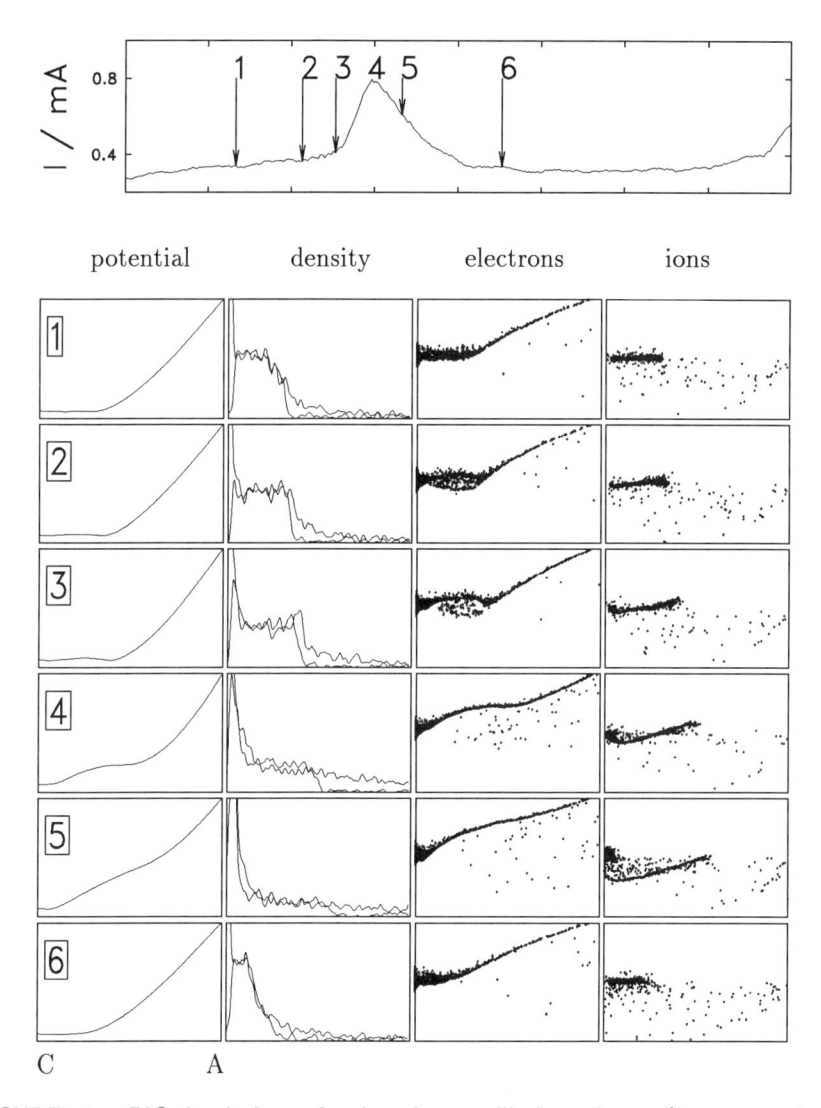

FIGURE 4: PIC-simulation of relaxation oscillations in a filament cathode discharge. The top trace shows a single period of the current oscillation. The diagrams below give the potential, electron and ion density, electron phase space, and ion phase space. Note the the plasma expansion from 1-3 and the triggering of the relaxation by formation of a transient double layer at 4.

The cathodic plasma is inherently unstable because of the free energy in the streaming electrons. The system can excite a Pierce-Buneman instability [29, 64, 75]. The simulation shows that the potential relaxation is triggered by the formation of a positive potential hump (panels 2-3) that leads to the formation of a vortex in electron phase space. This hump transforms into a transient double layer (panel 4). The increased potential allows that ions, which were trapped in the potential well of the virtual cathode before, now escape towards the cathode, where they efficiently neutralise the electron space charge and lead to the current increase. The streaming ion component appears clearly in panel 5 at negative velocities. The current peak decays by the loss ions and the original potential shape is reestablished.

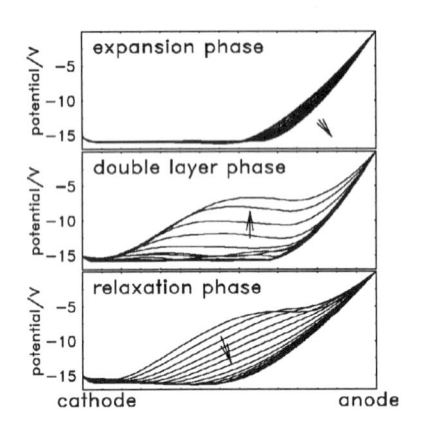

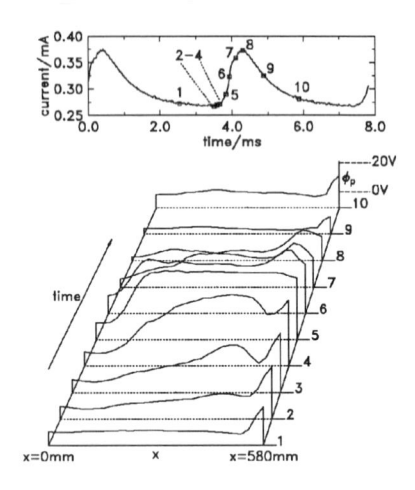

FIGURE 5: *left:* The evolution of the potential distribution in the PIC-simulation; *right:* Measured potential distributions from emissive probes for different phases of the relaxation oscillation.

The oscillation cycle consists of three distinct phases [29] (see Fig. 5): (a) the expansion phase, (b) the double layer phase, and (c) the relaxation phase. This typical sequence of events could be verified by potential measurements with emissive probes [30].

Summarizing, the relaxation oscillations are determined by the rate at which ions are trapped in the virtual cathode, which depends on discharge voltage and gas pressure. The external modulation creates additional ions and results in phase modulation. The trigger threshold for the Pierce-Buneman instability depends on ion density and size of the cathodic plasma. Hence, the backfolding of the chaotic phase space attractor that leads to period doubling can be understood by the periodic loss of ions in the relaxation process and the necessary time to meet the trigger criterion again.

Since undriven chaos was usually observed in discharges with many filament cathodes, we conjecture that such systems consist of a set of coupled oscillators, one of which is acting as driver. This may explain the similarity of the observed phenomena in both cases.

Synchronization

The synchronization regimes of externally driven filament cathode discharges have the shape of Arnol'd tongues [30] as in the circle map model. In the latter, a saddle-node bifurcation from a quasiperiodic to a locked state is found.

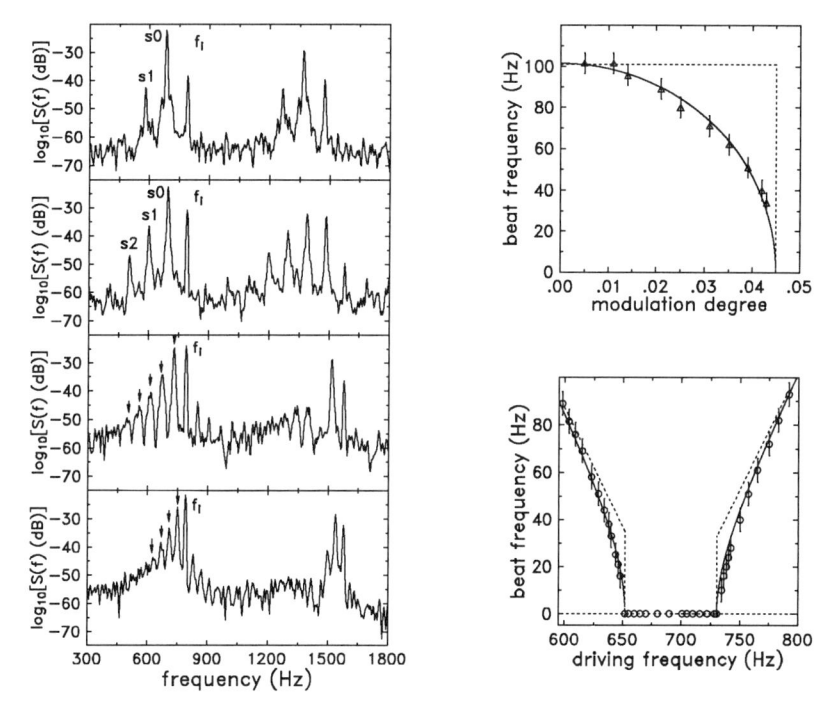

FIGURE 6: *left:* Measured spectra of driven relaxation oscillations in the filament cathode discharge. With increasing driver amplitude the spacing within the sideband gets narrower. *right, top:* Decrease of the average beat frequency with driver amplitude. *right, bottom:* Average beat frequency vs. driver frequency in comparison with the circle map (dashed line) and the van der Pol model (solid line).

The transition found in the experiment, however, takes an intermediate step of *incomplete synchronization*. This phenomenon is accompanied by periodic pulling of the system's frequency towards the driver [76, 77].

Periodic pulling can be identified by typical asymmetric spectra (Fig. 6), which are caused by simultaneous AM and FM modulation that suppresses the sidebands on the side of the driver frequency. Similar spectra were found in a gas discharge [14]. Periodic pulling can be modelled by the driven van der Pol equation:

$$\ddot{x} - \epsilon(1 - \beta x^2)\dot{x} + \omega_0^2 x = \omega_0^2 E \cos(\omega_i t) \tag{3}$$

Whereas the circle map only allows for nonlinear mixing of injection frequency ω_i and system frequency, the van der Pol model is capable to describe the detuning of the average beat frequency, which closely agrees with the experiment (see Fig. 6). Similar to the behavior of the circle map beyond the critical line, the van der Pol oscillator exhibits period doubling sequences for large driving amplitudes [78].

CONTROLLING CHAOS

A chaotic phase space attractor contains a dense set of unstable periodic orbits (UPOs). Ott, Grebogy and Yorke (OGY) [79] have suggested a technique to apply well defined changes to an accessible control parameter to stabilize UPO's. The OGY method, which has been successfully applied to many other physical systems, is now applied to plasma chaos. The OGY approach is different from mode suppression [80-83] as known from van der Pol systems, where a parameter of the system is periodically modulated and gives rise to parametric stabilization.

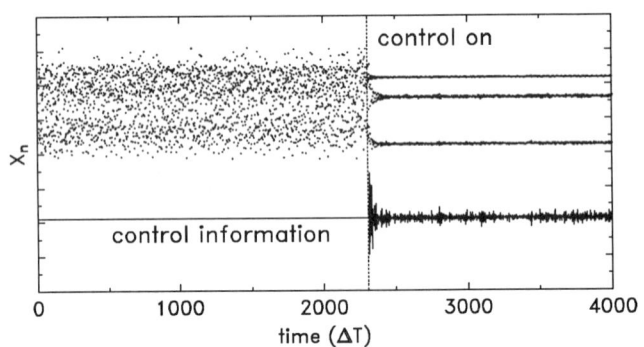

FIGURE 7: Stabilization of an unstable period-3 orbit for the chaotic state of ionization waves by the OGY-technique.

A general scheme for parametric stabilization of chaotic states was also discussed on the example of the Duffing-Holmes oscillator [84]. Apart from phase-space methods, other feedback techniques that are based on a detailed knowledge of the involved nonlinear modes have been successfully applied, e.g., to stabilize trapped particle instabilities in mirror machines [85, 86, 87].

We have applied the OGY technique to stabilize UPO's up to period 16 in a gas discharge system [88] (Fig. 7). Starting from a chaotic state, a signal proportional to the deviation from the actual fixed point of period p is fed back every p-th period.

THE ONSET OF DRIFT-WAVE TURBULENCE

We study the onset of drift-wave turbulence in the magnetized central section of the triple plasma device KIWI. The spatio-temporal structure of the drift-waves is observed by means of a 64-probe array, that maps the circumference of the magnetized plasma column (Fig. 8) and allows snapshots of 2 ms length with a 64-channel digitizer [53].

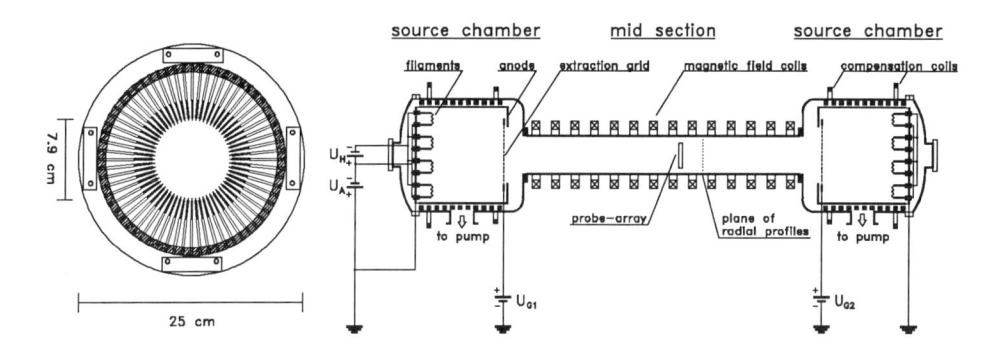

FIGURE 8: *left:* Arrangement of 64 Langmuir probes to study drift-wave turbulence on a magnetized plasma column. *right:* The probe array is placed in the center section of the triple plasma device KIWI. Drift waves are destabilized by varying the bias of grid 1.

The sequence of Hopf bifurcations is obvious from the frequency spectra in the second row of Fig. 9. The sequence shows (a) destabilization of a single mode frequency and its harmonics, (b) a second Hopf-bifurcation to a system with two modes of incommensurate frequency, (d) a third Hopf-bifurcation to a turbulent state. In between, (c) shows spatial mode-locking of mode $m = 1$ and $m = 2$. The wavenumber analysis shows that the two modes in (b) are $m = 1$ and $m = 3$. The nonlinear coupling of the modes can be read from the power spectrum $S(k,\omega)$.

The transition to chaos through subsequent Hopf-bifurcations can also be described by the evolution of Lyapunov exponents and dimensions (see Table 1). Increasing the applied voltage along the plasma column creates sets of Lyapunov exponents of the type (0,-,-), (0,0,-), and (+,0,-), which represent a 1-torus, a 2-torus, and a strange attractor. The turbulent state at 8V

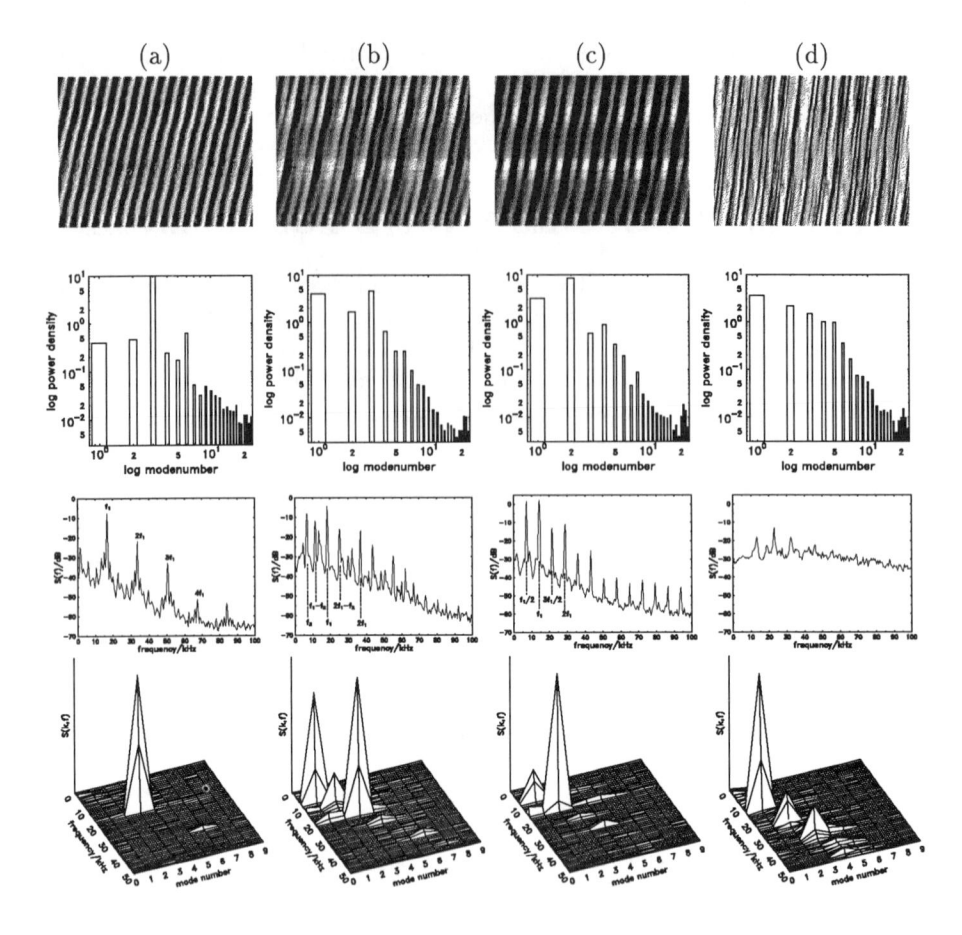

FIGURE 9: Analysis of regular, transitional and turbulent drift waves as recorded by the 64-probe array. 1st row: wave patterns of density fluctuations (horizontal: $t = 0 \ldots 1024\mu s$, vertical: angle $\phi = 0 \ldots 2\pi$), 2nd row: computed wave number spectrum, 3rd row: frequency spectrum from FFT-analyser, 4th row: $S(k,f)$ power spectrum. (a) regular $m = 3$-mode ($U_{g1} = 4.8$V), (b) quasiperiodic state with dominant $m = 1$ and $m = 3$ modes ($U_{g1} = 5.1$V), (c) spatial mode-lock between $m = 2$ and $m = 1$ ($U_{g1} = 5.9$V), (d) turbulent state ($U_{g1} = 8.0$V).

U_{g1}/V	λ_1	λ_2	λ_3	D_L	D_2
4.8	-0.9	-5.9	-17.7	1.0	1.05 ± 0.2
5.1	1.4	-1.2	-4.2	2.1	2.15 ± 0.2
7.0	15.6	0.8	-19.4	2.8	2.55 ± 0.2
8.0	(26.7)	(6.4)	(-9.5)	-	-

TABLE 1: The largest Lyapunov exponents (in bits/ms) and dimensions. The uncertainty of the λ_i is ± 2.

shows no saturation for the correlation dimension D_2 and there are two positive Lyapunov exponents.

DISCUSSION AND CONCLUSIONS

The relaxation oscillations of filament cathode discharges can be modeled by the driven van der Pol equation, which explains quasiperiodic, locked and periodic pulling states. The observed period doubling route is found at strong forcing inside a frequency locked state in accordance with the driven van der Pol oscillator. Unstable periodic orbits embedded in the chaotic phase space attractor can be stabilized by applying the OGY-method. This example demonstrates that a plasma system can be regarded as a simple oscillator.

This simplicity, which is expressed in the low dimensionality of phase space attractors and in an appropriate description by van der Pol's equation, results from the two well separated time scales involved: the slow time scale studied here is defined by the ion production and ion transit time, which determine the oscillation frequency and the frequency change by modulation. The fast time scale of the electron dynamics, that is involved in the Pierce-Buneman instability and double-layer formation, is practically invisible in the low-frequency signals, but appears as a trigger event that depends on plasma extension and density. The high frequency signals associated with the double layer phase have been verified [89].

Different from simple oscillator behavior, the onset of drift-wave turbulence is characterized by a series of three Hopf-bifurcations, which lead to a low dimensional chaotic attractor. This is a typical example for a Ruelle-Takens-Newhouse scenario [58]. It can be summarized that oscillator systems are mostly found to perform a period-doubling route to chaos, whereas wave-like systems show a preference for a sequence of Hopf bifurcations.

ACKNOWLEDGEMENT

Fruitful discussions with Dr. C. Wilke are gratefully acknowledged. This work was supported by DFG-grants SFB198/A8 and Pi185/6.

References

[1] T. Braun, J. A. Lisboa, R. E. Francke, and J. A. C. Gallas, Phys. Rev. Lett. **59**, 613 (1987).

[2] P. Y. Cheung and A. Y. Wong, Phys. Rev. Lett. **59**, 551 (1987).

[3] K. Ohe and H. Tanaka, J. Phys. D: Appl. Phys. **21**, 1391 (1988).

[4] C. Wilke and H. Deutsch, Experimentelle Technik der Physik **37**, 289 (1989).

[5] C. Wilke, R. W. Leven, and H. Deutsch, Phys. Lett. **136A**, 114 (1989).

[6] B. Albrecht, H. Deutsch, R. W. Leven, and C. Wilke, in *Proc. of XIXth ICPIG, Invited Papers*, edited by V. J. Žigman (Studio Plus, Belgrade, Jugoslavia, 1989), p. 150.

[7] K. Ohe, in *Proc. of XIXth ICPIG, Invited Papers*, edited by V. J. Žigman (Studio Plus, Belgrade, Jugoslavia, 1989), p. 150.

[8] C. Wilke, H. Deutsch, and R. W. Leven, Contrib. Plasma Phys. **30**, 659 (1990).

[9] J. A. Johnson III, L. E. Johnson, and Y. Hong, Phys. Lett. A **158**, 144 (1991).

[10] V. O. Papanyan and Y. Grigorian, Phys. Lett. A **164**, 43 (1992).

[11] T. Braun, J. Lisboa, and J. A. C. Gallas, Phys. Rev. Lett. **68**, 2770 (1992).

[12] B. Albrecht, H. Deutsch, R. W. Leven, and C. Wilke, Physica Scripta **47**, 196 (1993).

[13] K. D. Weltmann, H. Deutsch, H. Unger, and C. Wilke, Contrib. Plasma Phys. **33**, 73 (1993).

[14] T. Klinger, A. Piel, F. Seddighi, and C. Wilke, Phys. Lett. A **182**, 312 (1993).

[15] P. S. R. Prasad, R. Singh, and R. K. Thareja, IEEE Trans. Plasma Sci. **22**, 224 (1994).

[16] P. Y. Cheung, S. Donovan, and A. Y. Wong, Phys. Rev. Lett. **61**, 1360 (1988).

[17] P. Y. Cheung, S. Donovan, and A. Y. Wong, in *Proceedings of the International Conference on Plasma Physics* (PUBLISHER, New Delhi, India, 1989), pp. 793–796.

[18] Y. Jiang, H. Wang, and C. X. Yu, Chin. Phys. Lett. **5**, 489 (1988).

[19] J. Qin et al., Phys. Rev. Lett. **63**, 163 (1989).

[20] J. Qin and L. Wang, Phys. Lett. A **156**, 81 (1991).

[21] A. Piel and R. Timm, in *Proc. of XXth ICPIG, Contrib. Papers 1*, edited by V. Palleschi et al. (Instituto di Fisica Atomica e Moleculare, Pisa, Italien, 1991), p. 259.

[22] S. H. Fan et al., Phys. Lett. A **164**, 295 (1992).

[23] T. Klinger and A. Piel, in *Proc. of the 1992 ICPP*, edited by W. Freysinger et al. (EPS conf. abstr., Innsbruck, Austria, 1992).

[24] R. Timm and A. Piel, Contrib. Plasma Phys. **32**, 599 (1992).

[25] F. Greiner, T. Klinger, H. Klostermann, and A. Piel, Phys. Rev. Lett. **70**, 3071 (1993).

[26] F. Greiner, T. Klinger, A. Piel, and R. Timm, in *Proceedings of the Fourth Symposium on Double Layers and Other Nonlinear Phenomena in Plasmas*, edited by R. W. Schrittwieser (World Scientific, Innsbruck, Austria, 1993), pp. 208–213.

[27] W. X. Ding, H. Wei, X. Wang, and C. X. Yu, Phys. Rev. Lett. **70**, 170 (1993).

[28] H. Klostermann, F. Greiner, T. Klinger, and A. Piel, Plasma Sources Sci. Technol. **3**, 134 (1994).

[29] F. Greiner, T. Klinger, and A. Piel, Phys. Plasmas **2**, 1810 .

[30] T. Klinger, F. Greiner, A. Rohde, and A. Piel, Phys. Plasmas **2**, 1822 (1995).

[31] T. Klinger *et al.*, submitted to Phys. Rev. E (1995).

[32] W. X. Ding *et al.*, submitted to Phys. Rev. E (1995).

[33] W. Arter and D. N. Edwards, Phys. Lett. **114A**, 84 (1986).

[34] H. J. Barkley *et al.*, Plasma Phys. Controlled Fusion **30**, 217 (1988).

[35] C. P. C. Prado and N. Fiedler-Ferrari, Plasma Phys. Controlled Fusion **33**, 493 (1991).

[36] M. L. Sawley, W. Simm, and A. Pochelon, Phys. Fluids **30**, 129 (1987).

[37] V. Budaev, G. Fuchs, R. Ivanov, and U. Samm, Plasma Phys. Controlled Fusion **35**, 429 (1993).

[38] G. Ströhlein and A. Piel, Phys. Fluids B **1**, 1168 (1989).

[39] T. Klinger and A. Piel, Phys. Fluids B **4**, 3990 (1992).

[40] N. Ohno, M. Tanaka, A. Komori, and Y. Kawai, J. Phys. Soc. Jpn **58**, 28 (1989).

[41] M. Kono and A. Komori, Phys. Fluids B **4**, 3569 (1992).

[42] A. Komori, N. Ohno, T. Yamaura, and Y. Kawai, Phys. Lett. A **170**, 439 (1992).

[43] A. Komori, M. Kono, T. Norimine, and Y. Kawai, Phys. Fluids B **4**, 3573 (1992).

[44] A. Buragohain, J. Chutia, and Y. Nakamura, Phys. Lett. A **163**, 425 (1992).

[45] A. Buragohain *et al.*, Chaos, Solitons and Fractals **4**, 677 (1992).

[46] B.K.Sarma, A. Buragohain, and J. Chutia, Int. J. Bif. Chaos **3**, 455 (1993).

[47] L. I and M.-S. Wu, Phys. Lett. **124A**, 271 (1987).

[48] J. H. Chu and L. I, Phys. Rev. A **39**, 233 (1989).

[49] C.-S. Chern and L. I, Phys. Rev. A **43**, 1994 (1991).

[50] L. I and J.-M. Liu, Phys. Rev. Lett. **74**, 3161 (1995).

[51] K. H. Spatschek, in *Proc. XXIth ICPIG*, edited by G. Ecker et al. (APP, Ruhr-Universität, Bochum, 1993), p. 50.

[52] K. Ohe and T. Kimura, J. Phys. D: Appl. Phys. **22**, 266 (1989).

[53] A. Latten, T. Klinger, A. Piel, and T. Pierre, Rev. Sci. Instrum. **66**, 3254 (1995).

[54] R. M. May, Nature **261**, 459 (1976).

[55] M. H. Jensen, P. Bak, and T. Bor, Phys. Rev. A **30**, 1960 (1984).

[56] J. A. Glazier and A. Libchaber, IEEE Trans. Circ. Sys. **35**, 299 (1988).

[57] Y. Pomeau and P. Manneville, Commun. Math Phys. **74**, 189 (1980).

[58] S. E. Newhouse, D. Ruelle, and F. Takens, Commun. Math. Phys **26**, 35 (1978).

[59] P. Burger, J. Appl. Phys. **36**, 1938 (1965).

[60] W. Ott, Z. Naturforsch. A **22**, 1057 (1967).

[61] V. I. Kuznetsov and A. Y. Énder, Sov. Phys.-Tech. Phys. **22**, 1295 (1977).

[62] S. Kuhn, Plasma Phys. **23**, 881 (1981).

[63] G. Musa, in *Proceedings of 3rd International Symposium on Double Layers* (Analele Stiintifice ale Universitatii Al. I. Cuza, Iasi, Romania, 1987).

[64] S. Kuhn, in *Proceedings of 3rd International Symposium on Double Layers* (Analele Stiintifice ale Universitatii Al.I. Cuza, Iasi, Romania, 1987).

[65] F. Bauer and H. Schamel, Physica D **54**, 235 (1992).

[66] R. L. Merlino and S. L. Cartier, Appl. Phys. Lett. **44**, 33 (1984).

[67] S. L. Cartier and R. L. Merlino, IEEE Trans. Plasma Sci. **12**, 14 (1984).

[68] R. A. Bosch and R. L. Merlino, Contrib. Plasma Phys. **26**, 1 (1986).

[69] Y. Jiang, H. Wang, and C. X. Yu, Chin. Phys. Lett. **5**, 201 (1988).

[70] G. Knorr, Plasma Phys. Controlled Fusion **26**, 949 (1984).

[71] H. Sun, L. Ma, and L. Wang, Phys. Rev. E **151**, 3475 (1995).

[72] L. Malter, E. O. Johnson, and W. M. Webster, RCA Rev. **12**, 415 (1951).

[73] W. M. Webster, E. O. Johnson, and L. Malter, RCA Rev. **13**, 163 (1952).

[74] M. Carrère, *thesis*, Université de Provence, Aix-Marseille I (1994).

[75] H. Kolinsky and H. Schamel, preprint, to be published .

[76] H. Lashinsky, in *Symposium on Turbulence of Fluids and Plasmas* (PUBLISHER, Polytechnic Institute of Brooklyn, April 16-18, 1968), pp. 29–46.

[77] M. E. Koepke and D. M. Hartley, Phys. Rev. A **44**, 6877 (1991).

[78] R. Mettin, U. Parlitz, and W. Lauterborn, Int. J. Bifurc. Chaos **3**, 1529 (1993).

[79] E. Ott, C. Grebogi, and J. A. Yorke, Phys. Rev. Lett. **64**, 1196 (1990).

[80] B. E. Keen and W. H. W. Fletcher, Phys. Rev. Lett. **23**, 760 (1969).

[81] P. Michelsen, H. L. Pécseli, J. J. Rasmussen, and R. Schrittwieser, Plasma Phys. **21**, 61 (1979).

[82] H. Amemiya, Plasma Phys. **25**, 735 (1983).

[83] W. X. Ding, H. Q. She, W. Huang, and C. X. Yu, Phys. Rev. Lett. **72**, 96 (1994).

[84] R. Lima and M. Pettini, Phys. Rev. A **41**, 726 (1990).

[85] A. Sekiguchi, A. K. Sen, and G. A. Navratil, IEEE Trans. Plasma Sci. **18**, 685 (1990).

[86] P. Tham and A. K. Sen, Phys. Fluids B **4**, 3058 (1992).

[87] P. Tham and A. K. Sen, IEEE Trans. Plasma Sci. **21**, 588 (1993).

[88] K. D. Weltmann, T. Klinger, and C. Wilke, submitted to Phys. Rev. E (1995), submitted to Phys. Rev. E.

[89] A. Piel *et al.*, in *Dusty and Dirty Plasmas, Noise, and Chaos in Space and in the Laboratory*, edited by H. Kikuchi (Plenum, New York, 1994), pp. 501–521.

H.F. emission related to particle beams injected into ionospheric plasma from spacecraft.

Zbigniew Kłos

*Space Research Centre, Polish Academy of Sciences,
Bartycka 18 A, 00-716 Warsaw, Poland.*

During active injection of an electron beam, and charged neutral beams consisting of ions and electrons into the ionosphere from a spacecraft a broad spectrum of waves is generated. The characteristics of these emissions change substantially with altitude, beam current and ratio of ambient plasma to electron cyclotron frequencies. Many of rocket and satellite experiments performed up to now, have clarified the nature and features of these waves. This paper is concern with a particularly interesting aspect of these experiments, namely the high frequency (h.f.) wave generation.

INTRODUCTION.

Particle beams have been artificially injected into the ionosphere from rockets and satellites for many purposes. One is to provide an opportunity for reproducing naturally occuring phenomena in a more controlled environment. The artificially created aurora (1) was the first active experiment with an electron beam in space.

Charge-neutral (or plasma) beams consisting of ions and electrons injected from a spacecraft present another type of modelling of natural processes such as: solar wind-near Earth plasma interaction, modelling of the artificial comet, and plasma -magnetic field interaction.

The charge particle beam-plasma interaction in the ionosphere may simulate physical processes occurring elsewhere in the universe. It can help to understand the elementary interactions in beam-plasma system in application to astrophysical objects, and the structure and geophysical processes in the near-Earth plasma. On the other hand the near-Earth space can be used as a plasma laboratory without walls and plasma characteristics changing with altitude Fig.1. It opens up a very wide area of investigation of collective interaction of beam particles with plasma, namely the body neutralisation, beam dynamics, energy loss mechanism, and the most important active processes including beam plasma instabilities, wave-particle interaction and associated wave emission, strong turbulence, electron acceleration and radiation.

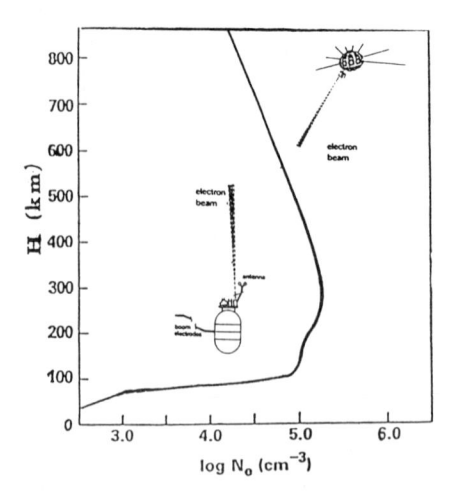

FIGURE 1. Altitude dependence of electron concentration in the ionosphere - the nearest natural plasma laboratory.

A wide range of active experiments has been rewieved in the past (2-8). The current research front is regularly presented in *Advances of Space Research [Active Experiments] COSPAR Proceedings Series*.

This paper is concerned with a particularly interesting aspect of these experiments, namely the h.f. wave generation and is mainly based on the experimental data from G60S rocket and APEX satellite experiments.

SPACE EXPERIMENT WITH INJECTION OF AN ELECTRON BEAM.

Electron beams experiments are one of the major active experiments in the ionosphere, as they provide an opportunity for the following studies: 1. controlled experiments which can test naturally occurring beam-plasma processes, 2. the electron beam-plasma interaction and wave excitation in free space, 3. the electron beam-plasma discharge phenomena, 4. the electron beam-atmosphere interaction, 5. the spacecraft charge neutralisation processes, 6. the idea that pulsed beam may replace conventional antennas in space for communication.

Injection of energetic electrons from an isolated body immersed in plasma creates a set of complex phenomena, the main obstacles for electron beam- space plasma experiment are: spacecraft charging, evolution of initially dense beam, restriction for space time diagnostics, technical malfunction. Generally experimental conditions are significantly different from idealised beam-plasma system, however many experiments have been performed up to now. They are summarised in Tab.1 and 2.

TABLE 1. Electron beam in Space. (rockets)

Experiment	Altitude H[km]	Electron Beam I [A]	E₀[kev]
1. **Hess - Artificial Aurora** (Aerobee - 1969, Strypit 1972)	269	0.49	9.7
2. **ZARNICA 1, 2** (1973, 1975)	109-151	~0.3	9.5
3. **ARAKS 1, 2** (1975)	110-196	~0.5	27 / 13
4. **Japanase K: 1 - 5** (1975)	178-320	~0.001 / 0.35	0.3 / 5
5. **ECHO 1- 7** (1970-1988)	215-350	0.1	30-40
6. **AFGL (Exede)** (1974, 1976)	106-135	1-10	3
7. **ELECTRON 1,2** (1978)	192	0.1	10
8. **Polar (Series)** (1976)	220	0.1	10
9. **G-60 -S** (1981)	1500	0.5	8
10. **MAIMIK** (1985)	400	0.8	8
11. **CHARGE 1, 2** (1983-1985)	240	0.04	1
12. **SCEX** (1982,1990)	211	0.05	6

TABLE 2. Electron beam in Space. (satellites)

Experiment	Orbit H[km]	Electron Beam I[A]	E₀[Kev]
1. **EXOS -B** (1978)	400- 6 R	0.001	0.2
2. **SCATHA**(1979)	5-8 Re	0.001	0.5-3
3. **VCAP - POP**(1982) (ST-3)	250	0.1	1
4. **PICPAB**(1983) (Spacelab-1)	250	0.1	8
5. **SEPAC** (Spacelab-2 1985)	250	0.1 modulated 1.22kHz	5
6. **APEX**(1991)	400-3000	0.15 modulated 40kHZ	10

Boundary condition for the electron beam emitted from spacecraft.

The isolated conducted sphere (which can model a spacecraft) of radius a [m] continuously emitting negative current I [A] into the vacuum will charge up to the positive potential at the rate:

$$\frac{d\phi}{dt} = 9\frac{I}{a} \, kv/\sec$$

For a = 1m and I = 1A during 1 µsec the body potential can reach the value of 9 kV and the escape of the electron beam with energy of some kiloelectronovolts will be protected. The simple one-dimensional analytical calculation (9,10) shows that the dynamics of escape electrons is determined by the time when the first emitted electron is stopped at some distance from the body. This stagnation time - t_S is determined as:

$$t_S \approx \omega_{pb}^{-1} \quad \text{here } \omega_{pb} \text{ - is electron beam plasma frequency}$$

When the injection is into the ambient plasma, plasma can provide a return (neutralisation) current to the body which (without ambient magnetic field) can be estimated as $I_o = \frac{1}{4} eNp \, V_{Te} \, A$; here A - body collecting area, e- electron charge and Np - electron plasma density, V_{Te} - electron thermal velocity which for the inosphere can reach - 10^6 cm^{-3}, and 10^7 cm/sec respectively. Thus, in ionospheric conditions for A = 1 m^2, I_o < 10 mA and it is impossible even with ambient plasma to compensate the spacecraft charge.

In the body emitted electron beam-plasma system, plasma response can be characterised by plasma response time :

$$t_R \approx \omega_{pe}^{-1} \quad \text{here } \omega_{pe} \text{ - ambient plasma electron frequency.}$$

After a more detailed computer simulation of the electron beam dynamics (11) it has become clear that the injection efficiency depends on the relation between these two times namely for :

$t_R > t_S$ - beam electrons are confined in the vicinity of the spacecraft
$t_R < t_S$ - beam is able to escape from the near spacecraft environment

To present the characteristics of the electron beam, ambient plasma and spacecraft interaction it is useful to refer to a one - dimensional case (without ambient magnetic field), which is described by the equations of motion for the particles

$$\frac{dU_n}{dt} = \frac{q}{m_n} \cdot E(z,t) \qquad n=b,e,n \qquad (1)$$

Un= velocity of beam electrons, plasma electrons and ions respectively ; q, m_n charge and mass of particles, E electric field of the charged body.

The Poisson's equation in the integrated form

$$E(z,t) = E(o,t) - 4\Pi e \int n(x't) dx \qquad (2)$$

and boundary condition at the body.

$$E(0,t) = 4\Pi \int_0^t [jb(0,t') - je(0,t') - jbr(0,t') + jp(0,t')] dt' \qquad (3)$$

here j $_b$, is the beam current j $_{br}$ - is the beam return current and j$_e$, p the ambient neutralisation current, E (x $_f$, 0) = 0 at the beam front.

The results of numerical calculations are presented in Figs.2-4 (12). They are in one-dimensional geometry, in which ambient plasma electron and ion currents flow into the beam front behind which they are exposed to electric charge on the body like emitted beam electrons. It is evident that for such strong beam injection the beam electrons group at some stagnation point (also called virtual cathode); separated from the body Fig.2. The beam concentration at this point oscillating in time reaches minimum values when beam particles return to the body or escape.

Thus even when the beam is injected continuously, its motion occurs impulsively, the result is not obvious at all. The plasma ions are repelled and build up a positively charged sheet at this distance from the body. Fig.3. The plasma electrons give fluctuating neutralisation current to the body, being trapped locally at some distance from the body in local potential wells. The electric potential and field on the body which oscillate with time have a strong spatial cut off at the distance at which an ion sheet is built up.

In case of an escaping beam in the presence of ambient magnetic field the beam particle dynamic as presented by many two-dimensional and three-dimensional computer simulations is determined by the magnetic field, only close to the body.

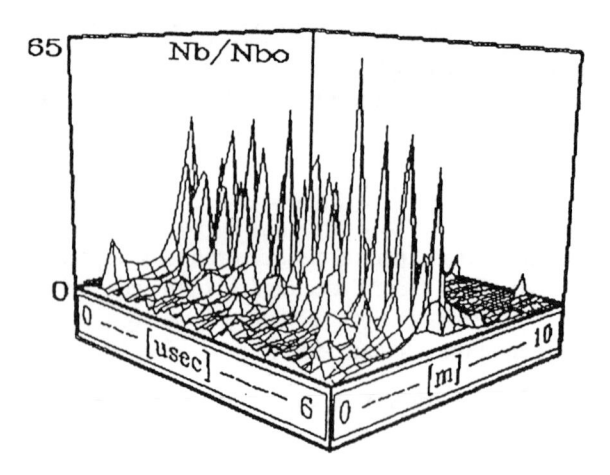

FIGURE 2. The electron beam density as function of time and distance from the body for continuous injection. The pulsing returns of beam particles to the body can be seen.

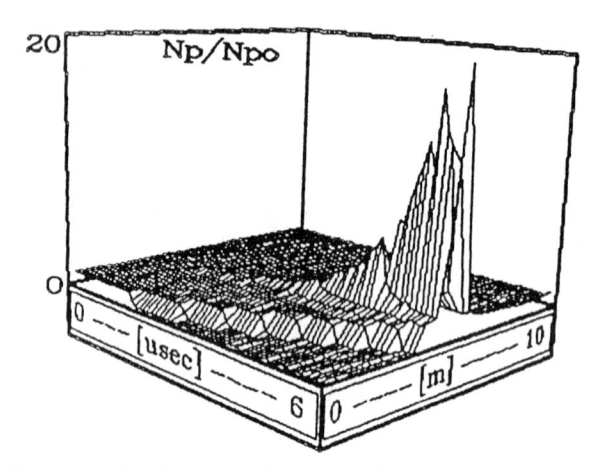

FIGURE 3. The plasma ion density as function of time and distance. The build up of ion sheet at some distance can be seen.

After some Larmor periods the collective plasma interaction forms a hollow cylinder (3,13,14,15), and the strong electrostatic repulsive field in the virtual cathode region causes the radial expansion (16).

One of the main goals of many experiments with electron beam in space was to avoid beam stagnation. Increase of current collection area, and of ambient plasma density close to the spacecraft by parallel plasma injection were used to increase neutralisation current (2).

During injection of the pulsed electron beam the period of modulation-exactly the ratio of beam on time to beam off time-is another characteristic parameter which determines the payload charging and in result the beam dynamics. When the beam on time is of the order of plasma response time and the beam of time is much longer, the ambient plasma has sufficient time to produce charge neutralisation of the spacecraft. In result a sequence of package of beam electrons can escape from the spacecraft vicinity, and injection is efficient (17).

The H.F.plasma waves excited by the electron beam in the ionosphere.

The unescaping electron beam .

In case of creation of stagnation point beam electrons return to the body and decrease its electric potential. The first electron is followed by other electrons returning to the body, and it can happen that many electrons reach the body at the same time. In this case a sharp decrease of the body potential occurs. Thus we obtain a return electron beam loop in the vicinity of the body. When the potential barrier drops below the injected beam energy, the loop breaks down and the potential starts to increase again. Thus the state of periodic charging and discharging of the body is obtained . The period of this process depends among others on plasma concentration, beam current and beam energy.
The negative space charge of the electron beam loop and the positive charge of the body by varying their relative configurations with time create the dynamic electric multipole which can radiate broad frequency range.

The rocket experiment Gruziya-60- SPURT (18,19,20) was one, in which a strong beam (I_b = 0.5 A ; E_b = 6.2 keV) was injected from the payload up to 1500 km of altitude, the plasma environment parameters ranging from these presented in the lower ionosphere to these in the thermosphere Fig.1. The waves related to the injection were measured on board in 0.1-10 MHz and 20-80 MHz range.

The observed emission in frequency range above 20 MHz appeared to be one of the unexpected phenomena and :
- At high altitude (lower ambient plasma density) this emission was considerable and decreased towards the end of the range, its structure being irregular.
- The emission appeared simultaneously with the increasing beam current density.
- For higher ambient plasma frequency the emission shifted to lower frequency and practically disappeared at the altitude of maximum local plasma frequency.

This experiment shows how much the process is controlled by the ratio t_S/t_R. The example of measured spectra is presented in Fig.4. The model calculations based on interaction of the electron with the charged payload are also presented . The correspondence of measured and calculated spectra is very good.

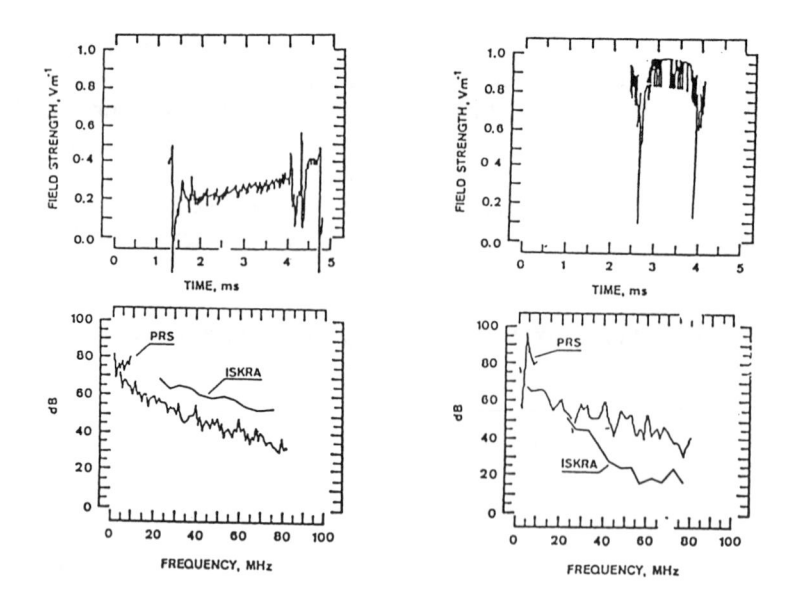

FIGURE 4. Time histories of electric field and the correspondence power spectra due to the electron beam loop configuration, calculated for the vicinity to the H.F. aerial and for condition at 1494 km(left) and 305 km(right) of altitude. The frequency spectrum from the model is shown together with the observation.

Similar h.f. emission (well beyond the ambient plasma frequency) was registered during mother-daughter rocket experiment CHARGE-2. In this experiment electron beam with energy of 1 keV and currents up to 40 mA was injected up to 262 km of altitude and h.f. noises were observed from separated payload (21).These two experiments have shown that the spacecraft-beam, plasma interaction generate fluctuating electric field in a broad frequency range.

The escaping electron beam.

When $t_R < t_s$ the electron beam escapes from the payload and collectively interacts with ambient plasma pervaded by the geomagnetic field. In this case the broad spectrum of plasma mode can be excited.

The Echo-7 rocket experiment results in which wave emission in frequency range 0-8 MHz related to the quasicontinous emission of electron beam (I_o=180 mA, E_o=36 keV) at different pitch angles, displays the rich nature of the excited spectra. The observations have been performed on the payload separated from the gun (22).

The Echo-7 observed spectra show the broad- band waves below the electron cyclotron frequency and narrow-banded emission near the harmonics of the electron cyclotron frequency. The observation yielded results similar to those obtained with electron beam on Spacelab1 in PICPAB experiment. Here the h.f. emission related to the electron beam (I_o =100 mA, E_o=8 keV) displays, successive electron cyclotron harmonics gathered together in periodic packet with maximum at the local plasma frequency and its harmonics.[23]

Electron beam generated waves near plasma frequency, its harmonics and upper hybrid frequency have been observed in SCEX series of rocket experiments (8) as well.

Another case occurred during the APEX satellite experiment. The APEX spacecraft was launched into an orbit with apogee of 3080 km, perigee 440 km; plasma density was in the range $10^5 - 10^3$ cm^{-3}. The electron gun operated in an unsynchronized mode sending 2 μs long pulses at the rate of 40 Khz. The pulse current did not exceed 0.15A and beam energy was 10 keV - with this parameter the initial beam density n_b was of the order of 10^8 cm^{-3}.

The r.f emission was registered on the same payload where the electron gun was located. The essential features of spectra excited during pulsed beam injection were observed.(24). The examples are presented in Fig 5-7.

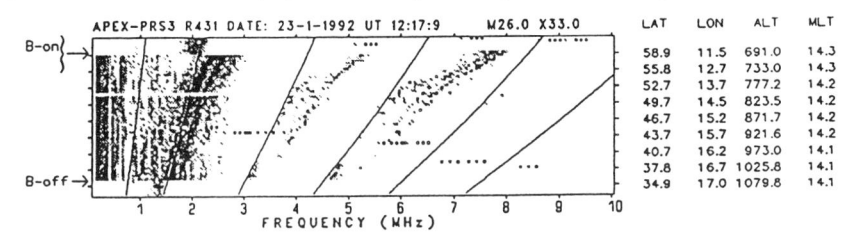

FIGURE 5. The sequence of h.f. spectra observed along the APEX satellite orbit. Intensity increases from white colour to black colour [26-33db] [1μv] respectively. Beam on and beam off are marked, the solid lines indicate the cyclotron frequency and its even harmonics.

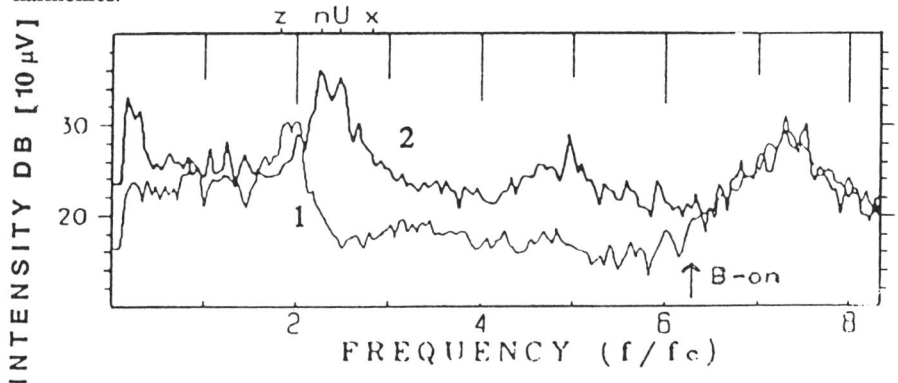

FIGURE 6. Two consecutive APEX Spectra (1- background and beam-on; 2- beam -on) showing dramatic change in the vicinity of local frequency due to beam injection. U - upper hybrid, n - plasma frequency and z and x, are z- mode and x- mode cut off frequency respectively. The harmonics of (f_p, f_u) band are visible.

The ambient spectra, prior to beam injection have pronounced structure which corresponds to the satellite electrical system noise. After injection of the electron beam a very prominent doublet emerges at frequencies corresponding to the plasma and upper hybrid frequency (f_p, f_u) band of the ambient plasma. This is illustrated in greater detail in Figure 6. The doublet structure evolves slowly along the orbit both in its form and frequency. These changes are in reasonable accordance with the expected changes of local plasma frequency. Assignment of the doublet to waves excited in frequency band (f_p, f_u) of the ambient plasma seems to be justified. The harmonics of this structure are also excited. Initially only harmonics of the higher frequency doublet member (f_u) are present. As the lower frequency member (f_p) approaches $2 f_c$ its harmonics also appear. The data presented suggest that intense emission emanates from places where resonance frequency of beam-plasma system is close to ambient plasma frequency f_p and that the ratio of f_p/f_c strongly controls the overall spectral structure. Another crucial factor for spectra interpretation comes from the observation that in many cases, spectral peaks exhibit fine structure in the vicinity of computed gyroharmonics or below them. These characteristics can be seen in Fig. 7, where spectra in order of increasing value of f_p/f_c are presented. The radiation slightly below 2fc and its harmonics, coherent with upper hybrid and its harmonics, and coupled upper hybrid- cyclotron beam emissions are presented respectively. On the other hand the excitation of Z-mode is presented in Fig.8.

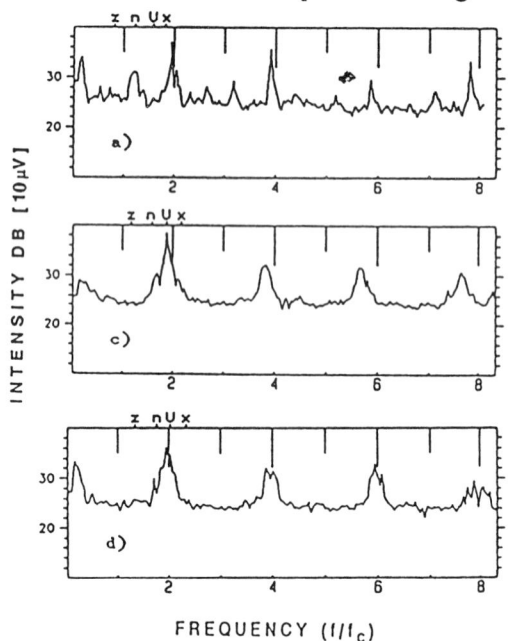

FIGURE 7. Selected spectra illustrating diversity of radiation generated by the electron beam in APEX experiment, n, U, z, x means f_p, f_u, z - mode and x-mode cut off frequency, respectively.

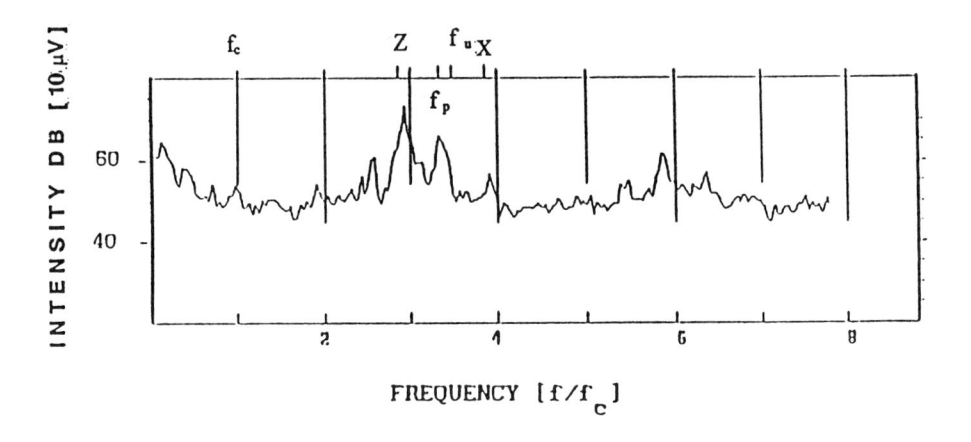

FIGURE 8. The Z - mode generated by the electron beam and observed in APEX experiment. the z, x cut off frequency, as well as the upper hybrid f_u and plasma f_p frequencies are marked.

Computer simulations were relevant in understanding of beam dynamics and possible wave generation mechanism. First of all the 1D electrostatic simulation (10), semiempirical one (17) and hydrodynamic modelling (25) clarified the physics of the processes. Oscillation of spacecraft potential for quasicontinous beam injection was the essential result. Much more realistic the 2D electromagnetic simulation modelling clarified not only the electron beam particles dynamics but also the behaviour of electric field outside and inside the beam (22). The wave field was determined selfconistently with the beam injection spacecraft charging and the return current system. It was confirmed, which was observed in experiments that the generation mechanism strongly depends on the f_p/f_c ratio. The detailed 2D simulation (22) showed that: for strongly magnetised plasma ($f_p/f_c <1$):-Whistlers waves-are driven by the parallel beam energy, fundamental Z mode and $2f_c$ of x mode -are generated by the electron maser instability driven by perpendicular beam energy or ExB drifts. It correlates with Echo -7, while Apex observations shows occurrence of these modes for $f_p/f_c \approx 3$. For weakly magnetised plasma ($f_p/f_c >1$) the 2D computer simulation (22) clarified that : nfc - maser emission is identified up to n = 4 and can operate near the spacecraft. First Bernstein mode and mode near $2f_u$ (i.e. 1.5 and 2.5 fc for ionospheric condition) are identified away from the spacecraft. These modes of generation have been observed in the APEX experiment.

The 2D simulation did not represent strictly the physical condition of the injection but the prediction of the essential h.f. characteristic like the Z mode generated by maser instability is numerous.

The selfconsistent 3D simulation performed by Pritchett (15) confirmed the oscillating character of the electric field, both inside the beam and outside, up to h.f. range. The result is presented in Fig.9.

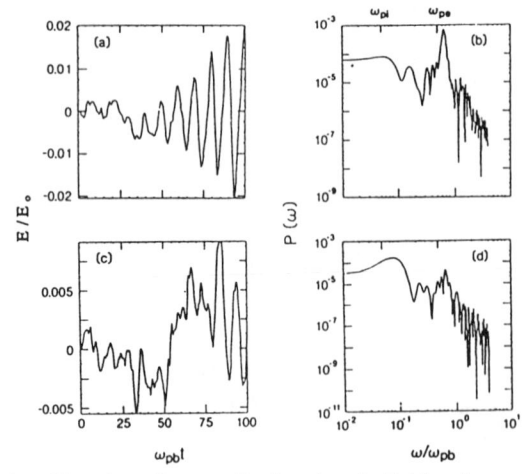

FIGURE 9. The time histories of perpendicular electric field and corresponding spectra outside (a,b) and inside (c,d) the electron beam cylinder obtained from 3D computer simulation (courtesy dr.P.L. Pritchett) Eo - the reference electric field value used in the simulation. [15]

The semiempirical modelling of the Apex spacecraft charging in the case of pulsating beam injection has showed that the spacecraft charging potential is periodic, depends on the ambient plasma frequency, and can even reach negative values (17). In APEX case the efficient escape of beam takes place.

Finally it can be concluded that the h.f. emission observed in Echo-7, SCEX rockets, Spacelab-1 and Apex experiments suggest that these plasma waves modes are generated by beam-plasma interaction, propagating in the medium and:
- The emission is controlled by the ratio f_p/f_c as well as the pitch angle of electron beam.
- For quasi perpendicular injection, the z mode, $2 f_c$ of x-mode and its harmonics, as well as Bernstein mode close, and upper hybrid mode and its harmonics are generated by perpendicular beam energy.
- For quasi parallel injection whistlers are generated as well as waves at $1.5 f_c$ and at $2.5 f_c$.
- Intense emission emanates from places where resonance frequency of beam - plasma system is close to ambient plasma frequency fp and the spectral structure is controlled by ratio f_p/f_c.
- Spectral peaks exhibited fine structure in the vicinity of nfc harmonics or slightly below them and harmonics of upper hybrid band.

H.F. WAVES RELATED TO THE PLASMA BEAM.

The charge neutral beams injected from the spacecraft have the advantage that they are not so strongly affected by the charging as electron beams. Several rocket and satellite experiments have been performed. They are summarised in Tab.3.

TABLE 3. Ion beam in Space (rocket and satellite)

Experiment	Plasma beam characteristic	Altitude (km)
Porcupine (Rocket)1979	Xe^+, E_i =200eV E_e= 2-3eV	400
Meteor (satellite)1977, 79	Xe^+, E_i= 130eV 1-3eV	800
ARCS 1,2,3,4 (rocket)	Ar^+ E_o 100 eV, 200 eV E_e = 1-2 eV	400
AELITA 1,2(rocket) 1978,79	Li^+, E_i 8.3 eV 2-3 eV	145
PLAZMA (rocket) 1985	Li^+F_i 8.3eV, E_e=2-3eV	150
SEPAC (Spacelab-1) (1985)	Ar^+ E_i =80eV E_e=3eV	250
APEX (satellite) 1991	Xe^+, E_i =200eV E_e= 2-3eV	400-2800

Ions play the crucial role in these beams dynamics and associated plasma heating. Ions from the beam have in the ionosphere the gyroradii of a few hundred meters and closely to the spacecraft can propagate across the geomagnetic field line. The neutralised electrons are pulled from the gun by ions but as their gyroradii is much smaller their motion is fieldaligned. Due to charges separation in the area inside the beam a strong electric field is built up. As was observed in PORCUPINE (26,27) these polarisation fields cause the gun electron motion with beam ions. These electrons are left behind and escape up and down the geomagnetic field. Together with ambient electrons they create the neutralised current of the magnitude of ion beam current. This beam particle dynamics has been confirmed in 2D computer simulation (28,29).

These currents are energy source for current-driven instability responsible for wave growth near the spacecraft (30). The current driven instability can excite not only low frequency waves but also the high frequency ones.

Observation of h.f. plasma beam related waves.

A limited number of active experiments with the injection of plasma beam in which H.F. waves were observed has been carried out in the upper ionosphere. From the rocket experiments it has been concluded that:

- the enhanced and unstructured spectrum in the 0.1-10 MHz frequency range is generated with amplitude increase by factor 2 (31).
- the H.F. broadband structure occurs in a very limited spatial region around the beam and it is electrostatic with narrow - frequency band emission around harmonics of f_c (30).
- the H.F. broadband structure is electrostatic in nature and is created only in the region penetrated by the ion beam (32).

Recently a series of experiments have been performed on the APEX spacecraft with xenon plasma beam. The h.f. waves were registered with swept frequency analyser 0.1.-10 MHz in the spacecraft vicinity.

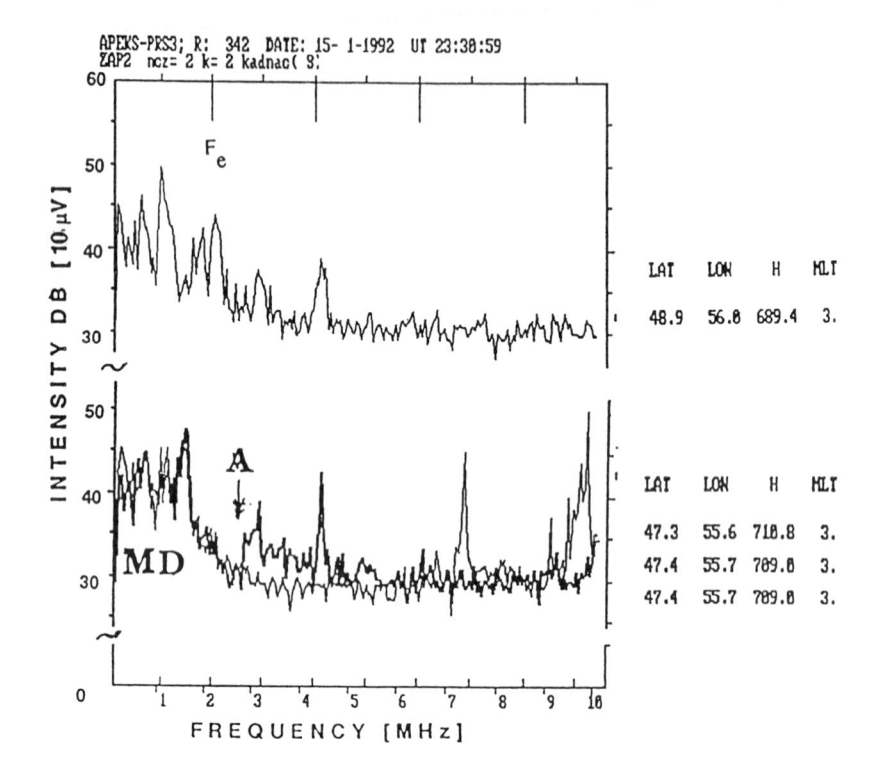

FIGURE 10. The spectra associated with the beam plasma(Xe^+,e) injection in APEX experiment. MD-gun heated cathode only, A - beginning of Xe^+ acceleration, respectively At the top the fully developed spectrum with spikes at electron cyclotron harmonics F_e is presented.

The observation along the satellite orbit shows that broad spectra of enhancement over background emission are generated in 0.1-10 MHz range (33). The process of wave excitation is strongly related to beam ion acceleration. It can be seen in Fig.10. in which the gun heated cathode, and accelerated ion related spectra are shown respectively. The electron cyclotron harmonics in the case of fully accelerated ion beam are very well manifested. This type of harmonics emission is similar to that observed in Porcupine, as well as in Ampte project in the solar wind plasma.

Thus it is evident that neutralised plasma beams (cloud) moving in the ambient plasma pervaded by the geomagnetic field generate h.f. emission at electron cyclotron harmonics. The unstructured broad spectrum is the result of ambient plasma heating in the plasma beam region which was observed in the experiment (31) and was concluded from computer simulation (28). The instability at electron cyclotron harmonics in current driven plasma is the most appropriate candidate to account for this emission.

CONCLUSION

Generation of h.f. waves related to electron beam injected from a spacecraft depends on the beam and ambient plasma parameters. In the case when the electron beam cannot escape from the spacecraft vicinity, the h.f. field is not classical beam-plasma mode interaction, but rather electron beam-charging spacecraft interaction. For the escaping electron beam the broad range of waves (cyclotron harmonics, plasma and upper hybrid harmonics, Z mode) are excited. All these experiments make detailed analysis of the generation mechanism possible.

The h.f. emission related to plasma beam is not so rich. It has electrostatic nature and unstructured spectra which occupy a brad frequency range. The appearing fine electron harmonics structures are related to the neutralised currents driven by the beam plasma expansion. The analysis of growth of the harmonics makes the estimation of the neutralisation current build up time possible.

REFERENCES

1. Hess,W.N., Trickel M.C., Davis T.N., Boggs W.C., Kraft G.E., Stassinopoulos E., and Maier E.J., J. *Geophys Res.* **76**, 6067-1971
2. Winckler, J.R., Rev. *Geophys Space Phys.* **18**, 659-682,(1980)
3. Grandal, B., ed., *Artifical Particle Beams in Space Plasma Studies*, Plenum Publ. 1982.
4. Kawashima, N., J.*Geomag Geolectr.* **40**, 1269-1281, (1988).
5. Shawhan, S.D., Murphy, G.B., Banks, P.N. Williamson, P.R. and Raitt, W.J., Radio Science, **19** 471- 1984.
6. Szuszczewicz, E.P., J. *Atm. Terr.Phys.*, **47**,1189-1210, 1985.
7. Papadopoulos, K. and Szuszczewicz, E.P., *Adv. Space Res.* **8** (1) 101- (1) 110,1988.
8. Kellog, P.J., *Adv. Space Res.* **12**, (12)15-(12)28, 1992.
9. Parks, D.E., Wilson, A.R. and Katz, I. IEEE Transactions on Nueclear Science, Ns - 22, 2368-2373,1975.

10. Winglee, R.M. and Pritchett P.L., J. Geophys Res., 6114-6126,1987.
11. Pritchett, P.L. and Winglee , R.M. , J.*Geomag. Geoelectr.* **40** ,1235-1256,1988.
12. Klos,Z., Zbyszynski,Z. Managadze, G.G., Gagua, T.I. and Leonov, N.A. , *Adv.Space Res.,* **10** ,7(155)-7(158),1990.
13. Ashour- Abdalla , M. and Okuda , H. , *Adv.Space Res.* **8** 1(137)-1(149),1988.
14. Pritchett, P.L. and Winglee R.M. J. *Geophys. Res.* , **92** , 763-7688,1987.
15. Pritchett , P.L. J. *Geophys.Res.*, **96**, 13781-13793,1991.
16. Koga, J. Lin, C.S., J. *Geophys Res.,* **30**, 3971-3983, 1994.
17. Zbyszyński, Z. and Klos,Z. *Adv. Space Res.* **15**, (12)29-(12)32, 1995.
18. Managadze, G.G., Balebanov, V.M., Burchudladze, A.A., Gagua, T.I., Leonov,N.A., Lyakhov, S.B., Martinson, A.A., Mayorov, A.D., Riedler, W.K., Friedrich, M.F., Torkar, K.M., Laliashvili, A.N.,Klos.Z., Zbyszynski.Z., *Planet.Space Sci.* **36**, 399- 410,1988.
19. Klos, Z., Zbyszynski,Z. Managadze, G.G., Lyakhov, S.B., Gagua, T.T., Tarkar, K.M., Friedrich, M. and Riedler, W. **8** , (1) 119-(1)122,1988.
20. Klos, Z.,*H.F.Emission in the Topside Ionosphere , in Proceedings of the International Workshop - The Solar Wind - Magnetosphere System.* Graz, Austria 1992 pp 319-334.
21. Neubert,T., Sasaki,S.,Gilchrist, B.E.,Banks P.M., Williamson, P.R., Fraser-Smith,A.C. and Raitt,W.J., J. *Geophys.Res.* **96**, 9639-9654,1991.
22. Winglee, R.M. Kellog, P.J., J .*Geophys. Res.* **95**, 6167-6191,1990.
23. Mourenas, D and Begin,C. *Radio Sci.,* **26** , 469-479,1991.
24. Kiraga ,A. Kłos , Z. Oraevsky, V.N., Dokukin V.C and Pulinets, S.A , Adv. Space Res. vol. 15 (12) 21- (12) 24, 1995.
25. Oraevsky, V.Na., Ruzhin, Ya, Ya , and Dokukin, U.S., *Adv.Space Res.* **12**, (12)43-(12)47, 1992.
26. Hausler, B. Treumann, R.A., Bauer, R., Carlson, C.W., Theile, B. Kelloy. M.C. , Dokukin , V.S., Ruzhin, Yu.Ya., J.*Geophys. Res.* **91**, 287-303,1986.
27. Hearendel , G. and Sagdeev, R.Z., *Adv.Space Res.* **1**, 29-46,1981.
28. Winglee, R.M. and Pritchett , P.L. J.*Geophys. Res.* **92**, 7689-7704,1987.
29. Papadopoulos, K., Mankofsky, A, Davidson, R.C., Drobot, A.T., Phys. Fuids B3, 1075-1090,1991.
30. Pottelette, R., Illiano, J.M., Bauer, O.H. and Treumann, R., J.*Geophys.Res.* **89**, 2324-2334,1984.
31. Erlandson, R.E., Cahill, L.J. , Pollock, G.J., Arnoldy, R.L., Scales, W.A. and Kinter , P.M. , J.*Geophys.Res.*, **92**, 4601-4616,1987.
32. Klos, Z. Zbyszynski. Z, Agafonov, V.F., Managadze, G.G. and Magarov , A.D., *Adv. Space. Res*, **13**, (10)91-(10)94, 1993.
33. Klos, Z. Kiraga, A., Zbyszynski, Z., Dokukin, V.C., Oraevsky, V.N., Pulinets, S.A., Sauer, K., Baumgartel, K., *H.F. emission observed during particles injection from APEX satellite,* 30th COSPAR Scientific Assembly, Hamburg , Germany 1994.

Different Aspects of Self-Consistent Modeling of Non-Equilibrium Discharges

Mario Capitelli, Claudine Gorse,
Daniela Iasillo and Savino Longo

Centro di Studio per la Chimica dei Plasmi del CNR
Department of Chemistry, Bari University, 70126 Bari, Italy

Abstract. Self-consistent models for homogeneous and one dimensional DC and RF discharges are illustrated for different chemical systems. In particular the coupling between vibrational kinetics, electronic kinetics and Boltzmann equation for the electron energy distribution function is illustrated in the case of nitrogen plasmas. Different self-consistent models including the homogeneous solution of Boltzmann equation, fluid model and PIC-MCC (particle in cell with Monte Carlo collisions) are then used for understanding RF discharges. In all cases it is emphasized the role of the coupling between heavy particle kinetics and free electrons dynamics.

INTRODUCTION

Modeling of non-equilibrium discharges has received considerable interest in the last 15 years due to the conviction that plasma technology can not progress without a serious understanding of the medium in which the relevant plasma processes occur. Therefore we have assisted to the development of a large number of sophisticated models improving both the included elementary processes as well as the treatment of electrical quantities. Prevalence of one aspect on the other one was mainly due to the back-ground of the different research groups working on the subject. In particular researchers coming from the chemical physics community favored the fundamental aspects of the processes while researchers coming from mechanical and electrical engineering groups were more interested in electrical and fluid dynamic aspects of the discharge.

This situation is also reflected on the used vocabulary. In particular the adjective 'self-consistent' is employed by the different groups for phrasing different thoughts.

As an example the chemical physics community introduced it for indicating a "self-consistent set of cross sections", a "self-consistent solution of Boltzmann equation for the electron energy distribution function and of vibrational and electronic master equation", while engineering groups mainly adopted it in a "self-consistent description of the electrical characteristics of a discharge".

Of course it is quite difficult to review the efforts made by both communities in the development of the different self-consistent models so that we are forced to favor the aspects we are more familiar with. In particular we will consider DC and

© 1996 American Institute of Physics

RF discharges preferring the cases where state to state kinetics plays a significant role.

COUPLING OF THE STATE TO STATE KINETICS WITH THE BOLTZMANN EQUATION

We start our analysis with the nitrogen system that can be considered one of the most studied system in molecular plasmas.

A serious kinetics for this system should include

- 1) a state to state vibrational kinetics of the ground electronic state $X^1\Sigma_g^+$,
- 2) a dissociation kinetics,
- 3) a collisional radiative model for the electronic states,
- 4) a description of the electron energy distribution function (eedf) by an adequate Boltzmann equation,
- 5) an ionization kinetics,
- 6) a self-consistent calculation of the reduced electric (E/N) field necessary to sustain the discharge.

All these points have been separately analyzed in the past by using uncoupled models. More recently it appeared clear that only a self-consistent treatment of all kinetics can realistically describe the properties of the system.

Let us however start the analysis of points 1 to 6 using uncoupled or partially coupled models.

The first attempts to study the vibrational kinetics of N_2 under DC electric discharges were made by Polak (1) and Capitelli et al. (2). Both groups solved a system of vibrational master equations including the following processes

$$e + N_2(v) \rightarrow e + N_2(w) \qquad \text{(e-V)}$$

$$N_2(v) + N_2(w) \rightarrow N_2(v+1) + N_2(w-1) \qquad \text{(V-V)}$$

$$N_2(v) + N_2 \rightarrow N_2(v-1) + N_2 \qquad \text{(V-T)}$$

Dissociation was also allowed either in the so called pure vibrational mechanism (PVM) or by direct electron mechanism (DEM).

According to PVM the vibrational quanta that cross a pseudo-level (v' + 1) located just above the last bound level of the molecule (v') can dissociate, i.e.

$$N_2(v') + N_2 \rightarrow N_2(v'+1) + N_2(v) \rightarrow 2N + N_2 \qquad \text{(VT)}$$

$$N_2(v') + N_2(v) \rightarrow N_2(v'+1) + N_2(v-1) \rightarrow 2N + N \qquad \text{(VV)}$$

On the other hand dissociation can occur also through direct impact with free electrons

$$e + N_2(v) \rightarrow e + 2N \qquad \text{(e-D)}$$

Polak et al. and Capitelli et al. using essentially the same model were able to show that PVM mechanism can prevail on DEM specially for low reduced electric fields and low translational temperatures. The calculations of the two groups were performed using a Maxwell distribution function for the electron energy distribution function (eedf). The main difference occurred for the vibrational kinetics, Polak used a stationary vibrational kinetics while Capitelli et al. used a time dependent approach.

Later on Cacciatore et al. (3) developed a new model that couples the Boltzmann equation with the system of vibrational master equation, the coupling occurring through inelastic processes involving vibrational levels as well as through superelastic vibrational collisions (SVC).

This new model, while confirming the previous results on the non-Boltzmann character of vibrational distribution function (vdf) and on the importance of PVM mechanisms in the dissociation and ionization of N_2, also showed a strong coupling between eedf and vdf. This coupling coming mainly from the process

$$e + N_2(v) \rightarrow e + N_2(w) \qquad v>w$$

which is able to return energy to electrons partially compensating that one they lost in the corresponding inelastic process. Practically the role of SVC is such to enlarge the bulk of eedf increasing the rate coefficients of electron-molecule processes with high energy threshold.

A similar situation was found for the ionization mechanism. In this case the electron impact mechanism, i.e., the process

$$e + N_2(v) \rightarrow e + N_2^+ + e$$

is unable to explain the ionization degree of a nitrogen plasma at low E/N (E/N < 60 Td) so that associative ionization process involving vibrationally excited molecules and metastable electronic states were invoked to explain the ionization process.

Further work performed by Loureiro and Ferreira (4) and by Boeuf and Kunhardt (5) essentially confirmed the results of our group.

It should be noted that the approach previously reported was able to reproduce the experimental CARS vibrational distributions of N_2 in flowing DC discharges (6) as well as the vibrational distributions of electronic states of N_2 (7-8).

Moreover a similar treatment was also used successfully to explain the highly non-equilibrium vibrational distributions of CO pumped by RF flowing He-CO discharges (9) as well as by nitrogen post discharges (10).

Apparently the model developed in reference 3, which essentially solves a system of vibrational master equations and the Boltzmann equation, contained many microscopic details able to reproduce several experimental observations.

However Paniccia et al. (11) showed that not only superelastic vibrational collisions but also superelastic electronic collisions (SEC) can affect eedf and the related quantities. The model developed in reference 3 was therefore enriched by a collisional-radiative model involving the most important electronic states of N_2. In particular we developed a kinetics for A, B and C states of N_2 coupled to the vibrational kinetics as well as to the Boltzmann equation (12). These new kinetics were coupled with the other ones not only through superelastic electronic collisions but also through the possibility of an inter-exchange of vibrational and electronic energy, a problem that is receiving increasing attention (see reference 13).

A sample of eedf calculated with this model either in discharge or post discharge conditions is illustrated in figures. 1-2. From both figures we can appreciate the coupling between vibrationally and electronically excited molecules and the eedf, the first affecting eedf in the energy range < 5eV the second having appreciable effects for energies> 6eV.

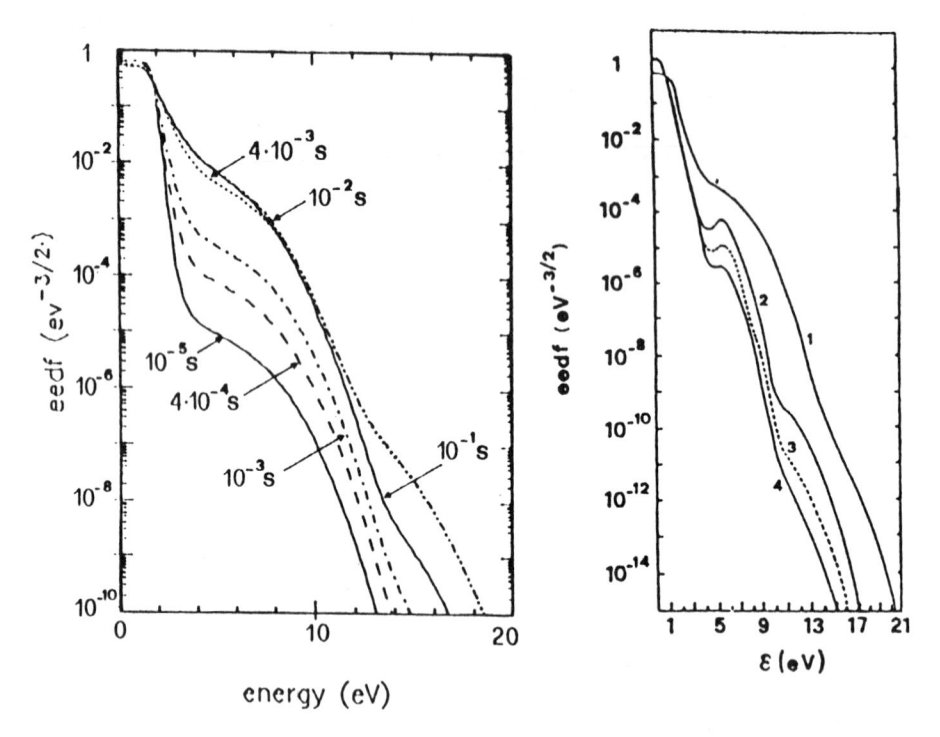

FIGURE 1. Time evolution of the eedf in a nitrogen discharge
($p=5$ torr, $n_e=10^{11}$ cm^{-3}, E/N=30 Td)

FIGURE 2. Time evolution of the eedf in a nitrogen afterglow showing the effect of excited states. (t_{pd}: residence time in the post-discharge, C_A = $N_2(A^3\Sigma_u^+)/N_{tot}$, $C_B = N_2(B^3\Pi_g)/N_{tot}$). curve 1: $t_{pd}=0$s, $C_A=1.4\ 10^{-5}$, $C_B=1.7\ 10^{-6}$; curve 2: $t_{pd}=4\ 10^{-7}$s, $C_A=1.4\ 10^{-5}$, $C_B=1.7\ 10^{-6}$; curve 3: $t_{pd}=10$ms, $C_A=1.4\ 10^{-5}$, $C_B=1.7\ 10^{-6}$; curve 4: $t_{pd}=37$ms, $C_A=1.4\ 10^{-5}$, $C_B=1.7\ 10^{-6}$.

Similar results were obtained by Nagpal and Ghosh (14) who, however, considered a complete collisional-radiative model for singlet and triplet states of N_2.

Apparently one could claim that our knowledge on the elementary processes occurring in N_2 discharges is high enough to now insert all the kinetics in more accurate models giving also the electrical characteristics of a discharge. As an example one could add an ionization kinetics for obtaining the electron density able to sustain a discharge at fixed values of E/N. This approach was followed by Gorse (15). Her results however, which are shown in figure 3, indicate a non stationary behavior of eedf, for a cylindrical discharge of 1 cm radius at E/N = 80 Td and p = 2 torr, in contrast to experimental results. Note however that the temporal evolution of eedf again contains all superelastic effects even though at longer times eedf converges toward the cold gas approximation since the abrupt decrease of electron concentration is not any more able to sustain considerable concentrations of vibrationally and electronically excited states.

This behavior is probably due to the use of crude rate coefficients for some of the numerous processes inserted in the whole kinetics.

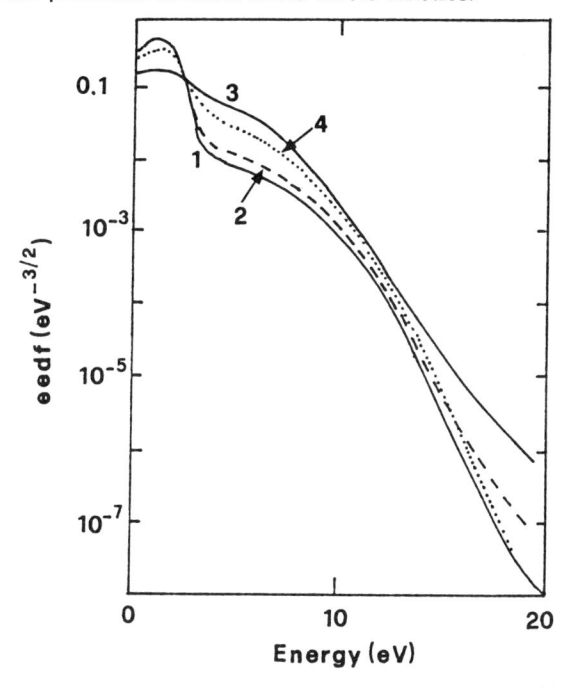

FIGURE 3. Temporal evolution of the N_2 eedf ($ne_{(t=0s)}$ =10^{11}cm^{-3}, E/N = 80 Td, p = 2 torr, R = 1 cm), the different curves correspond to 1: t = 4.5 10^{-8}s, 2: t = 4.3 10^{-4}s, 3: t = 1.6 10^{-3}s, 4: t = 1.0 10^{-2}s.

This means that improvement in the modeling of N_2 discharges can be achieved by a corresponding improvement in the relevant cross sections. An example in this direction is the use of more reliable rate coefficients for vibrational deactivation of $N_2(v)$ by atomic nitrogen, i.e., the process

$$N_2(v) + N \rightarrow N_2(w) + N$$

Insertion of the recent trajectory study of Laganá et al (16) for the corresponding rates strongly decreases the plateau in the vibrational distribution of N_2 under electrical discharges (17), decreasing also the importance of vibrational mechanisms in the dissociation and ionization processes. Of course an improvement of the potential energy surface for $N-N_2$ system urges in order to completely assess the dynamical calculations of Laganá et al.

Note however that improvement of the dissociation model used in our calculations could also favor PVM mechanisms in electrical discharges. Our approach is based on the so called "ladder climbing model", i.e., a model that considers dissociation as occurring only from the last bound level (v') of the molecule. Inclusion of dissociation processes from all vibrational levels (specially from those belonging to the plateau of the distribution) should strongly increase the dissociation by PVM mechanism (see for example reference 18).

Two other points should be addressed in future studies of the nitrogen system. The first one concerns the ultimate fate of energy in the recombination process of atomic nitrogen either in gas phase or on the surface.

In gas phase several mechanisms have been proposed for the recombination process, most describe it as occurring on electronically excited states. As a final state of this degradation one could obtain the metastable state $A^3\Sigma_u^+$ which could transfer its energy to the vibrationally excited states of ground state $X^1\Sigma_g^+$ through the following processes

$$N_2\,(A^3\Sigma_u^+) \;+\; N_2 \;\rightarrow\; N_2 \;+\; N_2\,(v=25)$$

$$N_2\,(A^3\Sigma_u^+) \;+\; N_2 \;\rightarrow\; N_2\,(v=12) \;+\; N_2\,(v=12)$$

In both cases the recombination energy should act as an additional pumping of vibrational energy with strong consequences on the vibrational kinetics. This point has been recently analyzed by Armenise et al. (19) in a completely different context. These authors studied the non equilibrium vibrational distributions of N_2 in the boundary layer of a body invested by a hypersonic flow of vibrationally excited molecules and atoms. Under these conditions highly non-equilibrium vibrational distributions develop as a result of the selective pumping of vibrational levels by recombination followed by a redistribution of the introduced quanta through V-T and V-V energy exchange processes. The consequences of such process have not been completely investigated for plasmas.

A similar situation could occur during the recombination of atoms on the surfaces that therefore could not only act on the relaxation of vibrational molecules but also on their production. This point should push the chemical-physics community toward a better understanding of the behavior of atoms and vibrationally excited molecules on surfaces. Study in this direction is in progress as recently pointed out by the recent review by Billing (20) on the subject.

RF PLASMAS

Modeling of RF plasmas has reached quite high levels of sophistication due to the large use of these plasmas in technological applications. Self-consistent studies of this kind of plasmas have been presented by different authors using completely different approaches. Here we will limit our review to plasmas produced by capacitively coupled RF reactors. To this aim we will illustrate three different self-consistent models, developed by Winkler et al (21-22), by Boeuf et al (23-24) and by Longo et al. (25-26).

RF Bulk Plasmas

The method of Winkler et al. was developed to understand the modulation of eedf in RF bulk plasmas, i.e., a time dependent Boltzmann equation was solved in a sinusoidal electrical field of the form $E = E_0 \cos \omega t$. Self-consistency was reached in the value of amplitude E_0 by imposing that the gain of electrons by the ionization process was balanced in the RF period by attachment and loss of electrons by ambipolar diffusion. This approach that does not consider the regions near to the electrodes (sheaths) can describe only the bulk plasma and was applied to various electronegative plasmas including SF_6 and SiH_4 plasmas. A typical example of these calculations is shown on figure 4 where the eedf in the SiH_4

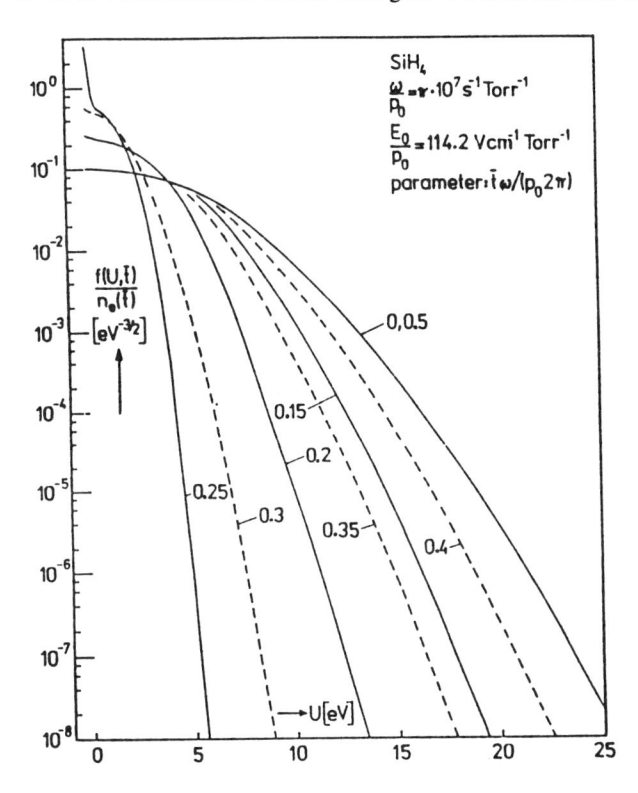

FIGURE 4 Periodic behavior of the isotropic distribution at $\omega/p_0 = \pi \, 10^7$

system is reported at selected times of the periodic evolution. Keeping in mind that the reduced time $t' = t\omega/(p_0 2\pi)$ represents for $t' = 0.25$ the passage of the field at the zero value and for $t' = 0$; 0.5 the maximum absolute value of the field we can notice, for the reported frequency ω ($\omega/p_0 = \pi 10^7 s^{-1} Torr^{-1}$), a large modulation of eedf following the sinusoidal field variation. The modulation is however not complete in the sense that eedf does not follow in a quasi-stationary way the field. This point can be understood by noting that, at $t' = 0$, eedf should be represented by a $\delta(\varepsilon)$ function centered at zero energy if a quasi-stationary approximation holds for eedf. Moreover large differences in eedf are present for the reduced times $t' = 0.2$ and 0.3, when the electrical field assumes the same absolute value. In the same figure we have also reported the E_0 value ($E_0/p_0 = 114.2$ V cm^{-1} Torr^{-1}) obtained with the self-consistent criterion above discussed. Note that this value is not so far from the experimental amplitude measured by Bohm and Perrin (27) for similar discharge conditions.

It is worth noting that the self-consistent approach completely disregards any linking with the kinetics of vibrational and electronic states. We note in this connection that the value of E_0 is strongly affected by the presence of excited states. As an example the presence of electronically and vibrationally excited SiH_4 molecules decreases E_0 up to a value of 88.2 V cm^{-1} Torr^{-1}, this effect depending strongly on the concentration of excited states.

Fluid Model Approach

The second approach for describing RF capacitively coupled discharges was essentially developed by Boeuf et al.who describe the macroscopic behavior of the discharge by using a fluid model approach. Macroscopic continuity equations for electrons and ions were therefore coupled to the Poisson equation for the electric field obtaining a self-consistent description of the discharge. In particular the electric field profile as a function of the distance, in a capacitively coupled RF discharge, was obtained emphasizing the importance of the sheath regions in accelerating electrons. The possibility of an electron beam accelerated by the sheath potential was also inserted in the model, this beam having a particular role under specific experimental conditions. Note that in the first approaches the rate coefficients for ionization and attachment were calculated by imposing an equilibrium assumption for eedf, i.e., they were calculated in any point of space by solving the uniform stationary problem with the local value of E/N field. Improvement was also obtained by calculating the rates at the average energy rather than at the instantaneous E/N value. Of course the macroscopic fluid model approach can give useful informations on the electrical profile, on the electron and ion densities as a function of the inter-electrode distance.

Fluid models can be also enriched by coupling them with a state to state kinetics. This was done by Capriati et al. (28) who coupled the fluid model with a non-local kinetics for excited states. The general scheme has been reported in figure 5 where the coupling between fluid model, chemical kinetics and Boltzmann equation is shown. The used fluid model considers two groups of electrons, the beam or γ electrons and the bulk or α electrons. A stationary continuity equation and a stationary energy equation are considered to describe beam electrons, while a second order fluid model (continuity and energy) is used to describe bulk electrons. The ion kinetics is described by a first order fluid model. The electric field space and time evolutions are obtained by solving the Poisson equation while a stationary Boltzmann equation is assumed taking into account also superelastic electronic collisions. The chemical kinetics includes three metastable states that are linked to the ground one through inelastic and superelastic collisions. Ionization from the ground as well as from metastable states is considered.

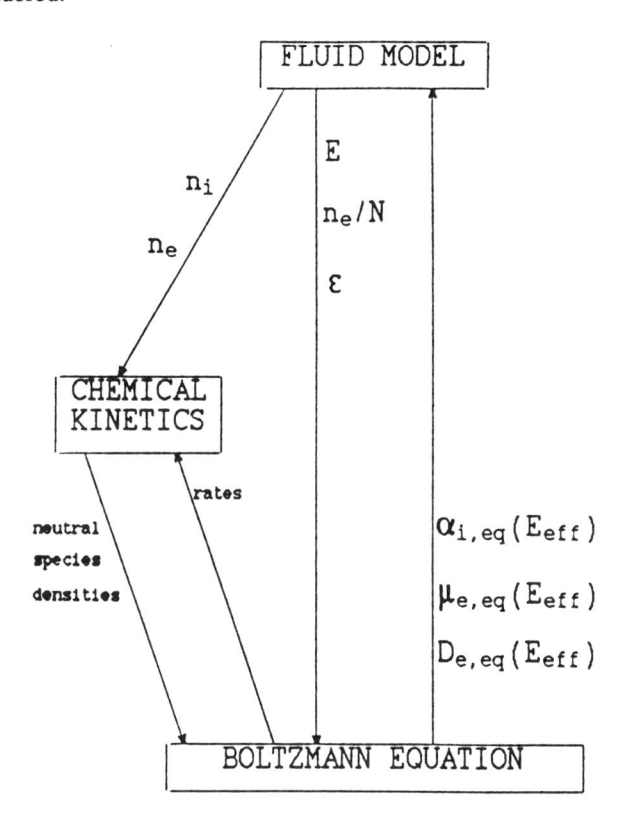

FIGURE 5. Scheme of the one-dimensional self-consistent RF discharge model discussed in the present work.

A sample of results is reported in figures 6-7 where we compare to experimental results theoretical ones for the electron temperature and density of bulk electrons obtained by including or neglecting the excited state kinetics. We can see that inclusion of excited state kinetics increases up to a factor 3 the bulk electron density having a minor role on the electron temperature. Note that this last quantity is in good agreement with the experimental results in the so called α regime underestimating the experimental results in the γ regime.

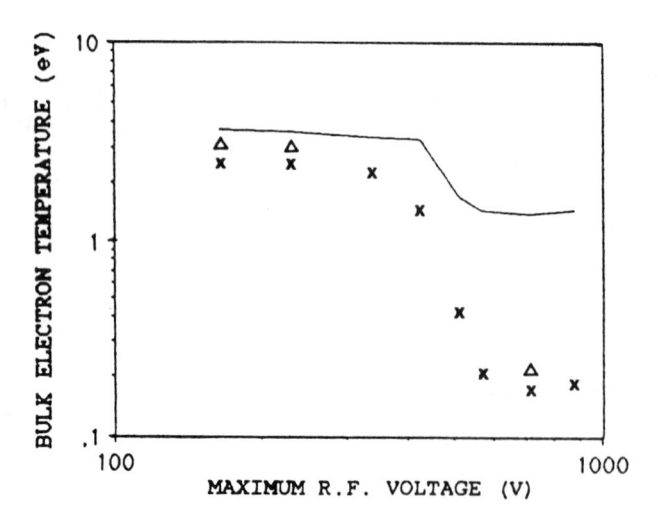

FIGURE 6. Bulk electron temperature as a function of the maximum RF voltage under the following condition: pressure = 1 torr, frequency = 3.2 MHz, secondary emission coefficient = 0.08. Comparison between experimental and theoretical values: full line, experimental values; x, theoretical values obtained by the fluid model without neutral kinetics; Δ, theoretical values obtained by the complete model.

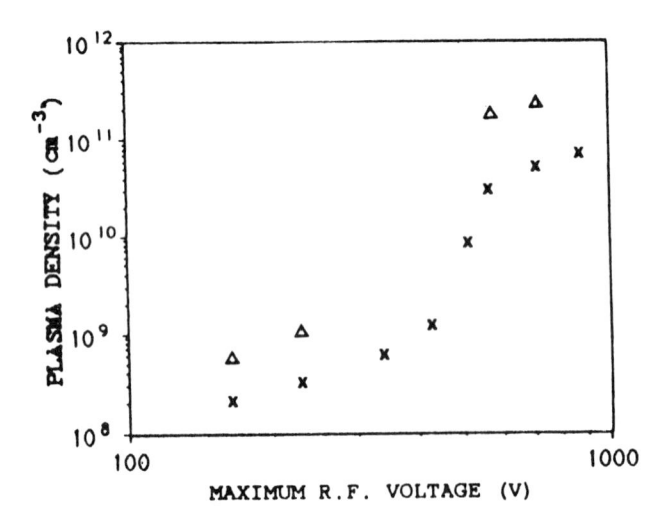

FIGURE 7. Bulk electron density as a function of the maximum RF voltage under the following condition: pressure = 1 torr, frequency = 3.2 MHz, secondary emission coefficient = 0.08: x, theoretical values obtained by the fluid model without neutral kinetics; Δ, theoretical values obtained by the complete model.

PIC/MCC and Vibrational Kinetics

Particle in cell method with Monte Carlo collisions (PIC/MCC) is one of the most powerful methods used to describe an RF discharge (30). We can note that we are following, by using first principles, the electron and ion dynamics which due to their different speeds can form space charge fields. Assuming instantaneous electron-neutral collisions, a Monte Carlo method is used to calculate the time to next collision for any charged particle, as well as the particle energy and velocity, after any collision event. The electric field is calculated self-consistently as in the fluid model, by solving the Poisson equation. The development of this classical PIC/MCC, realized in our laboratory, consists in considering a simplified vibrational kinetics coupled to the PIC-MCC method to study the influence of this coupling on the electrical characteristics of the RF discharge itself. The block diagram of the method is illustrated in figure 8.

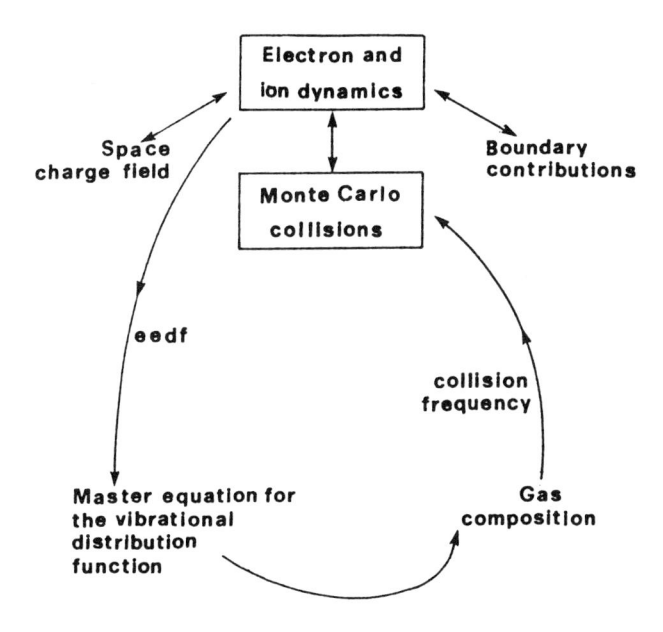

FIGURE 8. A schematic representation of PIC/MCC coupled to vibrational kinetics.

A very simplified vibrational kinetics (25) (specially if compared with our previous studies) is considered in this scheme: practically we insert only e-V (inelastic and superelastic processes) while all vibrational levels (8 in the present study) are submitted to a phenomenological deactivation rate f_d (sec^{-1}) which is considered in the present study as a parameter. This deactivation should represent V-T terms (by molecules and atoms) and energy exchange processes with surfaces, V-V energy exchange being neglected for the low pressure cases under study. Note that for $f_d = 0$ practically the vibrational distribution of N_2 is represented by a Boltzmann law at the average electron temperature T_e while large f_d values imply that all molecules are in the ground vibrational level.

We apply the model to the description of a parallel plate reactor at p = 0.1 torr (inter-electrode distance: 4 cm), under a sinusoidal applied potential V = V_{RF} sinωt with V_{RF} = 200V and a frequency of 13.56 MHz. These conditions have been studied both experimentally and theoretically by Turner and Hopkins (31), the difference with the present approach being our inclusion of vibrational kinetics.

Electron molecule collisions include elastic, inelastic, superelastic and ionization processes while for ions we consider elastic and charge exchange processes.

Due to the different relaxation times of the relevant kinetics the vibrational kinetics is solved using the quasi-stationary rate coefficients given by PIC-MCC. Then starts the dynamics of electrons and ions that includes superelastic vibrational collisions using the concentrations given in the vibrational kinetics. An iterative self-consistent method is therefore generated up to the achievement of a stationary condition.

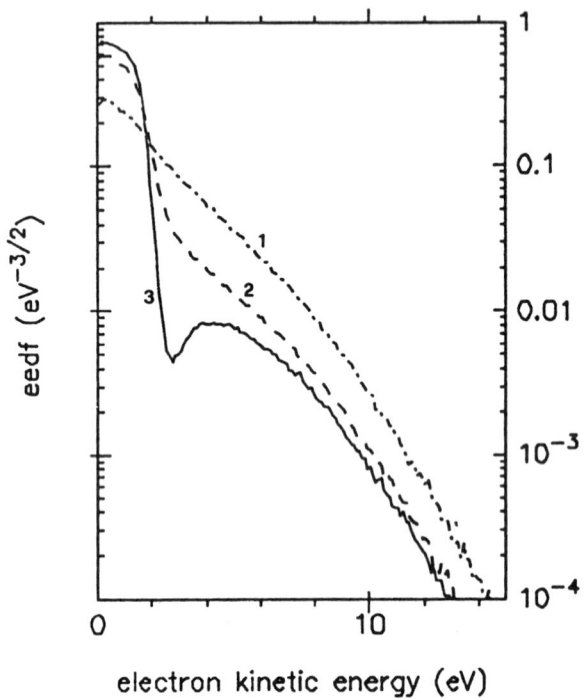

FIGURE 9. Electron energy distribution functions calculated in the center of the plasma at different f_d values. (1: f_d = 1s^{-1}, 2: f_d = 10^{1}s^{-1}, 3: f_d = 10^{3}s^{-1})

A sample of results, averaged on time at a given position, have been reported in figures 9-10. In particular figure 9 represents eedf in the center of the plasma for different f_d values. We note that the role of superelastic vibrational collisions is such to fill the hole in eedf generated by the corresponding inelastic vibrational losses that present a very strong peak in the relevant cross sections at 2 eV. This influence is as higher as lower is the f_d value, the case $f_d = 10^3$ sec^{-1} being very close to the cold gas approximation.

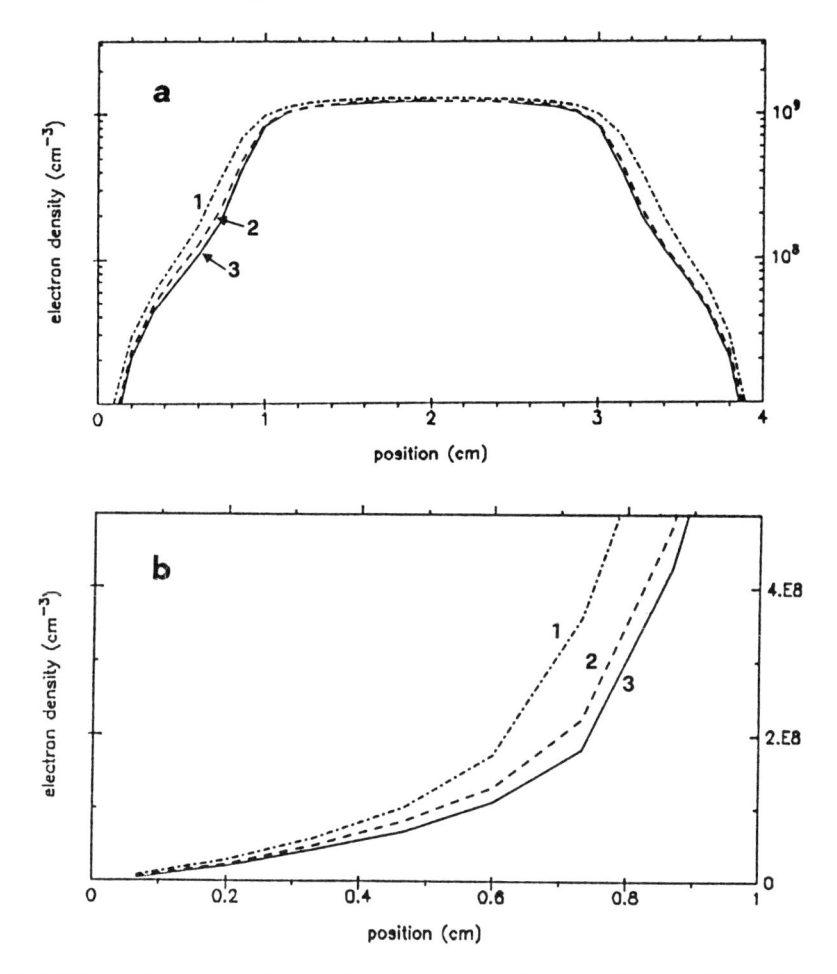

FIGURE 10. Electron density as a function as a function of position at different f_d values.
(1: $f_d = 1s^{-1}$, 2: $f_d = 10^1 s^{-1}$, 3: $f_d = 10^3 s^{-1}$)

The effect of such a situation on the electrical characteristics is not so dramatic (but not completely negligible) as can be seen by inspection of the profile of electron density reported in figure 10 a-b. We can see that only near the sheath the electron density profile depends on the hypothesis made in the vibrational kinetics. In particular a factor 2 in the electron concentration can occur depending on the selected f_d value. Of course the coupling of heavy particle kinetics with the PIC-MCC models is only at its infancy. Interesting should be also the study of the effect of superelastic electronic collisions on eedf as well as the insertion a more detailed kinetics in the PIC-MCC models.

CONCLUSIONS

In the present review we have reported several examples of the coupling of chemical kinetics, Boltzmann equation, electron and ion dynamics resulting in different self-consistent coupling schemes. Of course a true self-consistent approach is still far from the present possibilities specially in view of the lack of experimental and theoretical state to state cross sections (see ref. 32 for a review).

Chemical kinetics and electrical characteristics of a discharge should be in any case handled as a unique problem because of their conjoint activity.

However we are aware of the fact that a three-dimensional description of a discharge to be solved in feasible computer times should include only a limited number of elementary processes. This statement however should not separate engineering and chemical physics communities since in many cases only a state to state description of the chemical kinetics can properly describe the process to be used for the technological applications. This point, apart from all the examples previously reported, can be better understood by considering other two plasma applications: the formation of negative H^-/D^- beams for fusion applications (33) and the development of powerful excimer lasers (34).

In the first case the negative ions can be formed by dissociative attachment from vibrationally excited molecules and from Rydberg electronically excited states

$$e + H_2(v) \rightarrow H + H^-; \quad e + H_2^* \rightarrow H + H^-$$

so that a state to state description of vibrationally and electronically excited states of H_2 should be performed together with a description of the electrical characteristics of the discharge.

The second example "the formation of excimer molecules" follows the same lines: in fact in this case the excimer molecule is formed by recombination of Xe^+ and Cl^- in turn formed by dissociative attachment of vibrationally excited HCl molecules and by ionization of metastable Xe^*, i.e., by the mechanism

$$e + HCl(v) \rightarrow H + Cl^-$$

$$e + Xe^* \rightarrow e + Xe^+ + e$$

$$Xe^+ + Cl^- + Ne \rightarrow XeCl + Ne$$

As a conclusion we can say that future progress in understanding electrical discharges can be achieved only by using self-consistent models, the degree of s ophistication of which depends on the particular application under consideration.

ACKNOWLEDGMENT

This work has been partially supported by MURST (93)

REFERENCES

1. Polak, L., *Pure Appl. Chem.* **39**, 307 (1974)
2. Capitelli, M. and Dilonardo, M., *Rev. Phys. Appl.* (Paris) **13**, 115-123 (1978)
3. Cacciatore, M., Capitelli, M. and Gorse, C., *Chem. Phys.* **66**, 141-151 (1982)
4. Loureiro J. and Ferreira, C. M., *J. Phys. D:Appl. Phys.* **19**, 17-35 (1986)
5. Boeuf, J. P. and Kunhardt, E., *J. Appl. Phys.* **60**, 915-923 (1986)
6. Massabieaux, B., Gousset, G., Lefebvre, M. and Pealat, M., *J. Physique* **48**, 1939-49 (1987)
7. Massabieaux, B., Plain, A., Ricard, A., Capitelli, M. and Gorse, C., *J. Phys. B: At. Mol. Phys.* **16**, 1863-1874 (1983)
8. Plain, A., Gorse, C., Cacciatore, M., Capitelli, M., Massabieaux, B. and Ricard, A., *J. Phys. B:At. Mol. Phys.* **18**, 843-849 (1985)
9. De Benedictis, S., Capitelli, M., Cramarossa, F., d'Agostino, R., Gorse C. and Brechignac, P., *Optics Comm.* **47**, 107-110 (1983)
10. De Benedictis, S., Capitelli, M., Cramarossa, F., Gorse, C., *Chem. Phys.* **111**, 361-70 (1987)
11. Paniccia, F., Gorse, C., Bretagne, J. and Capitelli, M., *J. Appl. Phys.* **59**, 4004-4006 (1986)
12. Gorse, C., Cacciatore, M., Capitelli, M., De Benedictis S. and Dilecce, G., *Chem. Phys.* **119**, 63-70 (1988)
13. Wallaart,. H. L., Piar, B., Perrin M. Y. and Martin, J. P., *Chem. Phys.* **196**, 149-170 (1995)
14. Nagpal, R. and Ghosh, P. K., *J. Phys. D. : Appl. Phys.* **23**, 1663-1670 (1990)
15. Gorse, C., "Non equilibrium plasma modeling" in *Proceedings III of XXI International Conference on Phenomena in Ionized Gases*, G. Ecker, U. Arendt and J. Boseler eds., 1993, pp 141-148
16. Laganá, A., Garcia, E. and Ciccarelli, L., *J. Phys. Chem.* **91**, 312- 314(1987)
17. Armenise, I., Capitelli, M., Garcia, E., Gorse, C., Laganá, A. and Longo, S., *Chem. Phys. Lett.* **200**, 597-604 (1992)
18. Capitelli, M., *Molecular Physics and Hypersonic Flows*, Kluwer, 1996 to be published
19. Armenise, I., Capitelli, M., Celiberto, R., Colonna, R., Gorse, C. and Laganá, *Chem. Phys. Lett.* **227**, 157-163 (1994)
20. Capitelli, M., *Molecular Physics And Hypersonic Flows*, Kluwer, 1996, to be published
21. Winkler, R., Capitelli, M., Gorse, C. and Wilhelm, J., *Plasma Chem. Plasma Proc.* **10**, 419-442 (1990)
22. Winkler, R., Dilonardo, M., Capitelli, M., and Wilhelm, J., *Plasma Chem. Plasma Proc.* **7**, 245 (1987)
23. Boeuf, J. P., *Phys. Rev.* **A36**, 2782-2792(1987)
24. Boeuf, J. P. and Belenguer, P., *J. Appl. Phys.* **71**, 4751-54 (1992)
25. Longo, S. and Capitelli, M., *Phys. Rev.* **E49**, 2302-2306 (1994)
26. Longo, S., Gorse, C. and Capitelli, M., "Coupling vibrational kinetics of nitrogen molecules and electron dynamics in a PIC-MCC model of a parallel plate RF discharge" in *Proceedings of 12th ESCAMPIG*, 1994, pp. 119-120
27. Bohm, C. and Perrin, J., *J. Phys. D:Appl. Phys.* **24**, 865-881 (1991)
28. Capriati, G., Boeuf, J. P. and Capitelli, M., *Plasma Chem. Plasma Proc.* **13**, 499-519 (1993)
29. Godyac, V.A. and Khanneth, A.S., *IEEE Trans. Plasma Sci.*, **PS-14**, 112-123 (1986)
30. Birdsall, C. K. and Langdon, A. B., *Plasma Physiscs via computer simulation,* New York: Mc Graw-Hill (1985); Birsdall, C. K., *IEEE Trans. on Plasma Science* **19**, 65-85 (1991)
31. Turner,M.M. and Hopkins, M. B, *Phys. Rev. Lett.* **69**, 3511-3514 (1992)
32. M. Inokuti, *Advances in Atomic, Molecular and Optical Physics: Cross Section Data*, Boston : Acad. Press. 1994, ch. 9, pp. 322-366
33. Gorse, C., Celiberto, R., Cacciatore, M., Laganá, A. and Capitelli, M., *Chem. Phys.* **161**, 211-227 (1992)
34. Longo, S., Capitelli, M., Gorse, C., Demyanov, A. V., Kochetov, I. V. and Napartovich, A. P., *Appl. Phys.* **B54,** 239-245 (1992)

Formation and Evolution of the Cathode Sheath on the Streamer Arrival

Mirko Černák

Institite of Physics Faculty of Mathematics and Physics, Comenius University, 842 15 Bratislava, Slovakia

Abstract. Dynamics of the cathode region formed by the streamer arrival is clarified on the basis of a computer simulation model and experimental investigations. Prebreakdown streamers in positive corona discharges, negative corona pulses, and "breakups" of cathode sheaths during glow-to-arc transition breakdown processes are discussed in this context.

INTRODUCTION

At near-atmospheric pressures, the sequence of events leading to breakdown consists of the bridging the gap by primary streamers, and the subsequent heating of the channel created by the streamers. The arrival of the primary streamer on the cathode, forming an active cathode region that produces the electrons and ions by direct impact ionisation within the cathode fall, marks an important turning point in the streamer-initiated breakdown process. There seems to be consensus that upon the streamer arrival the main electron generation must take place very near (~0.1 mm) to the cathode surface, and that substantial charge transfer and neutralisation occur on a time scale of 10^{-9} - 10^{-8} s. Nevertheless, the details of the evolution of the primary streamer head via a streamer-cathode interface to the cathode region remain somewhat obscure.

The incomplete theoretical understanding of the streamer-cathode interaction seems to result from the fact that, because of instabilities introduced by numerical differentiation, the recent computer simulations of streamer-initiated breakdown processes have only been continued to the point when the streamer approaches the cathode. Probably the only theoretical models to date which provide a detailed description of the cathode region development at near-atmospheric pressures are the one-dimensional fluid model by Belasri et al. (1) and the two-dimensional fluid model by Simon and Bötticher (2). The one-dimensional approximation, however, is not adequate for the streamer-to-arc transition, where the discharge has a small cross section, and in the model by Simon and Bötticher an external circuit and discharge current computations are not included because of the computing time restrictions.

Development of the theoretical models is hampered also by experimental constraints, particularly from technical difficulties of viewing the small streamer-cathode interface with nanosecond time resolution. In the most commonly used plane-plane geometry of electrodes, a difficulty of the electrical diagnostic is that the streamer arrival generates only a small current hump on the discharge current growth waveform (4,5).

It is the main purpose of the present work to show how, combining computer simulations with experimental investigations, some insight can be obtained into the processes taking place at the streamer arrival at the cathode. In addition, it will be illustrated that the understanding of the streamer-cathode interaction is fundamental to the understanding of effects of the cathode surface properties on the glow-to-arc transition and negative corona (Trichel) current pulses.

STREAMER - CATHODE INTERACTION

The streamer-cathode interaction will be discussed for the discharge in a short positive point-plane gap, where the well-pronounced current signal induced by the genuine primary streamer-cathode contact can be measured using a small central cathode probe. The signal is relatively insensitive to gas composition and is characterised by a fast rising (the rise time of ~0.5 ns at atmospheric pressure) current peak of amplitude in 0.0 - 0.1A scale (3-5). This is followed in some 10 - 50 ns by a current hump and, subsequently, by a current portion associated with the formation of a glow-discharge-type cathode spot (6,7).

1.5-D Fluid Model for a Short Positive Point-to-Plane Gap

The one and half dimensional equilibrium simulation model is based on the numerical solution of Poisson's equation in conjunction with the continuity equations for electrons, positive ions, and negative ions. The effects of ionisation, attachment, electron diffusion, and photoemission and ion secondary electron emission from the cathode are included.

The cathode probe current I_p was computed as:

$$I_p = \int_S (q(j_i - j_e) + \varepsilon_0 \cdot \partial E/\partial t)\, dS \qquad (1)$$

where j_e, j_i, and E are the electron and ions flux and the field intensity, respectively, taken at the cathode probe surface S. Anode current I_a was computed according Rama-Shockley theorem. Figs. 1-4, show results computed for the streamer-cathode interaction in a short positive point-plane gap in N_2 at a pressure of 26.7 kPa, gap spacing of 10 mm, gap voltage of 4 kV, values of secondary emission coefficients $\gamma_i = 2.10^{-3}$ and $\gamma_{ph} = 10^{-2}$, and the streamer channel radius expanding from 0.5 mm near the anode to 1.2 mm at the cathode.

The good agreement obtained between the discharge behaviour observed experimentally and those in our simulation model indicates that 1.5-D models provides an adequate physical picture of streamer arrival at the cathode and , taken together with experimental investigations, can serve as a basis for developing a 2-D model.

The reader is referred to (7) for more details. Here, for brevity, we shall restrict ourselves to the results in Figs. 1-4 and to discussion of principal conclusions.

Conclusions and Comparison with Experimental Observations

Since, for pressures above say 10 kPa, the streamer discharge behaviour prior to the glow-to-arc transition is very similar to that in air (8), the model is believed to qualitatively describe the streamer-cathode contact also in air at near-atmospheric pressures, where the vast majority of the experimental studies have been made. In addition, based on the results by Kennedy (Ref.10, see Fig.5.6b there) and by Martin et al. (11) which show striking similarities between the current signal induced in the cathode probe hit by the streamer in a short point-plane gap and that in a parallel-plane gap, we suppose that the model provides an insight into the streamer-cathode interaction also for this gap geometry

The computed time evolution of electron and ion densities near the cathode surface at the streamer arrival shown in Figs.1-2 illustrate the transformation of the streamer front structure to an abnormal cathode fall of roughly 1 kV, which is consistent with the experimental results by Cavenor and Mayer (12) indicating that cathode spot created by the streamer arrival operates in the abnormal glow regime. Also, the time evolution of the cathode sheath thickness indicated by Figs.1-3 is in fair agreement with that observed by Bertault et al. (6).

The results reveal that the dominant component of a sharp current spike induced by the streamer arrival in a cathode probe (see Fig.4) is the

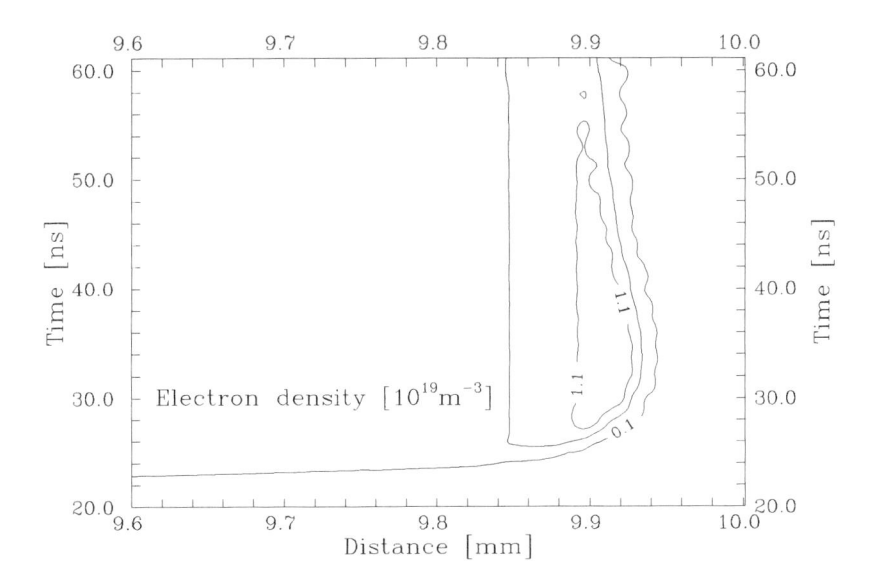

FIGURE1. Spatio-temporal development of electron density during streamer-cathode contact (the cathode is situated at the distance 10 mm).

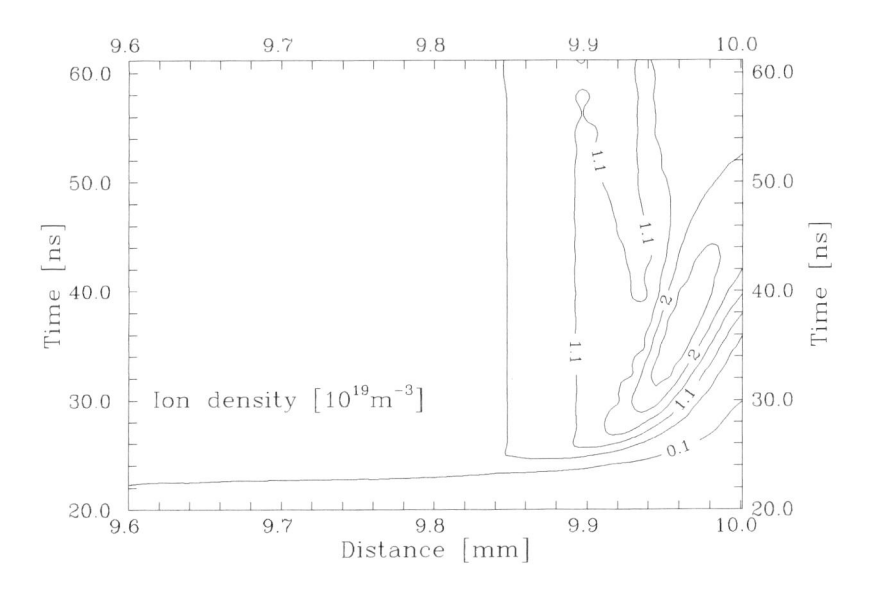

FIGURE 2. Spatio-temporal development of ion density during streamer-cathode contact

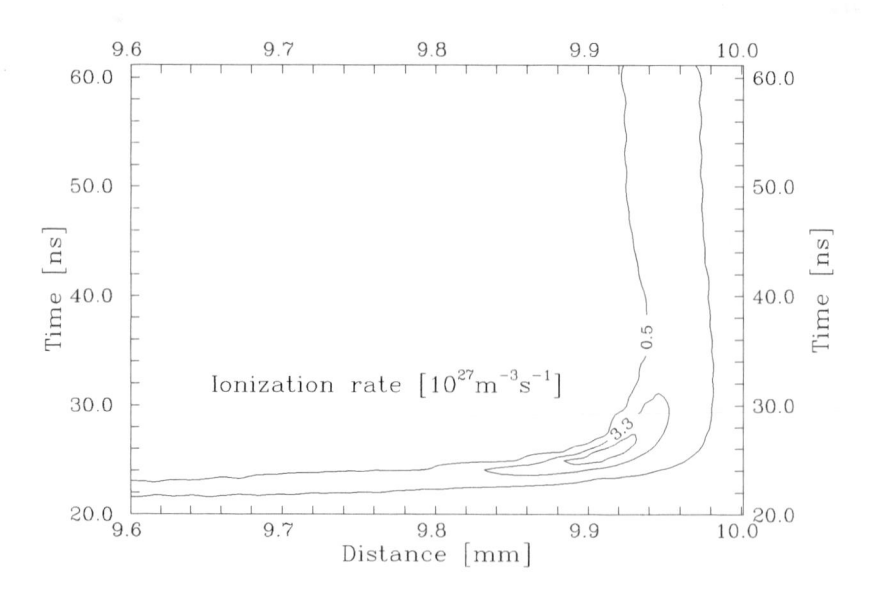

FIGURE 3. Spatio-temporal development of the ionisation rate during streamer-cathode interaction.

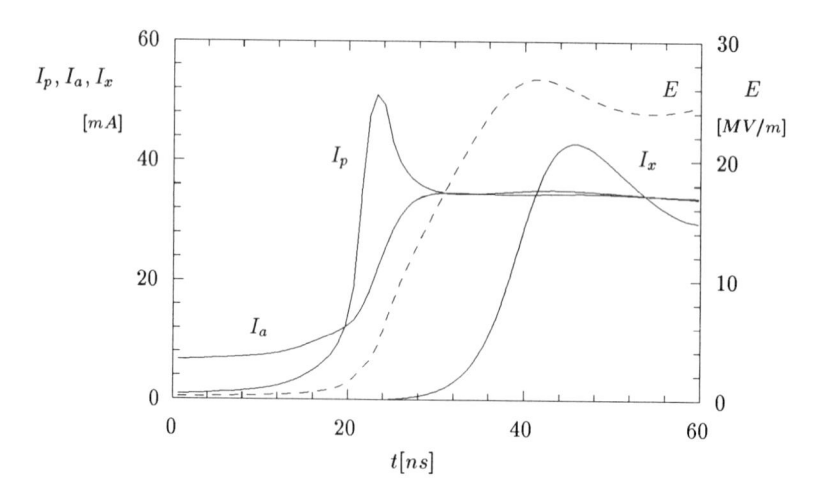

FIGURE 4. I_p, I_a, I_x, are the total probe, anode and conductive probe currents, respectively; E - intensity of electric field at the axis of the probe.

displacement current and that this current signal is not very sensitive to cathode emission.

Also, the sudden increase in ionisation activity near the cathode surface seen in Fig.3 (and observed experimentally as the bright flash of light as the streamer hit the cathode (6)), is not sensitive to secondary electron emission. The conductive current due electron emission processes and to positive ion collection by the cathode becomes the dominant part of the cathode probe current some 10 -20 ns after the streamer arrival marked by the cathode-probe current spike. This is in contrast to the commonly held belief (5,13) that the streamer arrival at the cathode is associated with a sudden burst of electrons leading to the neutralisation of the positive charge in the streamer head.

It is interesting to refer here to several implications of the model, which were tested experimentally and can be used as a tool for further studies on the cathode sheath evolution:

i) An implication of the model is that an increase in secondary photoemission coefficient has little effect on the current spike induced at the streamer arrival, but results in a reduction in the time lapse between this spike and the following current hump due to incoming positive ions (see Fig.6 in Ref.7). This is in conformity with the results in Ref.14, Fig.17, where values of the photoemission coefficient were increased using CuI-coated cathode surfaces.

ii) The results in Fig.4 indicate that the field at the cathode surface reaches its maximum several ns before the current hump due to incoming positive ions. This is in striking conformity with the fact that "streamer-like-instabilities of the cathode sheath" discussed in the following Section (see Figs. 5 and 6) take place just at this moment (14), Figs. 15 and 22.

iii) Our computer simulations, and also the results by Morrow (15), Fig.11, indicate that under certain conditions the discharge current can exhibit damped oscillations with period of the order of 10 ns reflecting the back and forth motion of the electric field in the cathode sheath. This phenomenon was experimentally observed in (5), Fig.5a; (14), Figs. 5 and 9, and (16), Fig.3.

Finally, it may be noted that a transition from capacitive to resistive behaviour of the cathode sheath seen in Fig.4 is closely analogous to that in Ref.1.

STREAMER FORMATION IN THE CATHODE SHEATH

It is generally accepted that ionisation instabilities in the cathode region ("breakups" of the cathode sheath (17)) play a crucial role in limiting the duration of the high-pressure space glow discharge used, for example, in gas lasers excited by transverse discharges (17-19). The ionisation instabilities result in the "hot spots" formation near the cathode, which trigger filaments of strongly enhanced current densities, leading to the arcing. Up to now no adequate model of the ionisation instabilities in the cathode sheath of high-pressure glow discharges has been developed. The most applicable model to the ionisation instabilities seems to be that of Bityurin et al. (20), which predicts the possibility of a cathode-directed ionising wave driven cathode-sheath instability with a repetition period ~ 25 ns in N_2 at atmospheric pressure. In (1) streamers in the cathode fall have been simulated, which propagate from the cathode surface towards the positive column. This is, however, in contrast to the experimental observations (19) indicating that the 'streamers' bridging the cathode sheath begin from the positive column side.

It is believed that the formation of ionisation instabilities in the cathode sheath can be at least partially understood in terms of the "positive-streamer-like instabilities of the cathode sheath" discussed below.

In Fig. 5 the I_p waveform measured in CO_2 at a pressure of 13.33 kPa, gap spacing of 10 mm, and gap voltage of 3kV, using the conditioned cathode surface (waveform 1) is compared with that measured using unconditioned (i.e., freshly polished) cathode surface under the same experimental conditions.

The interpretation of the waveform 1 in terms of our computer model is clear: The initial sharp current spike corresponds to the displacement current generated at the streamer arrival at the cathode, the subsequent current hump is due to incoming positive ions, and the following current portion corresponds to a filamentary (abnormal) glow discharge. By comparing the spikes denoted by X, which are seen on the waveform measured using unconditioned cathode surface, with the 1, it can be seen that they are remarkably similar in shape to the Ip waveform generated by the streamer arrival at the cathode (They are, practically identical to I_p waveforms taken at a reduced gap voltage, see (21)). For this reason they can be called 'positive-streamer-like'. The same phenomenon can be observed using point cathode (14, 21,22), where the ionisation is confined to a thin cathode region. This is why we use the term 'positive-streamer-like instabilities of the cathode sheath' (PSLI) to refer to the phenomenon. We refer to the papers (14, 21,22) for a more complete analysis and a possible explanation for the phenomenon.

Despite of its incomplete understanding the PSLI could well have practical importance. Figs. 5 and 6 show that the appearance of the PSLI due to the use of unconditioned cathodes (Fig.5) and due to 'ageing' of the cathode surface (Fig.6) resulted in transition of the discharge to spark. This phenomenon can limit performance characteristics of pulsed corona devices (4) and wire chambers used as detectors for ionising particles. This is in contrast to most switching applications, where the glow discharge phase is undesirable because of its low

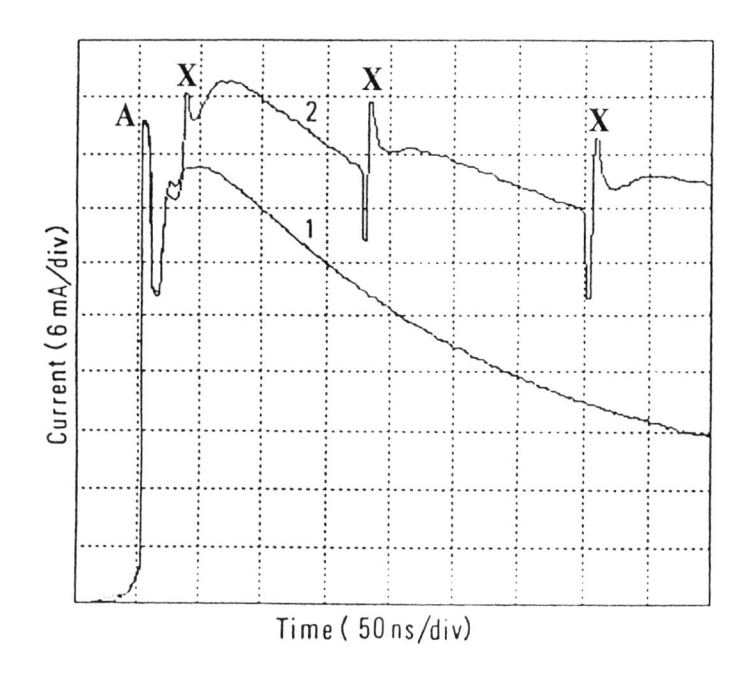

FIGURE 5. Current signal induced in the cathode probe hit by the streamer measured using the conditioned (1) and freshly polished (2) cathode surface.

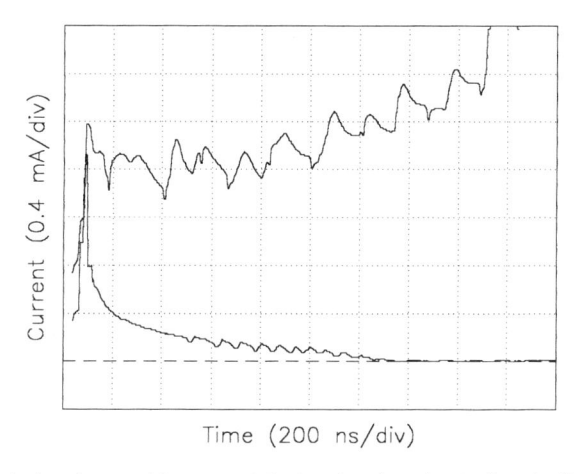

FIGURE 6. Spark development in a coaxial wire chamber due to "ageing". Conditions: Ar +10% CH_4, 100 kPa; anode and cathode diam.: 20 μm and 8 mm; gap voltage of 2500 V.

conductivity. Thus, for example , based on Ref. 11, one may speculate that an artificial ignition of PSLI, accelerating the arc formation, could improve operation characteristics of switching spark gaps. Also, we hypothesise (22,23) that, streamers resulting in the "hot spots" formation in high pressure glow discharges are due to the same phenomena.

POSITIVE-STREAMER MECHANISM FOR NEGATIVE CORONA CURRENT PULSES

Evidence has been accumulating over the past twenty years that the steep negative corona current rise is associated with the development of a positive streamer (14, 16, 22,24-28) and its arrival at the cathode. It is notable, however, that this mechanism does not seem to be generally accepted by the workers in the field of corona discharges. Perhaps, the most common objection raised against the positive-streamer mechanism for negative corona pulses, is that the time for positive ions to move to the cathode is much longer than the recorded pulse rise time (29). This objection is based on the unrealistic assumption that the streamer arrival at the cathode surface is associated with the immediate neutralisation of the positive ions at the cathode (see Fig.4).

A detailed discussion of the mechanisms for the negative corona pulse rise is beyond the scope of this paper. It is of interest, however, to note here broad similarities observed between the negative corona current pulses, current pulses generated at the arrival of positive streamers on the cathode, and a current signal corresponding to the positive-streamer-like instabilities of the cathode sheath, which is indicative of the same physical mechanisms (14, 21).

ACKNOWLEDGEMENTS

I would like to express my particular gratitude to Prof. T. Hosokawa and Dr. E. Marode for their substantial contributions to this work. The computer simulation model presented was the PhD thesis of I. Odrobina.

REFERENCES

1. Belasri A., Boeuf J.P., and Pitchford L.C, *J. Appl. Phys.* **74**, 1553-567 (1993)
2. Simon G. and Bötticher W., *J. Appl. Phys.* **76**, 5036-46 (1994)
3. Inoshima M., Černák M., and Hosokawa T., *Jpn. J. Apl. Phys.* **29**, 1165-72 (1990)

4. Černák M., van Veldhuizen E.M., Morva I., and Rutgers W.R., *J. Phys. D: Appl. Phys.* **28**, 1126-32 (1995)
5. Kondo K. and Ikuta I., *J. Phys. Soc. Jpn.* **59**, 3203-16 (1990)
6. Bertault P., Dupuy J., and Gilbert A., *J. Phys. D: Appl. Phys.* **10**, L219-L222 (1977)
7. Odrobina I. and Černák M., to be published in *J. Appl. Phys.* **78**(6), 15 September 1995
8. Odrobina I. and Černák M,"Formation of cathode region of filamentary high-pressure glow discharges" in *Contributed Papers of the 4th International Symposium on High Pressure Low Temperature Plasma Chemistry*, Bratislava 1993, pp.165-70.
9. Černák M., Marode E, and Odrobina I.,"Comparison of current waveforms induced by prebreakdown corona streamers in N_2 and air" *Proceedings of 21th International Conference on Phenomena in Ionized Gasses*, Bochum, 1993, pp. 399-400.
10. Kennedy J.T.,"Study of the avalanche to streamer transition in insulating gases" PhD.-Thesis, Eindhoven University of Technology 1995.
11. Martin T.H., Seamen J.F., Jobe D.O., and G.E.Pena,"Gaseous prebreakdown processes that are important for pulsed power switching" in *Proceesings of the 8th Pulse Power Conferernce*, San Diego 1991, pp.323-27.
12. Cavenor M.C. and Mayer J., *Austral. J. Phys.* **22**, 155-167 (1968)
13. Achat S., Teisseyre T., and Marode M., *J. Phys. D: Applied Phys.* **25**, 661-8 (1992).
14. Černák M. and Hosokawa T., *Aust. J. Phys.* **45**, 193-219 (1992)
15. Morrow R., *Phys. Rev.* **A 32**, 1799-809 (1985)
16. Černák M. and Hosokawa T., *Jpn. J. Appl. Phys.* **26**, L1721-L1723(1987)
17. Turner R., *J. Appl. Phys.* **52**, 681-92 (1981)
18. Bötticher W., "Modelling of discharge pumped XeCl lasers. Open questions" in *Abstracts of Invited Talks and Contributed Papers of the 10th ICPIG*, Vol.14E, Orleans 1990, pp.8-11.
19. Makarov M., *J. Phys. D: Appl. Phys.* **28**, 1083-93 (1985).
20. Bityurin V.A., Kulikovski A., and Lyubimov G.A. *Zh. Tekh. Fiz.* **59**, 50-63(1989)
21. Černák M. and Hosokawa T., "Positive-streamer-like instabilities of the cathode sheath of filamentary glow discharges" in *Contributed Papers 4 of XX ICPIG*, Il Ciocco 1991, pp.917-18.
22. Černák M., Hosokawa T., and Inoshima M., *Appl. Phys. Lett.* **57**, 339-40 (1990)
23. Hosokawa T. and Černák M., "Acceleration of filamentary glow-to-arc transition in CO due to cathode sheath instabilities" in *Proceedings of the 10th Int. Conf. on Gas Discharges and their Appl.*, Swansea 1992, pp.452-55.
24. Ikuta N. and Kondo K., *IEE Conf. Publ.* **143**, 227-30 (1976)
25. Golinski J. and Grudzinski J., *J. Phys. D: Appl. Phys.* **19**, 1497-505 (1986)
26. Černák M. and Hosokawa T., Phys. Rev. **A 43**, 1107-9(1991)
27. Černák M., Hosokawa T., and Odrobina I., J. Phys. D: Appl. Phys. **26**, 607-18(1993)
28. Liu J. and Govinda Raju R.R., *IEEE Trnas.on Dielectrics and El. Insul.* **1**, 520-29(1994)
29. Dancer P., Davidson R.C, Farish O., and Goldman M.,"A unified theory for the mechanism of the negative corona Trichel pulse" in *Proceedings of IEEE-IAS Conf. on Electrostatics*, Cleveland 1979, pp.87-90.

On the Use of the Plasma in III-V Semiconductor Processing

G. Bruno, P. Capezzuto, M. Losurdo

*C.N.R. - Centro di Studio per la Chimica dei Plasmi
Dipartimento di Chimica-Università di Bari
via Orabona,4- 70126 Bari, Italy*

Abstract. The manufacture of usable devices based on III-V semiconductor materials is a complex process requiring epilayer growth, anisotropic etching, defect passivation, surface oxidation and substrate preparation processes. The combination of plasma based methods with metalorganic chemical vapor deposition (MOCVD) offers some real advanteges: in situ production and preactivation of PH_3 and sample preparation using H-atom. The detailed understanding and use of the plasma (using mass spectrometry, optical emission spectroscopy, laser reflectance interferometry and spectroscopic ellipsometry) as applied to InP material is discussed.

INTRODUCTION

The processing of III-V semiconductor materials (InP, GaAs, ...) is assuming, day by day, a very complex configuration, in that a variety of processes for manufacturing the final devices is required. Among others, epilayer growth (1), anisotropic etching (2), defect passivation (3,4), surface oxidation (5) and substrate cleaning (6,7) are dry and/or wet processes performed, nowadays, by using various chemical, physical, electrochemical and photochemical techniques.

In these last years, there has been a considerable interest to develop "new" methods which may give, on one side, a further impulse to industrial development and, on the other side, an unvaluable tool for answering the many unsolved scientific problems.

The level of complexity that has been reached, at present, can be seen in metalorganic chemical vapor deposition (MOCVD) process (8), for example. In MOCVD the deposition process concept is simple, but it requires collateral processes, such as substrate cleaning and surface passivation, before performing a useful epilayer growth. Moreover, the MOCVD technique still needs developments in process control and gas precursors for its complete confirmation. With respect to this, scientific themes of investigation are related to the high toxicity of the commonly used V-group precursors (PH_3, AsH_3) and, also, to their relatively high thermal stability. This last point is particularly important as it often requires high process temperature to incorporate the V-group element and, because they are layer-by-layer deposition processes, the possibility to produce new multilayer devices is often limited by the occurance of interdiffusion processes.

© 1996 American Institute of Physics

Within the above matters, the development of "new" methods involving plasma is seen as one of the most promising as it will be helpful not only in improving the III-V processing, but also in integrating all processes in a unique technology.

We will present some plasma process studies we have performed using remote plasma-MOCVD (RP-MOCVD) technique. These includes:

- P_{red}-H_2 plasma, for the in-situ production of PH_3 by ablation of red-phosphorus with H-atoms;
- PH_3-H_2 plasma, for the PH_3 preactivation before interacting with hot deposition surface;
- H_2 plasma, for the production of H-atoms and of H atoms-InPsurfaces interactions, for native oxide removal, material defect passivation, and phosphorus ablation processes;
- deposition of InP epitaxial films from PH_3 and trimethylindium ($In(CH_3)_3$) under PH_3 plasma preactivation conditions;
- oxidation of InP surfaces in the O_2 plasma downstream region.

Emphasis is given to non-intrusive optical diagnostic techniques such as optical emission spectroscopy (OES), laser reflectance interferometry (LRI), and phase modulated spectroscopic ellipsometry (PMSE) to fingerprint the plasma phase, its downstream region and the interacting surface.

RP-MOCVD REACTOR ARCHITECTURE

The experimental apparatus, which can be classified as remote-plasma MOCVD (RP-MOCVD) reactor operating at low pressure (0.1-10 torr), is schematized in Fig.1. It consists of a quartz tube, where plasmas of PH_3-H_2, H_2 and O_2 are generated by applying a r.f. electric voltage (13.56 MHz) between two semicircular external electrodes, and a stainless-steel MOCVD reactor where the radiatively heated molybdenum susceptor is placed during the InP epilayer growth, and the InP substrate treatments.

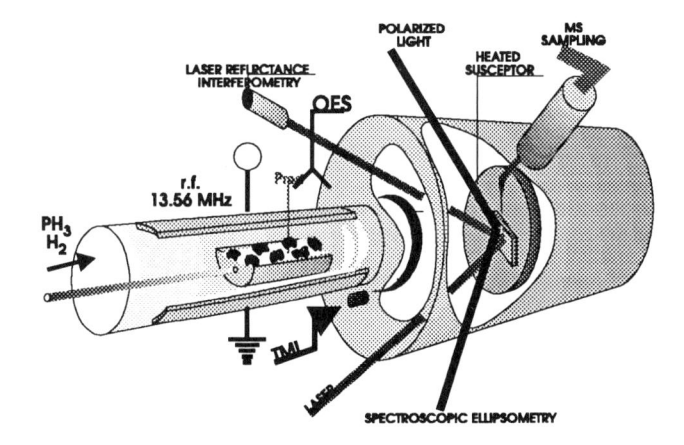

FIGURE 1. Schematic representation of the RP-MOCVD apparatus.

A molybdenum boat filled with red-phosphorus can be positioned in the quartz plasma tube for studying the in situ PH$_3$ formation by the ablation of P$_{red}$ with plasma produced H-atoms. Two load-lock chambers allow to charge the substrate and the red phosphorus in the MOCVD chamber and plasma tube, respectively. The system is evacuated by turbomolecular pumps to reach a base vacuum of 10^{-8} torr, and by roots-rotative pumps to mantain the process pressure constant.

Many "windows" on the reactor allow to see inside the processes by the following diagnostics:

- quadrupolar mass spectrometry (VG-SPX-Elite 600), to analyze gas phase composition in the hot boundary layer near the growth surface. The MS sampling is performed through a 100 μm orifice positioned in the afterglow region close to the substrate holder.
- phase modulated spectroscopic ellipsometry (UVISEL, Isa Jobin-Yvon), to study the surface kinetics (growth and etching) and the modification of the surface optical properties and morphology.
- laser reflectance interferometry (He-Ne laser), to measure the rates of the ongoing deposition and/or etching processes.
- optical emission spectroscopy (OMA, EG&G-PAR), to monitor the emitting species present in the plasma phase.

InP PLASMA ASSISTED DEPOSITION

Conventional MOCVD processes require high deposition temperature to supply the activation energy for both gas phase and surface reactions, and high hydride (PH$_3$ or AsH$_3$) flux to grow good quality III-V epitaxial layer. The plasma, as a secondary source of energy, offers low temperature (9) and low V/III ratio processing (10), mainly by the pre-cracking of the thermally stable PH$_3$. Plasma process also offers a valid alternative to the use of the highly toxic PH$_3$, by performing the in situ generation of PH$_3$ through the ablation of red phosphorus in H$_2$ plasma (11).

PH$_3$ in situ production

The PH$_3$ formation process by P$_{red}$ and H$_2$ plasma is a very simple one; it can be schematized as

$$P(s) + 3 H \rightarrow PH_3 \quad (1)$$

The amount of PH$_3$ produced by the above process depends on pressure, r.f. power and H$_2$ gas flow rate, in that all these parameters affect the H-atom density (12). However, the PH$_3$ production is limited by the partial chemical equilibrium (PCE) as shown in fig. 2. In the figure, the PH$_3$ density is plotted vs residence time for both PH$_3$ formation (lower half) and PH$_3$ decomposition processes (upper half). As indicated in Fig. 2, at very long residence time a steady state condition is reached and, hence, a sort of chemical equilibrium (PCE) is established:

$$PH_3 \rightleftarrows P + 3/2\ H_2 \quad (2)$$

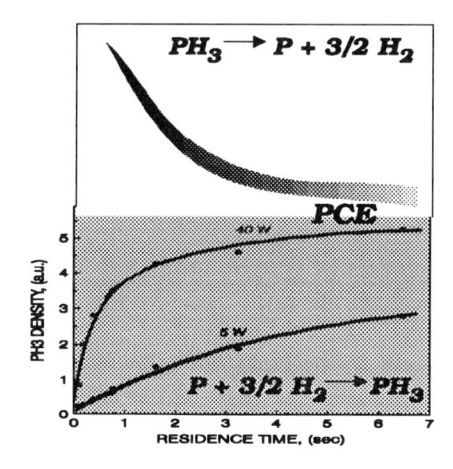

FIGURE 2. Plot showing the oncoming of the PH$_3$ partial chemical equilibrium (PCE) from both side of PH$_3$ formation and PH$_3$ decomposition processes.

Far from PCE, i.e. at short residence time, the PH$_3$ production rate is linearly related to H-atom density:

$$r_{PH3} = k\,[H] \qquad (3)$$

The validity of eq.3 is confirmed by the result shown in fig.3, where PH$_3$ production rate, as derived by MS analysis, is plotted vs the correspondent H-atom density values as obtained by actinometric OES (13).

Although PCE exists, PH$_3$ flow rate as high as 20 µmol/min have been measured. This amount could be enough for InP deposition under low pressure conditions.

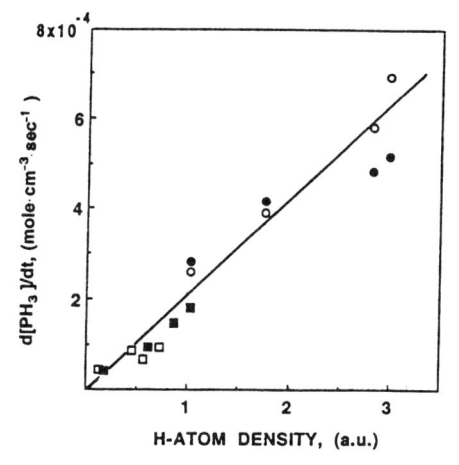

FIGURE 3. The PH$_3$ production rate vs H-atom density in P$_{red}$-H$_2$ plasmas for different r.f. power and P$_{red}$ amount.

PH3 plasma preactivation

The PH$_3$ plasma preactivation is an important process for MOCVD deposition of P-based III-V materials, as it is an efficient source of PH$_x$ radicals and H-atoms. Both these species are effective in assisting the material growth as PH$_x$ radicals promote the phosphorus deposition, whereas H-atoms favour the desorption of surface carbon coming from indium organometallic precursor (In(CH$_3$)$_3$).

The parametric study of the PH$_3$-H$_2$ plasma has revealed in r.f. power the main parameter affecting the PH$_3$ decomposition. The typical products distribution is shown in Fig.4.

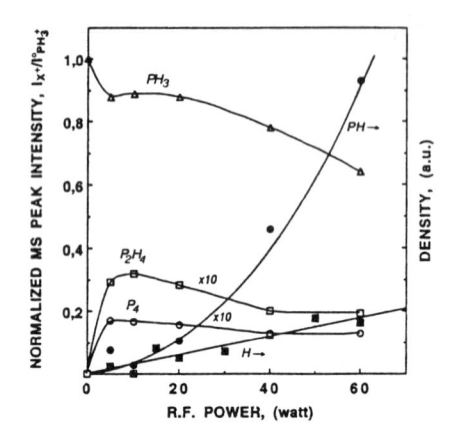

FIGURE 4. Effect of r.f. power on the H-atom and PH radical density (right hand side) and on the gaseous reaction product distribution (left hand side) in PH$_3$-H$_2$ plasmas (100 sccm PH$_3$/H$_2$ = 0.032, 1 torr).

Here, in addition to PH$_3$, P$_2$H$_4$ and P$_4$ density profiles, the trends of PH radicals and H-atoms, derived by OES, are shown. The very high density of PH radicals at high r.f. power can be read in the following reaction scheme:

$$PH_3 + e \rightarrow PH_x + (3-x) H + e \qquad (4)$$

$$H_2 + e \rightarrow H + H + e \qquad (5)$$

$$PH_3 + H \rightarrow PH_2 + H_2 \qquad (6)$$

in which reaction of eq.6 plays a determinant role for PH$_x$ radical formation.

However, the true values of PH$_x$ and H densities at the substrate position also depend on the disappearance processes in the downstream region. Here, the presence of active phosphorus has been demonstrated by the phosphorus film deposition simply measured by LRI.

InP epilayer deposition

From the InP deposition experiments we have obtained important informations regarding the chemistry near the surface, the growth kinetics and the material chemistry and morphology.

The effect of PH_3 plasma decomposition on the InP growth kinetics has been evaluated, at fixed depositon temperature, through the analysis of the growth rate profile as a function of PH_3 molar fraction in the gas feed. The resulting data are included in Fig.5.

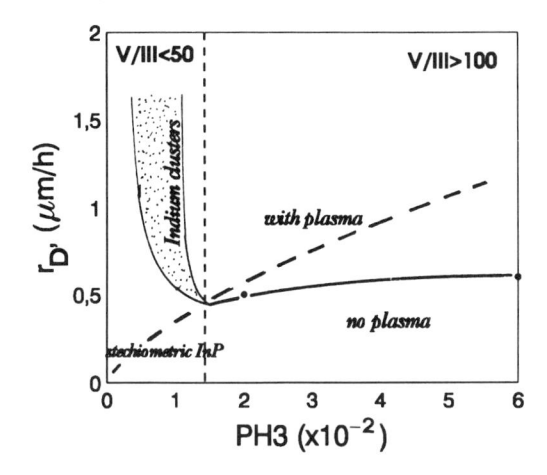

FIGURE 5. Deposition rate of InP films vs PH_3 molar fraction in the TMI-PH_3-H_2 gas mixture. Dashed line is for the effect of PH_3 plasma preactication. Shadowed area is for the deposition of InP films with metallic indium droplets (550°C, 2 torr, 500 sccm, 20 watt)

From the figure, it comes out that, at high V/III ratio (>150), the growth rate is almost independent on PH_3 partial pressure and is mainly determined by $In(CH_3)_3$ partial pressure. Under conditions of very low V/III ratio (<50), and without plasma preactivation, non stoichiometric InP epilayers are grown and, correspondently, the growth rate increases. The effect of the plasma is double: on one side, at low V/III ratio, it allows the deposition of stoichiometric InP epilayers; on the other side, at high V/III ratio, it induces an increase in the growth rate (14). Both these phenomena result from the much higher density of phosphorus active species present at the growth interface under plasma preactivation conditions.

Also, different chemical environments near the growth surface, depending on the V/III ratio, substrate temperature and on PH_3-H_2 plasma activation, have been revealed by MS analysis (15). As an example, large amounts of P_2H_4 and P_4 have been detected when under plasma preactivation conditions.

The effectiveness of the plasma in depositing "new" InP epilayers is evidenced by the PL spectra shown in fig.6. The "new" character of the plasma deposited InP epilayer can be read in the very high PL efficiency and in the very different spectrum features, when compared to those of conventional material.

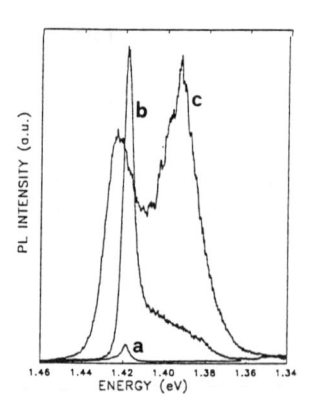

FIGURE 6. Photoluminescence spectra of (a) InP substrate, (b) InP film deposited at T=600°C, V/III=150 and (c) InP film deposited under PH$_3$ plasma preactivation at T=550°C, V/III=20

REMOTE PLASMA TREATMENT OF InP SUBSTRATE

Recently, there is a great interest in developing plasma dry processes for the treatment of III-V materials and, in particular, i) the substrate cleaning, to remove native oxide before epilayer growth, ii) the material passivation, to reduce defect density so improving the optoelectronic quality, and iii) the surface oxidation, to prepare a stable insulating layer for MOS-like devices.

Among the various plasma configurations, the use of remote plasmas seems the most appropriate since the ion bombardment process is avoided and damage-free surfaces are obtained.

H$_2$ plasma cleaning and passivation

In this section, we discuss the in situ cleaning and passivation of InP surfaces by a controlled atomic hydrogen flux present in the downstream region of a low pressure H$_2$ r.f. plasma. Thus, the plasma characterization and, in particular, the measurement of H-atom density in the remote plasma region, assumes noticeable importance. With respect to this, a P-target has been used as a H-atom density probe. The method is based on the LRI measurement of the P-etching rate, which is linearly related to H-atom density (see eq. 3). The calibration of the etching rate in terms of absolute density of H-atoms has been done on the basis of the following kinetic evidences:

- the H-atom formation kinetics is very fast and a plasma equilibrium condition is reached even at very high flow rates (1000 sccm). This results from the constancy of OES emission intensity ratio (I_{H^*}/I_{Ar^*}) at various H$_2$ gas flow rate (100-1000 sccm);
- the H-atom decay kinetics, in the downstream region, is controlled by the homogeneous recombination only (H + H + H$_2$ → H$_2$ + H$_2$), since the wall recombination process (H + wall → 1/2 H$_2$) is negligible.

The data of phosphorus etching rate (r_E) at different gas flow rate are shown in Fig.7. In the same figure the right hand scale gives the correspondent value of the H-atom density.

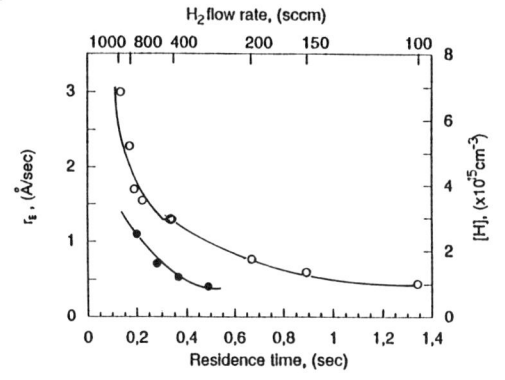

FIGURE 7. Phosphorus etching rate, r_E, and the corrisponding H-atom density, [H], vs rasidence time in the plasma downstream region (1 torr, 60 watt).

The plasma cleaning experiments have been performed under plasma conditions which assure an H-atom flux of about $5 \cdot 10^{20}$ at/cm²sec impinging on the InP substrate which is positioned at 10 cm from the plasma tube end.

The interaction of H atoms with as-received InP substrate implies the processes of:
- native oxide removal

$$InPO_x + 2x\ H \rightarrow InP + x\ H_2O \quad (7)$$

- selective phosphorus ablation

$$InP + 3\ H \rightarrow In + PH_3 \quad (8)$$

- and defect (d) passivation

$$InP(d) + H \rightarrow InP(H) \quad (9)$$

The occurrance of the above processes depends, besides H-atom density and exposure time, on the surface temperature. In fact, XPS analysis of InP substrates (16) treated at different temperatures, has shown that 230°C is the optimum for the complete oxide removal.

However, in order to discriminate the occurance of the processes of eqs. (7) and (8), real time SE measurements have been performed. Figure 8 shows a typical trace of the imaginary part of the dielectric function, ε_i, at 4.65 eV, recorded during the H-atom treatment. The higher the ε_i value, the lower the oxide thickness. From the figure the following considerations can be drawn: i) the oxide removal process at T=230°C is well discriminate with respect to phosphorus ablation, and of the process end point is well detected; ii) the phosphorus ablation process is very fast at room temperature as observed by the strong decrease of ε_i and iii) the annealing at T=500°C, in presence of PH$_3$, is able to reconstruct the damaged material.

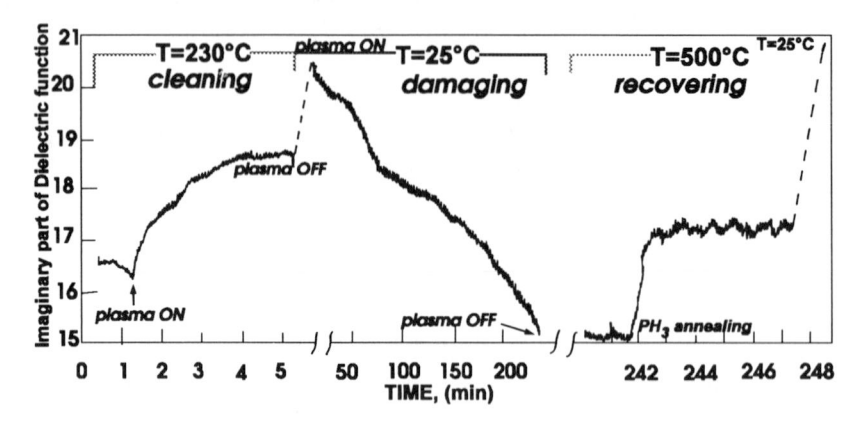

FIGURA 8. Trends of the imaginary part of dielectric function during the H-atom treatment for InP cleaning (oxide removal) and damaging (P-ablation). The material recovering by PH$_3$ annealing is also shown.

As for the material passivation, the ex situ photoluminescence (PL) measurements have shown for the H$_2$ plasma cleaned substrate, a PL efficiency two order of magnitude higher than that measured for the PH$_3$ annealed and untreated substrate (15). Also, the PL signal stability of the cleaned substrate remains high even after long time air exposure (3).

O$_2$ plasma oxidation

As regarding the InP oxidation to form a stable and defect-free oxide-InP interface, standard thermal oxidation methods result in non homogeneous oxides, and altered stoichiometry at the InP surface, i.e. a phosphorus enrichment at the oxide-InP interface (17). Also, the wet elecrochemical oxidation methods gives oxides with a large carbon contamination. Thus, the remote plasma oxidation of InP, because of the low temperature and low energy plasma processing, is expected to be a key technique in preparing new oxides layers with high quality electrical properties. Preliminar data on the InP remote plasma oxidation have been obtained by using low pressure (0.6 torr) O$_2$ plasma operating at 13.56 MHz and at the r.f. power of 90 watt. The InP surface is at 10 cm away from the plasma and is kept at room temperature. Under the above conditions, an oxide growth rate of about 0.3 A/min has been measured.

The oxidized surfaces have been in situ characterized by SE whose typical ε_i spectrum is shown in Fig.9. The main result is that the plasma grown oxide layer is a mixture of phosphates, InPO$_x$, and indium oxide, In$_2$O$_3$, as shown in the inset of Fig. 9.

In conclusion, the plasma oxidation method does not induce the formation of free phosphorus at the oxide-InP interface, which degrades its electrical properties.

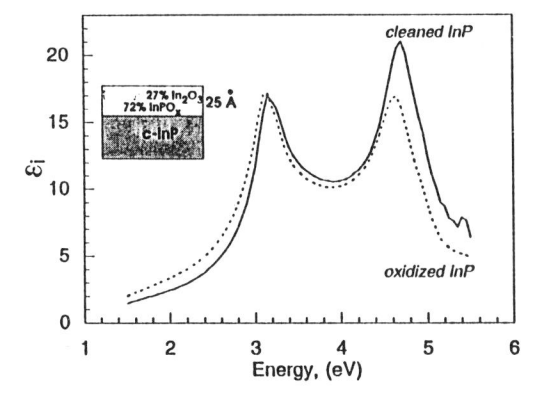

Figure 9. Spectrum of the imaginary part of dielectric function, ε_i, for plasma oxidized InP surface. For comparison the spectrum of a cleaned c-InP surface is shown. The inset shows the top layer structure as derived by BEMA model.

REFERENCES

1. M. Razeghi, *"The MOCVD Challenge, vol.1: A survay of GaInAsP-InP for photonic and electronic applications"*, Adam Hilger, 1989.
2. S.J. Pearton, F. Ren, T.R. Fullowan, A. Katz, W.S. Hobson, U.K. Chakrabarti, C.R. Abernathy, *Materials Chemistry and Physics*, **32**, 215 (1992).
3. E.S. Aydil, K. P. Giapis, R.A. Gottscho, V.M. Donnelly, E. Yoon, *J. Vac. Sci. Techn.B*, **11**(2), 195 (1993)
4. T. Sugino, H. Ninomiya, t. Yamada, J. Shirafuji, K. Matsuda *Appl. Phys. Lett*, **60**(10), 1226 (1992).
5. M.P. Besland, P. Louis, Y. Robach, J. Joseph, G. Hollinger, D. Gallet, P. Viktorovitch, *Appl. Surf. Science*, **56-58**, 846 (1992)
6. A. Guivarc'h, H. L'Haridon, G. Pelous, G. Hollinger, P. Pertosa, *J. Appl. Phys.*, **55**(4) 1139 (1984).
7. P.G. Hofstra, D.A. Thompson, B.J. Robinson, R.W. Streater, *J. Vac. Sci. Techn. B*, **11**(3) 985 (1993).
8. G. B. Stringfellow, *"Organometallic Vapor Phase Epitaxy: Theory and Practice"*, Academic Press Inc., 1989.
9. U. Sudarsan, N.W. Cody, T. Dosluogolu, R. Solanki, *J. Crys. Growth* **94**, 978 (1989).
10. M. Behet, A. Brauers, H. Luth, P. Balk, *J. Crys. Growth* **107**, 209 (1991).
11. M. Naitoh, T. Suga, T. Jimbo, M. Umeno, *J. Crys. Growth* **93**, 52 (1988).
12. G. Bruno, M. Losurdo, P. Capezzuto, *Appl. Phys. Lett*, **66**(26) 3573 (1995).
13. G. Bruno, M. Losurdo, P. Capezzuto, Jour. Vac. Sci. & Technol. A, **13** (9), 349 (1995).
14. G. Bruno, M. Losurdo, P. Capezzuto, *Jour. de Physique*, to be published.
15. G. Bruno, M. Losurdo, P. Capezzuto, *"In situ Mass Spectrometric Diagnostics During InP Deposition in a Remote Plasma- Enhanced MOCVD System"*, in Proceedings of the 6th International Conference on Indium Phosphide and Related Materials, Santa Barbara, California, 1994, pp 94.
16. M. Losurdo, P. Capezzuto, G. Bruno, J. Vac. Sci. Techn. b, to be published.
17. F. Schroder, W. Storm, M. Altebockwinkel, *J. Vac. Sci. Techn. B*, **10**(4) 1291 (1992).

Breakdown and Discharges in Dense Gases Governed by Runaway Electrons

Leonid P. Babich

Russia Federal Nuclear Center: VNIIEF. Arzamas-16, 607200, Region of N.Novgorod, Russia.
E-MAIL: otd4@expd.rfnc.nnov.su

Abstract. The phenomenon of runaway electrons (REs) at high values of the ratio field intensity /gas number density E/N and N up to the Loshmidt number $N_L \approx 2.7 \times 10^{19}$ cm^{-3} is described. REs are shown to govern the breakdown and discharges at such condition.

INTRODUCTION

The idea for charged particles to runaway from collisions with neutral species of dense plasma has been considered by C. Wilson thirty years earlier than the famous works of Drieser et. al. on runaway electrons (REs) were published at the outset of the era of thermonuclear research. Wilson suggested that strong electric fields such as those of thunderclouds can accelerate charged particles to very high energies in dense lower layers of the Earth's atmosphere (1). It seems that for the first time the runaway phenomenon was observed in 30-th during thunderstorm activity.

Table 1 allows to get a notion on the first laboratory experiments performed in 60-th to detect x-rays at $N \approx N_L$. For references see (2). Here U_{idle} and τ_{idle} are the maximum value and risetime of the idle-running high-voltage waveform (HVW) To detect x-rays Frenkel et.al. and Noggle et.al. ran 10,000 pulses. Stankevich et.al. ran 100 pulses at high overvoltage $\Delta = (U/U_{st}-1)$ over statics selfbreakdown value U_{st}. Subnanosecond τ_{idle} allowed for Tarasova et.al. to realize extremely

TABLE 1. First experiments on REs in dense gases

Authors	Frenkel et.al.	Noggle et.al.	Stankevich et.al.	Tarasova et.al.
Geometry	point-plane	point-plane	rod-plane	rod-plane
d,cm		5	0.04	0.5-8.5
Gas	He	He	air	air
τ_{idle}, ns		<10	<2	<0.6 (2)
U_{idle}, kV		300	40	300
mR/pulse	<0.04	<0.04	0.4	electrons (2)

high Δ and detect a penetrating radiation of very high intensity, which was shown later to be REs of extremely high energy(2). These pioneer experiments were followed by numerous studies. X-rays from nanosecond discharges in dense gases were detected by Asinovskii et.al., Byszewski and Reinhold, Dashuk et. al., Kremnev and Kurbatov. Bosamykin et.al. observed microsecond x-ray emission from the air gap (wire parallel to plane) at normal conditions (3). Loiko succeeded to detect directly REs with $\varepsilon \approx eU$ from the breakdown in the air at normal conditions ($d \approx 2$ cm, rod - plane, $U \approx 70$ kV) at relatively low $\Delta \approx 0.6$ (2).

Undiscrepant model of breakdown and discharges in dense gases should incorporate the nonlocal phenomenon of REs to be adequate at high values of E/N, especially at high Δ or in the case of high spatial inhomogeneity of E/N, which may be primordial or arise as a result of E or N distortion. Though fundamentals of the nonlocal model were formulated rather long ago by Babich and Stankevich [4] and Kunhardt and Byszewski [5], theoretical treatment, capable to describe the overall dynamics of discharges with REs, is still being developed. Experiments are by far more numerous and successful. So in the present lecture we shall be considering mainly experimental studies performed for the air up to normal density (2).

GENERAL TREATMENT OF RUNAWAY ELECTRONS

The phenomenon of REs is an effect of cross sections of electron-atomic particle interactions to be decreasing functions of electron energy ε in the range from about 100 eV up to 1 MeV. Adequate stochastic definition of RE was formulated by Kunhardt, Tzeng and Boeuf: "...an electron is runaway if it does not circulate through all energy states available to it at a given E/N, but on average moves towards high-energy states" (6). In the framework of the deterministic approach, based on the effective retarding force $F(\varepsilon)$, for electron to become runaway it should overcome a definite energy threshold depending on E/N. It is convenient to present $F(\varepsilon)$ in terms of energy loss function $F(\varepsilon) = eL(\varepsilon)P$. In the air $L(\varepsilon)$ hits maximum value $L_{max} \approx 356$ $eV/cm/Torr$ at $\varepsilon_{max} \approx 150$ eV, passes the minimum $L_{min} \approx 1.2$ $eV/cm/Torr$ in the vicinity of $\varepsilon_{min} \approx 1$ MeV and after all slowly increases on account of the relativistic rise of cross sections and radiative loss. According to Fig. 1 for $F(\varepsilon)$ the right hand side of the conventional energy balance equation

$$d\varepsilon/dt = f(\varepsilon) = -(eE\mu + F(\varepsilon)) \times p(\varepsilon) \, c^2/\varepsilon$$

is a characteristic function accounting at $\mu \in [-1, 0)$ for the dynamics of the simplest bistable element. Here μ is the cosine of the angle between E and electron momentum p. For $E \in (F_{min}/e|\mu|, F_{max}/e|\mu|)$ the equation $f(\varepsilon) = 0$ has three roots: $\varepsilon_1 < \varepsilon_{max}$, $\varepsilon_2 \in (\varepsilon_{max}, \varepsilon_{min})$ and $\varepsilon_3 > \varepsilon_{min}$, corresponding to stationary states of the

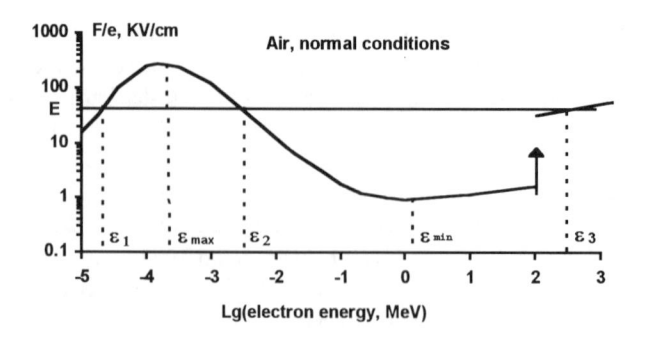

FIGURE 1. Energy dependence of the retarding force $F(\varepsilon)$.

tested electron. States ε_1 and ε_3 are stable, whereas ε_2 is absolutely unstable. In the range $(\varepsilon_1, \varepsilon_2)$ electron slows down to ε_1, where the conventional local approach is valid. Within $(\varepsilon_2, \varepsilon_3)$ it accelerates up to the relativistic state ε_3 with an adequate drift velocity. So ε_2 is the runaway threshold. In the domain $< \varepsilon_{max}$ the slow drift is superimposed over the fast chaotic motion. Contrary, in the range $> \varepsilon_{min}$ directed motion prevails on account of slow chaotization of p. In the case $E < F_{min}/e |\mu|$ there is only one conventional stationary state ε_1. Extremely strong field $E > F_{max}/e|\mu|$ also allows for one stationary state ε_3. The last case assumes zero runaway threshold, which means actually the initial energy $\varepsilon(0)$ of electron to be well below ε_{max}. This case is adequate to the self-breakdown in dense gases at very high Δ. For zero-energy threshold a critical field is reasonable to evaluate as $E_{cr} \approx L_{max}P$, $(\mu = -1)$. Kunhardt and Tzeng (7) calculated for N_2 the critical value to be 1.5 times higher owing to electron scattering was taken into account. The above simplest consideration infers that an electron assembly should generate energy distribution in the vicinity of ε_1 and ε_3. State ε_3 is realized if $\varepsilon(0) > \varepsilon_2$ or $E > F_{max}/e|\mu|$. High E/N along with applied voltage of the order of tens and hundreds of kV were sufficient to overcome ε_2 in laboratory at N of the order of N_L. To achieve ε_3 the voltage should be of the megavolt range. Presently ε_3 is possible to observe only within thunderstorm fields in lower layers of planet atmospheres.

DISCHARGES IN THE AIR AT VERY HIGH OVERVOLTAGES

Most of results on REs have been obtained along the line of experiments started by Tarasova et.al. with HVWs up to 300 kV and $\tau_{idle} < 0.6$ ns (2). The initial supply of electrons was 5×10^{13}. As a rule the breakdown occurred at the front of $U_{idle}(t)$ (Table 2 below, where $U(d) < U_{idle}$) At normal conditions and $d > 1$ cm the

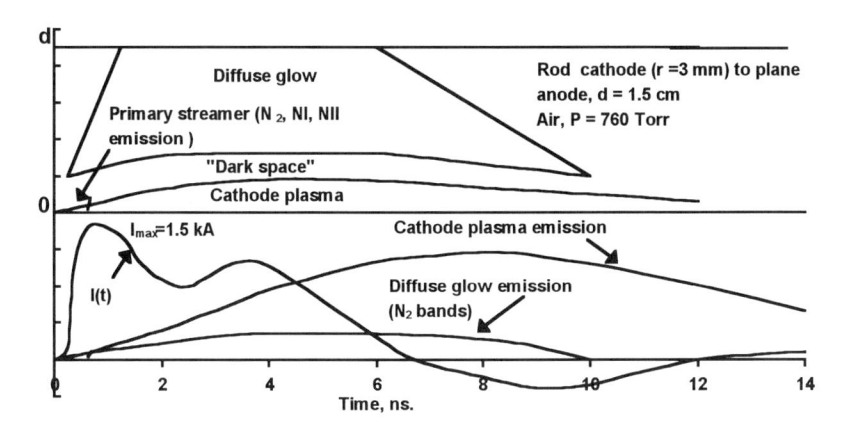

FIGURE 2. Current and photohronogram of the light emission.

breakdown developed into a diffuse glow discharge. The glow was separated by narrow (<< d) "dark space" from small ($l \approx 1$ mm $<< d$) one or several bright near-cathode plasmas with a crown. $U(t)$ and current $I(t)$ oscilloscope traces displayed almost lonely spikes. At lower d the plasma grew through the glow up to the anode to develop contracted channels. The current risetime was $\tau_I < 0.5$ ns for both cases. The near-cathode plasma emitted N_2 bands, NI, NII and lines of cathode ions and (or) atoms along with a broad continuum. The nitrogen emission began about 1 ns prior to the appearance of the continuum and metal lines. The diffuse glow of the main volume emitted only N_2 bands, the same as the near-cathode plasma emitted. The temporal evolution of the light emission and current $I(t)$ are illustrated in Fig.2. The glow propagation velocity was estimated to be less than 2×10^9 cm/s. The emission implies that over time shorter than τ_I, a faintly glowing streamer-like channel developed in the near-cathode region. It was rather power-consuming formation. Electrons accelerated in front of the streamer, became runaways and preionized the main volume. The spatial homogeneity of the plasma within it resulted from the overlapping of secondary avalanches. Within the "dark space" the energy of secondary electrons was too large to provide detectable light emission.

RUNAWAY ELECTRONS AT HIGH OVERVOLTAGES

Subsequent study of the angle distribution and absorption properties of the radiation observed by Tarasova et.al. outside the anode of the above discharges, have shown it to be an intensive flux of high-energy REs. Fig.3 presents the dependence of the RE number on the initial air pressure for two values of absorber thickness. The N_e decrease at low P infers that the generation of REs was volumetric process. Some high-energy portion of RE distribution rather weakly depended on P and consequently on $U(t)$. In that way REs of "anomalous energy"

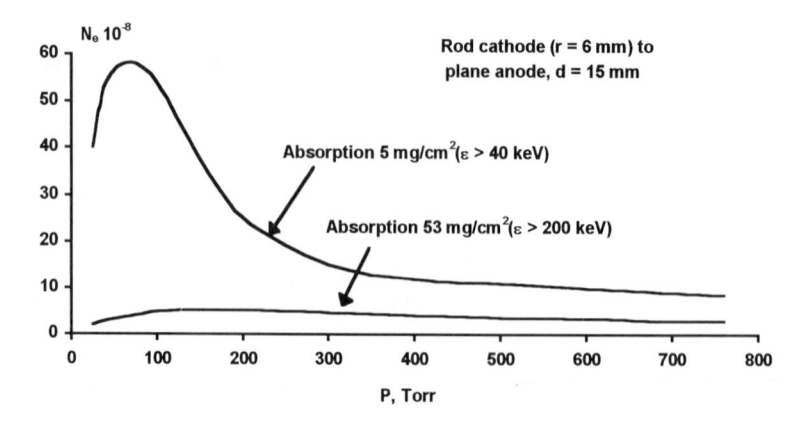

FIGURE 3. Dependence of the number of REs on the air pressure.

for the first time revealed themselves. In this sense absorption curves of REs occurred to be very illuminative. At *P* of the order of hundreds Torr the curves displayed features typical for monoenergetic electrons, i.e. the most steeply decreasing linear section in the middle and pronounced straggling at the end. RE energy was evaluated from the extrapolated range to be $> eU$ ("anomalous energy"). Babich and Loiko published energy distributions of REs (2) obtained by the method of magnetic spectroscopy (Fig.4). Discharges at high P generated distributions with a well-defined maximum, position ε_m of which increased along with *P* increase. Its value really was higher than *eU*. The intrinsic width of these distributions was essentially less than the measured value $W \approx 60$ *keV*, which did not change within the pressure range *200 - 760 Torr*. The lack of REs below $\varepsilon_m - W/2$ along with "anomalous" value of ε_m and $W \ll \varepsilon_m$ appealed for an acceleration mechanism unobserved earlier in discharges in dense gases. At $P < 200$ *Torr* the distributions expanded owing to the generation of a large amount of REs with lower energies. Table 2 shows the energy excess $\Delta\varepsilon = \varepsilon_m - eU$ was 100-

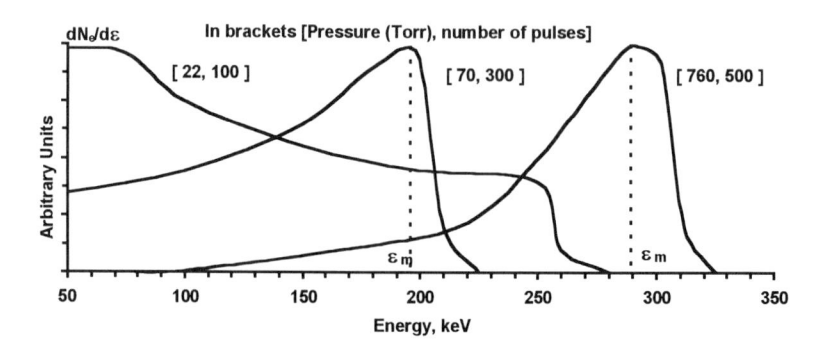

FIGURE 4. Energy distributions of REs. Air, *d* = 2 cm, cone to plane grid gap.

TABLE 2. Energy of REs

d, cm	0.5	1	2	3.5
U, kV	130	150	190	210
ε_m, keV	180	260	290	320

110 keV for $d \geq 1$ cm, while for lower d it was less pronounced. These results correlate with the form of the discharges: diffuse glow at $d > 1$ cm and bright contracted channel at $d < 1$ cm. It was of crucial importance to know spatial and temporal characteristics of REs in order to identify their origin and understand the way, by which they govern the breakdown and overall dynamics of the discharge. REs were generated within the cathode region. At normal conditions the pulse width of anomalous energy REs was measured to be $\Delta t \approx 0.5$ ns. As the gas density was reduced, Δt increased, to become at $P \approx 10$ Torr equal to the width ($\approx$ 2 ns) of the first oscillation of HVW. To identify the onset of REs was of particular importance. REs govern the breakdown provided that they are generated at the stage of primary avalanches and streamers. Otherwise, they are an effect rather than one of the crucial causes of breakdown. The anomalous energy REs were detected at the front of high-voltage pulse $U(t)$ used. By means of a barrier discharge, which delayed the cathode emission and gas ionization, it was proved the onset of anomalous REs to coincide with the start of the conductivity current. N_e was essentially independent on I in the range 0.2 - 1.5 kA. So the duration of a stage responsible for the generation of anomalous REs was reasonable to estimate as $(0.2/1.5)\tau_I < 0.1$ ns $<< \Delta t \approx 0.5$ ns. The problem arose to understand the nature of this phenomenon. The generation of such particles by low-temperature collision - dominated plasma appeals to extrapolate the Roether treatment of the breakdown phenomenology, incorporating localized fields of space charges, to the domain of extremely high fields. The proximity of the detected number of anomalous energy REs (Fig. 3) to the Roether critical avalanche $exp(20)$ is very significant.

SELFACCELERATION OF RUNAWAY ELECTRONS

From the very outset of the streamer model of the breakdown in dense gases the ionization rate of a gas, intensification of the field E_f in front of the anode-directed streamer and permanent increase of a local drift velocity $V_d\ (E_f)$ were considered to be selfconsistent. Actually this collective process one may treat in the framework of the conventional local approach as an electron drift self-acceleration by the edge field soliton, propagating owing to the ionization of a background gas by the swarm of electrons, trapped by the soliton, and polarization of the resultant plasma. The selfconsistency, if being incorporated with the nonlocal approach, infers that under certain conditions REs may also be trapped by the soliton, i.e. at sufficiently high external field intensity E_0 the joint motion of REs and the soliton, coupled due to the ionization of a gas, becomes in fact "pure" (not drift) self-acceleration in the

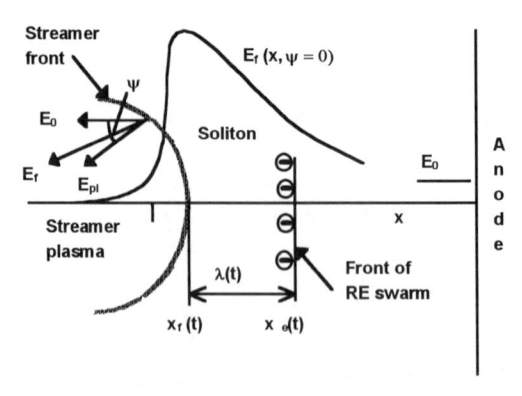

FIGURE 5. Selfacceleration of REs in front of anode-directed streamer.

laboratory frame of reference at least within some limited space-time domain of the overall breakdown process. Fig. 5 illustrates the self-acceleration mechanism. Here $x_f(t)$ is the apex of the streamer quasi-neutral plasma. The dimension $\lambda(t) = x_e(t) - x_f(t)$ of the active zone, where ionization, polarization, field displacement and RE acceleration take place, may be treated as the wave length of the soliton. For the motion of REs to be synchronized with the soliton propagation it is necessary for the characteristic time τ of the field to displace from the plasma trail into the soliton space domain to be of the order of the RE motion within the soliton. Have proceeded from the idea of two energy groups of electrons, proposed by Kunhardt and Byszewski (5), and believed that the field shielding and displacement were due the drift motion and Townsend α-avalanching of secondary slow electrons, initiated by the swarm of REs within the soliton, Babich has deduced the necessary condition $d\tau/dt < 0$ for the self-acceleration. It is valid for the air with $N \approx N_L$ at high E/N provided that $(E_f/E_e) \times \alpha\lambda_{ion} > 15$. Under this condition undamped electron avalanches, initiated by the swarm of REs, synchronize the motion of the swarm and the field soliton, which are in a sense coupled by secondary avalanche chains. Here E_e is the selfconsistent field of the RE swarm and λ_{ion} is the RE ionization free path. In the case of very strong E_f, at which all electrons within the soliton space domain run away, $E_f/E_e \to 1$ and $\alpha\lambda_{ion} \to 1$, so that the inequality may be violated. For zero avalanching ($\alpha = 0$) the condition $d\tau/dt < 0$ never is satisfied. Apparently λ should be $\ll l \approx 1$ mm, the length of the near - cathode streamer. To estimate λ Babich and Mozgovoi (8) took into account that N_e did not depend on I up to 200 A, so that $\lambda \approx eN_e c/2I \approx 100\ \mu \ll l$. Babich estimated that REs acquired $\Delta\varepsilon \approx 100\ keV$ (Table 2) along l provided that the ratio l/streamer width had realistic value 5. To satisfy $E_f > E_e$ and the necessary condition for the formation of the avalanche chains the number of REs should be within the interval $[4\times10^5, 4\times10^9]$. The number of anomalous REs $< 10^9$ was detected to be within it. Babich and Mozgovoi evaluated that the finite speed of the light c limited the energy increase by the value $(2/\sqrt{3} - 1)\ mc^2$, which is very similar to the measured excess $\Delta\varepsilon$ (Table 2). Apparently this fundamental restriction accounts for the failure of all attempts to increase $\Delta\varepsilon$. We hope that the above estimations of separate features of the self-acceleration, which agree rather well with experimental results, will be a basis for the continuous relativisitically invariant description, incorporating selfconsistent electromagnetic field and taking into account complex interplay of ionization and acceleration processes.

REs accelerated away from the moving front, either relax to the external field E_0 or become really runaway, i.e., escape from a gap, provided that $\varepsilon > \varepsilon_2$. Runaway criterion met first of all at the streamer apex $(x = x_f, \psi = 0)$, gradually spreads over all front surface, so that an "injection cone" of ψ angles (Fig. 5) develops with a maximum number of REs at the streamer axis (5). At low Δ REs within the cone have sufficiently low energies to initiate secondary avalanches and develop a narrow filament, which grows up to an anode. At very high Δ the breakdown develops as a short near-cathode channel with a "crown" at large ψ due to short-range REs and a diffuse glow at small ψ and far to an anode due to preionization of an ambient gas by long-range REs and overlapping of secondary avalanches (2).

DISPLACEMENT OF *U(Pd)* MINIMUM

Participation of REs in the breakdown should manifest itself as an adequate behavior of $U(Pd)$ curves. It is well known that the breakdown "on the left-hand branch" of the static Pashen's curve $U_{st}(Pd)$ is governed by REs up to $(Pd)_{min}$. The runaway criteria for zero-energy threshold (which is the case for the breakdown) $(eU_{st})_{min} > \varepsilon_{max}$ and $e(U_{st}/Pd)_{min} > L_{max}$ hold for many gases. $(U_{st}/Pd)_{min}$ differs not very much from L_{max} with the exception of strongly attached gases. This implies to go so far as to suggest that $e(U/Pd)_{min} \approx L_{max}$. From it follows that an increase of Δ

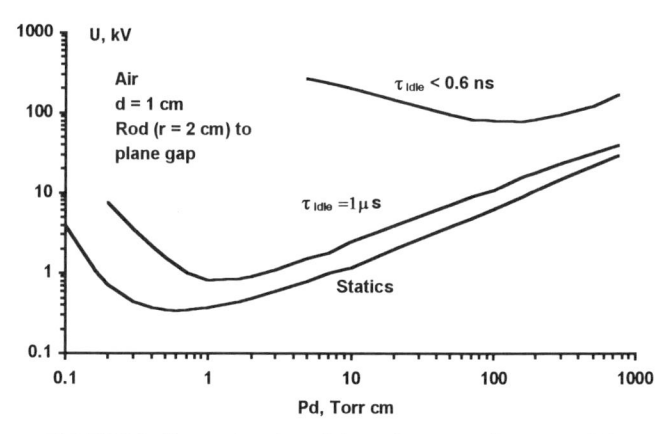

FIGURE 6. Dependencies of the voltage maximum on Pd.

should lead to the rightward shift of the *U(Pd)* minimum. Fig.6 shows curves *U(Pd)* measured for three τ_{idle} values. The shift is strongly pronounced and *(U/Pd)$_{min}$* values differ not very much from L_{max}. So the displacement predicted from the idea for the breakdown to govern by REs, do take place. In its turn this result supports the nonlocal conception of breakdown. In general, the dielectric strength of gases should be characterized by the family of curves *U(Pd, τ_{idle})*, with τ_{idle} being a parameter.

X- RAYS AND NEUTRONS FROM THUNDERSTORM

Production of penetrating radiation by thunderstorm is an intriguing problem. Adequate activity started by Schonland in 1933, never terminated. To search high-energy electrons or x-rays directly from lightning Whitmire placed dosimeters at TV tower for a long term during which it received 200 strokes (9). Pronounced anisotropy of the detected radiation to the north-east ruled out the influence of environmental permanent factors and the solar radiation. McCarthy and Parks carried out flight experiments to search x-rays within thunderstorm clouds (10). Along with active channel they used a passive one to control an electromagnetic environment. Their observations were as follows. (a) X-ray flux increased up to 3 orders of magnitude above the background throughout the tested energy range from 5 keV up to > 110 keV. (b) Space scale of the x-ray source was of the order of 1 km. (c) Elevated flux of x-rays preceded lightning by several seconds and ceased coincident with its onset. McCarthy and Parks concentrated on a model assuming that a large-scale thunderstorm electric field increased flux of high energy electrons produced by cosmic rays (11). The lightning was assumed to account for a collapse of the voltage and termination of the flux. Evidence that lightning generate neutrons was published by Shah et. al (12). Table 3. gives a number of neutron events with N_n recorded neutrons, correlated with lightning strokes. By means of the comparison with the cosmic ray statistics $N_n > 3$ events were put down to fusion reactions within the lightning. Multiple output of neutrons was recorded when trees were damaged by the lightning. The authors estimated total yield as 3×10^6 - 2×10^{12} neutrons per stroke. These time resolved results present a reliable evidence that local disturbance of *E/N* associated with the lightning, is capable of accelerating charged particles up to very high energies. They support Whitmire's belief, that he recorded penetrating radiation from lightning. Most likely two mechanisms accounts for high-energy REs and ions. The first is associated with overall thunderstorm activity, the other is governed by a local enhancement of electric field caused by lightning. Gurevuch, Milikh and Roussel-Dupre proposed a

TABLE 3. Neutrons from lightning.

N_n	1	2	3	6	9	60
Number of events	10,818	250	241	59	39	1

mechanism of lightning based on avalanching of relativistic REs permanently produced by cosmic rays and accelerated by thunderstorm electric field (13).

CONCLUSION

Runaway phenomenon in collision dominated plasmas of gas discharges was observed for a vast range of conditions: gas specimen, neutral density, geometry of a gap, HVWs. It revealed itself in a full measure at Δ several times over the static selfbreakdown voltage. New gas discharge phenomena and dependencies were discovered. (a) The subnanosecond pulse of REs with the energy $\varepsilon > eU$ is the most remarkable. The adequate mechanism is not exotic and is followed straightforward from the conventional Roether model. (b) Due to the preionization of a gas by high-energy REs discharges might acquire a high degree of spatial homogeneity even at very high N. The flux of REs of lower energies supports the homogeneity. (c) The displacement of the $U(Pd)$ minimum to high Pd was observed as Δ was increased from zero up to very high values. It is the most reliable evidence of REs to govern the breakdown. The flight and long term ground experiments provided solid evidences on high-energy phenomena associated with thunderstorm and lightning. These phenomena call for a lightning theory to incorporate nonlocal processes. A large work to study fundamentals of gas discharges, governed by nonlocal processes, and associated high-energy phenomena is still to be done by means of theory, computer simulations and experiments both in laboratory and in the nature. Presently these discharges are used at VNIIEF to generate ultra-short pulses of electrons, x-rays and neutrons, to initiate and pump high-power atmospheric pressure TEA lasers. They are studied to be applied in ecology to disintegrate exhausted industrial gases, to generate ultrasonic waves for non-destructive testing of solids, in mining to select minerals.

REFERENCES

1. Wilson, C.T.R., *Proc. Cambr. Phil. Soc.* **22**, 534 (1925).
2. Babich, L.P., Loiko, T.V., and Tsukerman, V.A., *Sov.Phys.Usp.* **33**, 521 (1990).
3. Bosamykin, V.S., Karelin, V.I., Pavlovskii ,A.I., et.al., *Pis'ma Zh. Tekh. Fiz.* **6**, 885 (1980).
4. Babich, L.P., and Stankevich, Yu.L., *Sov. Tech. Phys.* **17**, 1333 (1972).
5. Kunhardt, E.E., and Byszewski, W.W., *Phys.Rev.* **A 21**, 2069 (1980).
6. Kunhardt, E.E., Tzeng, Y., and Boeuf, J.P., *Phys. Rev.* **A. 34**, 440 (1986).
7. Kunhardt, E.E., and Tzeng, Y., *Phys. Rev.* **A 38**, 1410 (1988).
8. Babich, L.P., and Mozgovoi, A.L., "Relativistic restrictions on electron selfacceleration in dense gases at high overvoltages", in *Proc. of ICPIG XX*, Italy, **v.2**, p. 909-910 (1991).
9. Whitmire, D.P., *Let. Nuov. Cim.* **26**, 497 (1979).
10. McCarthy, M., and Parks, G.K., *J.Geophys. Res.* **97**, 5857 (1992).
11. McCarthy, M., and Parks, G.K., *Geophys. Res.Let.*, **12**, 393 (1985).
12. Shah, G.N., Razdan, H., Bhat, C.L., and Ali, Q.M., *Nature* **343**, 773 (1985).
13. Gurevich, A.V., Milikh, G., Roussel-Dupre, R.A. *Phys. Lett.* **A165**, 463 (1992).

Magnetic and Electric Probe Diagnostics in Inductive Plasmas

Valery Godyak, Robert Piejak and Benjamin Alexandrovich

OSRAM SYLVANIA INC.
71 Cherry Hill Drive, MA 01915, USA

Abstract. An abbreviated review of recent experiments performed in argon inductive discharges at **OSRAM SYLVANIA** Research Center is given in this presentation. Results of three different diagnostic methods (Langmuir probe, magnetic probe and measurement of the external electrical discharge characteristics) are presented here. The measurements were done in an axially symmetric, co-axial configuration with an internal exciting coil driven at 13.56 MHz. For the first time, the electron energy distribution function, plasma density and effective electron temperature and the radial distributions of rf electric field and discharge current density were measured altogether in an inductive discharge in argon ranging over three orders of magnitude in gas pressure. Some specifics of electrostatic and magnetic probe diagnostics in inductively coupled discharges are discussed.

INTRODUCTION

Low pressure inductive discharges offer new opportunities in plasma processing and light source technology because of the absence of electrodes and because of the high plasma density that can be achieved with low plasma dc and rf potential. Recognition of these attractive features intrinsic to inductively coupled discharges has served to stimulate a comprehensive study of basic plasma phenomena in such discharges. Increased understanding of these plasmas has led to an improvement in the control of the rf discharge characteristics upon which the importance of these discharges to modern technology continues to grow.

Experimental research of rf discharges include a wide variety of diagnostic techniques among which measurement of the external discharge parameters (rf voltage, current, power and impedance) together with measurement of the internal discharge parameters (plasma density, electron temperature, rf field and rf current density) are fundamental.

For meaningful diagnostics of the internal discharge characteristics, it is essential to properly define the experimental conditions. Definition of these conditions generally requires measurement of the external discharge characteristics. Defining the discharge conditions by merely measuring rf power delivered to an experimental device (which disturbingly is the case in many rf discharge experiments) can be misleading, since the power absorbed by an rf plasma is generally an unknown fraction of the total measured power.

Here we report on Langmuir probe and magnetic probe (B-dot) measurements together with measurement of external electrical characteristics in a low pressure inductive rf discharge with an internal coil configuration typical for electrodeless light sources. Based on measurements of the external electrical parameters it was possible to discriminate between power delivered to the plasma and that lost in the inductor-matcher circuitry, thus all internal discharge parameters given here are related to rf power dissipated in the plasma.

EXPERIMENTAL SYSTEM AND DIAGNOSTICS

Measurements were made in a discharge chamber with an internal exciting coil surrounded by an electrostatic shield to eliminate capacitive coupling and thus insure that the discharge was purely inductive. The vacuum chamber consisted of two coaxial glass cylinders, one with a 14.2 cm ID and another with a 3.8 cm OD, that were limited at their ends by two aluminum plates separated by 6.7 cm (see Fig. 1). Except for the inner glass tube, the chamber and vacuum-gas flow system were identical to those of our earlier experiments with capacitive rf discharges (1, 2). The screen was made up of eight parallel copper foil strips glued to the inner glass surface (outside the vacuum) with 1 mm gaps between each strip to make the screen transparent to the rf magnetic field.

The excitation inductor consisted of 10 turns with a length of 4.5 cm and a diameter of 3.2 cm and was centered near the chamber midplane. Rf power consumed by the plasma (discharge power) P_d was determined by subtracting coil and discharge chamber losses from the rf power transmitted (incident minus reflected power) to the matching network. In practice, this was done by first measuring the transmitted power without plasma as a function of rf coil current and subtracting that power from the transmitted power (at that same coil current) with the plasma present. This technique is described in detail in Ref. 3.

Rf plasma potential was measured with a capacitive probe formed by a 5 cm long, 2 mm diameter metal rod inserted into a glass capillary glass tube as shown in Fig. 1. The rf plasma potential was measured using an rf probe with two different values of input capacitance (4). This technique allows one to account for the capacitance between the capacitive probe and the plasma thus obtaining the true value of rf plasma potential.

Langmuir probe measurements were done with a probe station developed and used in our previous capacitive rf discharge studies and it is described in Ref. 5. The EEDF was obtained through analog double differentiation of the probe characteristic followed by digital averaging and processing. Noise suppression techniques with a fast probe voltage sweep (5) made it possible to significantly extend the dynamic range of reliable EEDF measurements (up to four orders of magnitude); this is especially important at the lowest gas pressure where the discharge manifests a low frequency instability. The plasma potential was taken to be the probe voltage corresponding to the zero crossing of the second derivative. The plasma density n and the effective electron temperature T_{eff} were obtained as integrals of the EEDF (see Ref. 5, for details).

Effective electrostatic screening of the plasma from the inductor coil in the present work resulted in negligibly low rf plasma potentials (compared to the electron temperature), thus there was no need for rf filtering in the probe circuit.

Note, that the use of self-resonant inductors as rf filters can bring about two problems specific to inductive discharge. One problem is induction an rf voltage on the filter inductor due to rf magnetic field that maintains the discharge; thus, instead of reducing the rf voltage on the probe sheath, the filter inductor may be a source of additional rf voltage! A second problem, generally encountered at the high plasma density typically found in inductively coupled plasmas, occurs when the dc resistance of the filtering inductor appears to be comparable to the dynamic resistance of the Langmuir probe (5). When this happens, depletion of the low energy part of the measured EEDF occurs. This effect can be seen in some recently published works on EEDF measurements in inductive rf discharges.

The B-dot probe is a small coil of wire that produces a voltage proportional to the time rate of change of magnetic flux enclosed by the coil. The probe used here consisted of six turns of #30 enamel covered magnet wire wound in a tight coil which was about 1.8 mm long and 2.5 mm in diameter. The probe rode within a glass cylinder (4 mm OD) radially directed in the mid plane of the discharge. Since the discharge current density was found in this work through spatial differentiation of the B-dot signal, special attention was paid to accuracy of the probe positioning. The probe was spatially driven by a micrometer based linear motion stage which slided the probe shaft radially inside the glass tube. The probe (together with the balun transformer) was calibrated by creating a known time varying magnetic field inside a long solenoid and measuring the voltage from output of the probe across a 50 ohm load. The B-dot probe was positioned so as to measure the axial component (z) of the magnetic flux density.

The magnitude and relative phase of the B-dot probe signal was measured with and without a discharge at 1 mm intervals along the discharge radius in the mid plane of the primary coil. Since the driving frequency of the discharge was fixed, the voltage measured with the B-dot probe was proportional to the magnitude of the magnetic flux density B.

Proper design and electrostatic shielding are important issues in B-dot probe construction. Generally, to attain a reasonable spatial resolution, the probe radius must be smaller than the scale of change in the electric field. In this experiment both the probe and the discharge system are designed to minimize capacitive pickup of the voltage on the primary induction coil. Capacitive pickup was minimized in this system by reducing the rf plasma potential with the electrostatic screen between plasma and primary coil and using the balun transformer to de-couple the common mode signal from the probe output voltage proportional to dB/dt.

Two vector voltmeters (each having two channels) were used in this experimental system. One was used to measure the primary coil rms voltage, rms current and relative phase shift. A second vector voltmeter was used to measure the magnitude and phase of the voltage from the dB/dt probe. The phase of the voltage from the dB/dt probe was referenced to that phase induced on the B-dot probe by the primary coil current in vacuum. Measurements reported in this work have been done in the mid plane of an axially symmetrical inductive discharge driven at 13.56 MHz in the wide range of argon pressure between 1 mTorr and 1 Torr.

As mentioned earlier all results given here are related to the rf power consumed solely by the plasma P_d and not to the total rf power dissipation P of the entire discharge device. To our knowledge all data in the literature concerning inductive rf discharges is related to P, however, we believe that plasma parameter measurements related to P_d provide a defined plasma condition rather than a condition of the particular discharge arrangement, which in large extent may depend on type and quality of induction coil, matching circuit elements and transmission line.

EXPERIMENTAL RESULTS AND DISCUSSION

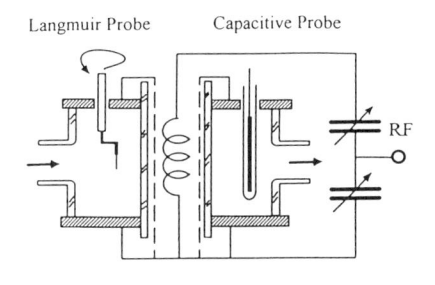

FIGURE 1. Schematic diagram of experimental setup.

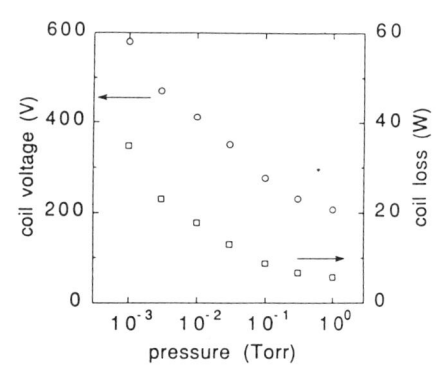

FIGURE 2.. Induction coil voltage and coil power dissipation versus argon pressure.

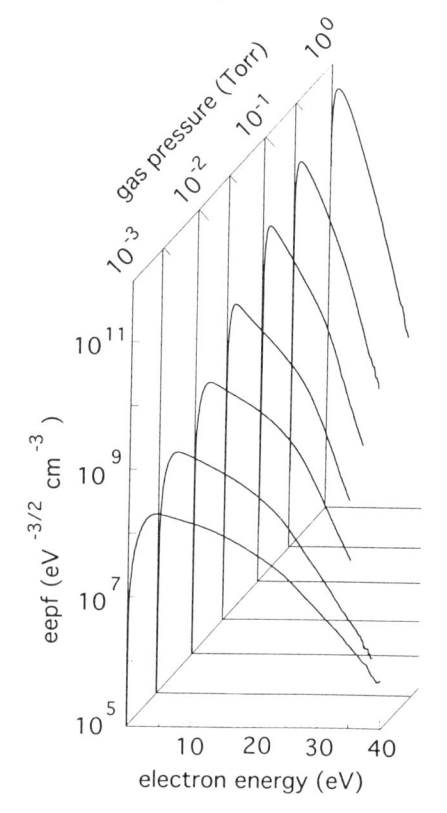

FIGURE 3. EEPF evolution with argon pressure.

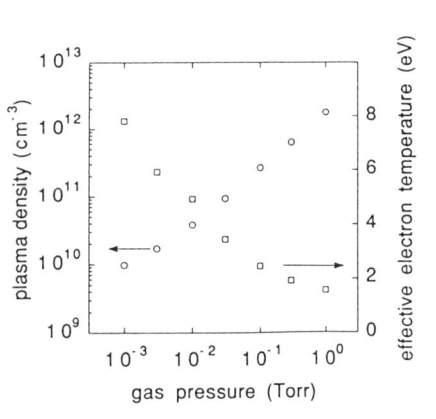

FIGURE 4. Plasma density and effective electron temperature versus argon pressure.

Discharge Electrical Charecteristics

The rms coil voltage and the power loss in the coil versus argon pressure are given in Fig. 2. With decreasing argon pressure, the coil voltage grows because the plasma rf field E_p required to maintain the discharge increases (6). The increase in E_p is mainly due to an increase in electron inertia which is accounted for by the factor ω/ν_e, where ν_e is electron collision frequency and ω is the driving frequency. Therefore, as gas pressure decreases a larger rf current in the coil is needed to maintain the discharge. A larger current results in higher coil voltage and larger power loss in the coil. Measurement of the coil loss P_c allows one to find the power transfer efficiency: $P_d/P = P_d/(P_d+P_c)$ which for $P_d = 50W$ increases from 60% at p = 1 mTorr to about 90% at p = 1 Torr.

Capacitive probe measurements of the plasma rf potential V_{rf} show that V_{rf} monotonically decreased with argon pressure from about 2 V_{rms} at p = 1 mTorr to about 30 mV at p = 1 Torr. For all gas pressures, V_{rf} was less than a quarter of the effective electron temperature T_{eff}. The small ratio of V_{rf} to T_{eff}, ($V_{rf}/T_{eff} < 0.5$) suggests that the probe measurements are undistorted even without rf compensation of the probe (4). The plasma dc potential found from the probe measurements was 21 V at p = 1 mTorr and about half that at p = 1 Torr.

EEDF and Plasma Parameters

The following Langmuir probe measurements were obtained in the mid-plane of the discharge at a radial distance of 5 mm from inner glass tube in the region of strong rf field. The evolution of the electron energy spectrum with argon pressure is given in Fig. 3 where the EEDF's are presented in terms of the electron energy probability functions (EEPF's) which are proportional to the second derivative of the probe characteristic.

Over the measured energy range the slopes of the EEPF's vary and so they cannot be interpreted as a Maxwellian corresponding to a single temperature. For all pressures they are nearly Maxwellian distributions only in the elastic energy range. At higher electron energy and lower gas pressure, the EEPF's clearly are depleted due to inelastic collisions. Depending on gas pressure, the depletion threshold occurs somewhere between the excitation ($\varepsilon^*=11.55eV$) and ionization ($\varepsilon_i = 15.8$ eV) energy. At high electron energy the EEPF decreases more rapidly and can be characterized by a distribution temperature $T_e = [d(\ln f(\varepsilon))/d\varepsilon]^{-1}$ which is lower than that in the elastic energy range. The two slope EEPF's (found here at the lower gas pressures) can be interpreted as "two-temperature" distributions typical of dc discharge and they differ dramatically from those non-Maxwellian distributions found in capacitive rf discharges in the same chamber using the same probe measurement system (5). The two-temperature EEPF's found here cannot be presented as a sum of two Maxwellian distributions with different temperatures as in the case of low pressure (p < 0.1 Torr) capacitive rf discharges. The fundamental difference between the EEPF in inductive and capacitive discharges is that in the former case the distribution temperature of low energy electrons T_{e1} is higher than that of high energy electrons T_{e2}, $T_{e1} > T_{e2}$ while in the latter case $T_{e1} < T_{e2}$. In contrast with the Druyvesteyn distribution which is typical for dc and rf discharges in a Ramsauer gas, the Maxwellian-like distribution function found in inductive plasmas in the elastic energy range is due to the relatively high plasma density of the inductive rf plasma. This trend towards a Maxwellian distribution occurs because the frequency of electron-electron interaction v_{e-e} grows with increasing plasma density n and falling electron temperature T_e following the relation: $v_{e-e} \propto nT_e^{-3/2}$.

From the coil towards a region of weaker induced rf field, probe measurements along the discharge radius showed an insignificant change in the EEPF shape corresponding to less than a 10% drop in the mean electron energy, however the peak plasma density was 2 to 4 times larger than that measured at 5 mm from the inner glass wall. The rather homogeneous spatial distribution of the mean electron energy (or effective electron temperature, $T_{eff} = 2/3<\varepsilon>$) is due to the large electron thermal conductivity. Over this pressure range the electron energy relaxation length λ_ε is estimated to be larger than the plasma size. At such a condition the EEPF is a homogeneous function of the total energy: $w = \varepsilon + e\phi$, where ϕ is the plasma ambipolar potential. The small change observed in EEPF shape was just a shift equal to the local ambipolar potential and since the EEPF for an overwhelming majority of electrons was Maxwellian this shift affected the effective electron temperature very little.

The plasma density and effective electron temperature versus argon pressure at a distance of 5 mm from the inner wall are shown in Fig. 4. As the gas pressure increased from 1 mTorr to 1 Torr, the plasma density increased more than two decades while the effective electron temperature dropped from 7.9 eV to 1.6 eV. Thus, in inductively coupled plasma over a large range of argon pressure EEDF's in the elastic energy range were found to be very close to Maxwellian due, as mentioned earlier, to the high density plasma and a correspondingly large v_{e-e}.

In the mTorr pressure range where EEDF measurements were extended to energies far beyond the excitation and ionization energies, the EEDF's were found to drop with electron energy faster than in the elastic energy range ($T_{e1} > T_{e2}$). Such two temperature EEDF's are essentially different from EEDF's found in low pressure capacitive rf discharges and in inductive rf discharges with a capacitive coupling (5). The differences in the electron temperature of the bulk electrons and the high energy tail electrons make it clear that extrapolation of the EEDF from the elastic energy range into the inelastic energy range may bring about a dramatic error in calculation of the excitation and ionization processes in non-Maxwellian plasma of inductive rf discharge.

RF Magnetic and Electric Field Distributions

Figure 5 shows the axial component of rf magnetic induction B normalized to the primary coil current I measured as a function of radial position with no plasma and at three gas pressure with $P_d = 50$ W . Fig. 5 shows that the presence of a plasma increases the magnetic flux density in the plasma near the primary coil. This enhancement of the rf magnetic induction which appears as a "plasma ferromagnetism" is a general feature of an inductive discharge and is a consequence of the superposition of B-fields created by the coil and by the plasma.

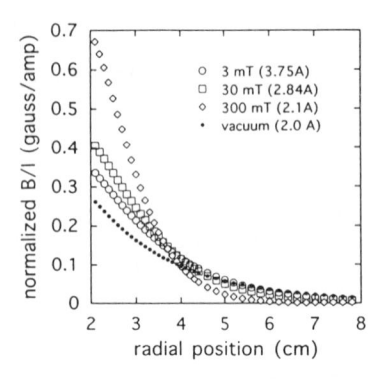

FIGURE 5. Radial distribution of the normalized induction field. The numbers in paranthesis indicate coil current.

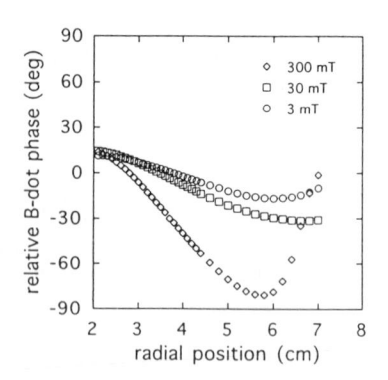

FIGURE 6. Relative phase of B-dot signal versus radial position.

To clarify this, notice that the magnetic flux outside the coil (in the plasma) has a phase that is opposite to that inside the coil. As predicted by Lenz's law, the induced discharge current is directed so as to create a magnetic flux that opposes the total magnetic flux it encloses (the majority of which is inside the coil). Thus, the discharge current creates a magnetic flux that reduces the flux within the coil and increases the flux in the region outside but near the coil.

As shown in Fig. 5, the presence of the plasma makes the B/I ratio decay faster than that in vacuum. Moving radially outward there always comes a point (crossing points in Fig. 5) where the B/I ratio in discharge becomes less than that in the vacuum case due to plasma shielding. Fig. 5 shows that B/I grows with gas pressure while the characteristic length of B/I decays ($\Lambda = [d\ln(B/I)/dr]^{-1}$ is falling). A similar trend is observed with growing the discharge power. In both cases a more pronounced shielding effect is associated with growth in plasma density and thus a reduction of the skin depth δ.

The relative phase of the dB/dt probe voltage with respect to the vacuum rf field created by the primary coil current is shown in Fig. 6. Note that the phase dependence of the B-dot signal must be incorporated in spatial integration or differentiation to infer rf field and current distributions.

The radial dependance of the plasma rf field E(r), found from Faraday's law as: $E(r) = -j\omega \int B(r)rdr$, is shown in Fig. 7. Since B(r) is complex, this equation has a real and an imaginary part and the magnitude of B(r) is simply the square root of the sum of the squares. The area of integration was between r and the chamber wall where the electric field was negligably small.

The decay rate of the rf field outwards from the induction coil depends on the plasma skin-depth δ but the characteristic length of rf field decay $\Lambda = [d\ln E/dr]^{-1}$ is always smaller than δ since for circular and/or multi-dimensional inductive plasmas (which is always the case) the space decay of electromagnetic field is governed not only by skin effect but also by the system geometry.

In textbooks skin depth is defined for a homogeneous half-space plasma as the characteristic length of rf field decay $\delta = [d\ln E/dx]^{-1}$. There is no reason to expect

FIGURE 7 Radial distribution of the rf electric field. Numbers in paranthesis indicates coil rf voltage.

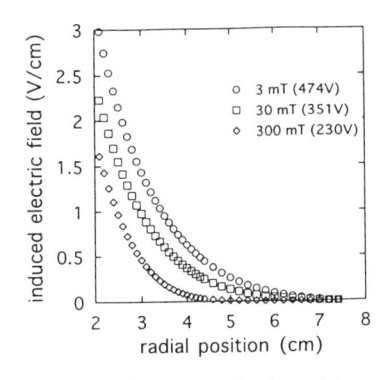

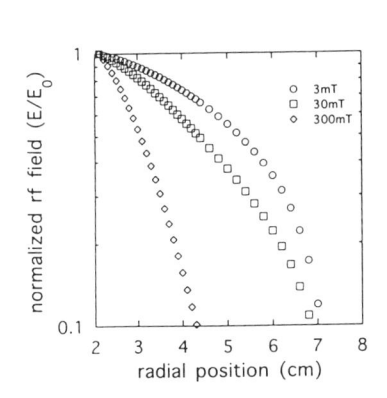

FIGURE 8. Normalised rf field in semi-logarithmic scale.

an exponential rf field decay corresponding to a one-dimensional flat model to be the same as for a two dimensional (at least) rf field pattern of an experimental device with an inhomogeneous plasma adjacent to the wall, and thus the field decay length found in this way is not equal to the skin depth.

To remove the "geometric aberration" and evaluate the skin effect due to plasma shielding we have plotted the rf electric field distribution twice normalized to that in vacuum and that at the plasma boundary. The results is shown in Fig. 8 where one can see that the normalized field decay is not an exponential function, i.e., the characteristic length Λ varies along rf field decay. The main reason for such behavior is the radial inhomogeneity in the plasma density.

The spatial dependence of the rf current density $J(r)$ was determined from Maxwell equation: $\text{rot}\mathbf{B} = \mu_0 \mathbf{J}$, neglecting displacement current since the driving frequency is much less than the plasma frequency. From the geometry of the experiment it is expected that the plasma current is axisymmetric and has only an azimuthal θ component, thus, the Maxwell equation yields: $\mu_0 J_\theta(r) = \partial B_r/\partial z|_d - \partial B_z/\partial r|_d$, where subscript d denotes the spatial derivative measured in the discharge (due to both primary coil current and plasma current). This equation may be written where each spatial derivative is given as a superposition of two terms as shown here: $\mu_0 J_\theta(r) = \partial B_r/\partial z|_v + \partial B_r/\partial z|_p - [\partial B_z/\partial r|_v + \partial B_z/\partial r|_p]$. Here the subscript v denotes the spatial derivatives due only to the primary coil current (in vacuum) and the subscript p denotes the spatial derivatives due only to the plasma current.

Unable to measure the B_r component along radial and axial directions (to find $\partial B_r/\partial z$ as a function of radial position) we calculated the $J_\theta(r)$ distribution in assumption that the radial component of the magnetic field due to the plasma current has no axial gradient in the discharge midplane ($\partial B_r/\partial z|_p \approx 0$). This assumption implies that $\partial B_r/\partial z|_d$ in the plasma is entirely due the coil current and is equal to $\partial B_r/\partial z|_v$ in vacuum. Since in vacuum $\partial B_r/\partial z|_v - \partial B_z/\partial r |_v = 0$, the discharge current density can be written: $J_\theta(r) = \mu_0^{-1}[\partial B_z/\partial r|_v - \partial B_z/\partial r|_d]$ which

FIGURE 9. Current density distribution for different argon pressure. $P_d = 50W$.

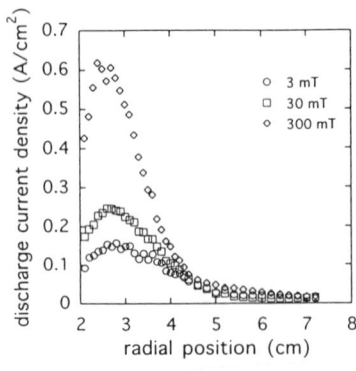

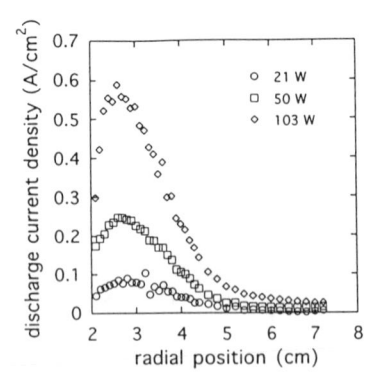

FIGURE 10. Current density distribution for different discharge power. $p = 30$ mT.

is composed of spatial derivatives that are easily measured with a simple one-dimensional B-dot probe. Since B_z is complex, J_θ also has real and imaginary parts, thus: $J_\theta = [\text{Re}^2(J_\theta) + \text{Im}^2(J_\theta)]^{1/2}$. Note, that assuming $\partial B_r/\partial z|_p \approx 0$ is equivalent to assuming a one-dimensional (1-D) plasma structure which is the more realistic the closer the plasma is to the coil, i.e., when the ratio of plasma length/diameter is large.

The azimuthal current density J_θ determined from the technique described above is shown in Fig. 9 for the three gas pressure at a fixed discharge power of 50W. As the gas pressure increases, the peak current density increases and appears to move slightly toward the axis. Fig. 10 shows the radial distribution of the discharge current density for three discharge powers at a constant gas pressure of 30 mTorr. The current density (and electron density) increases with gas pressure and discharge power and this results (due to skin-effect) in the peak of the current density moving closer to the coil.

It is interesting to compare the drift velocity derived from rf field measurements (which is independent of the 1-D assumption) and the drift velocity determined from coarsely measured discharge current density (which is based on the assumption of a 1-D topology). Having measured the plasma density n and the EEDF at r = 24 mm (Fig. 4) we can evaluate the rms value of the electron rf drift velocity v_{dr} for r = 24 mm using formulae: $v_{dr}(E) = eE/m(\omega^2+v_e^2)^{1/2}$ and $v_{dr}(J) = J/en$. In these formulae the rf field E, discharge current density J and plasma density n are measured, while the electron-atom collision frequency v_e is calculated using the collision cross section for argon and the EEDF's shown in Fig. 3. The values of $v_{dr}(E)$ and $v_{dr}(J)$ calculated for P_d = 50 W at different argon pressures are given in Table I together with other relevant measured and calculated parameters.

As one can see in Table I the drift velocities found in both ways are in reasonable agreement and that supports the 1-D assumption at discharge midplane near the coil. Observe that the absolute value of the drift velocity v_{dr} at 3 mTorr is small compared to the electron thermal velocity $v_{th} = (2T_e/m)^{1/2}$ also given in Table 1, $v_{dr}/v_{th} \leq 1/3$. The ratio v_{dr}/v_{th} defines the degree of anisotropy in the electron energy distribution regarding to Langmuir probe measurement.

TABLE 1. Plasma parameters measured at 5 mm from inner wall. Pd = 50 W

p (mT)	T_e (eV)	$n\,10^{-10}$ (cm^{-3})	J (A/cm^2)	E (V/cm)	$\nu_e\,10^{-8}$ (S^{-1})	$v_{th}\,10^{-7}$ (cm/S)	$v_{dr}(E)\,10^{-7}$ (cm/S)	$v_{dr}(J)\,10^{-7}$ (cm/S)
3	5.9	1.7	0.14	2.3	0.25	14.5	4.59	5.15
30	3.4	10	0.23	1.7	1.26	10.9	1.97	1.44
300	1.9	67	0.58	1.1	3.90	8.20	0.48	0.54

As shown recently by Sheridan and Goree (7) the probe characteristic of a cylindrical probe is insensitive to the electron drift at $v_{dr}/v_{th} \leq 1/3$. That justifies the application of the Langmuir probe at the condition of the present experiment.

Although the current density distributions shown in Fig. 9 and 10 appear to be reasonable, these results should be considered to be a qualitative approximation since the assumption that $\partial B_r/\partial z|_p$ due to the plasma current can be ignored remains unproven. A real two-dimensional B-dot measurement is needed to infer rf current density distribution in a two-dimensional inductive dischage.

REFERENCES

1. Godyak, V. A., Piejak, R. B., and Alexandrovich, B. .M., *Rev. Sci. Instrum.* **61**, 2401 (1990).
2. Godyak, V. A., Piejak, R. B., and Alexandrovich, B. M., IEEE Trans. Plasma Sci. **19**, 660 (1991).
3. Godyak, V. A., and Piejak R. B., J. Vac. Sci. Technol. **A8**, 3833 (1990).
4. Godyak, V. A., and Piejak, R. B., J. Appl. Phys. **68**, 3157 (1990).
5. Godyak, V. A., Piejak R. B., and Alexandrovich, B.M., Plasma Sources Sci. Technol. **1**, 36 (1992).
6. Piejak, R. B., Godyak, V. A., and Alexandrovich, B. M., Plasma Sources Sci. Technol. **1**, 179 (1992)
7. Sheridan T. E., and Goree J., Phys. Rev. E, **50**, 2991 (1994).

Numerical and Analytical Kinetic and Fluid Models for RF Discharges

Chwan-Hwa "John" Wu, F. Dai, C. Li[+], F.F. Young[*] and J. Tsai[&]

Department of Electrical Engineering
Auburn University, Alabama 36849-5201
[+]National I-Lan Institute of Agriculture and Technology, Taiwan, R. O. C.
[*]Feng Chia University, Taiwan, R. O. C.
[&]National Center for High-Performance Computing, Taiwan, R. O. C.

Abstract. A review of multidimensional radio-frequency (rf) discharge models for capacitively- and inductively-coupled plasma sources are presented in this paper. The models for capacitively-coupled discharges include single- and three-moment fluid models, direct solutions of the Boltzmann equation and particle simulations. The results from those capacitively-coupled models are compared for better understanding the physical insight. Analytical kinetic model is used to investigate the particle-wave interaction in a planar inductively-coupled plasma source. The comparison of electron energy distribution functions with experimental results is also presented.

INTRODUCTION

Low-pressure rf glow discharges are widely used for semiconductor and material processing in the past decade. It is important to quantitatively understand the complicated processes in rf discharges for better use of the plasma sources. Many researchers have devoted to this subject and a significant amount of work have been published. This paper cannot review such a large scope of work; hence, a limited amount of work is discussed here.

In this paper, we will discuss multidimensional capacitively- and inductively-coupled plasma sources. The fluid models for capacitively-coupled discharges, including single- and three-moment equations, will be discussed in the next section. Then, the direct solutions of the Boltzmann equation and the particle simulations for capacitively-coupled discharges are presented. Results from these models are compared to assess the validity of models. Furthermore, the inductively-coupled plasma sources are studied analytically to study the plasma-wave interaction.

MULTIDIMENSIONAL FLUID MODELS FOR CAPACITIVELY-COUPLED DISCHARGES

Multidimensional fluid models are obtained from taking the moments from the Boltzmann equation.[1] We have studied the single-, two-moment and three-moment fluid models.[2-8] A brief review of the single- and three-moment fluid models for a cylindrically symmetrical parallel-plate, two-dimensional (2D) geometry is given in the following.

Single-Moment Fluid Model

The single-moment fluid model contains the particle conservation equation (1) and the average velocity is a sum of the diffusion and drift velocities for both ions and electrons. Simulation results for the 2D model show a sheath close to the radial wall as well as a sheath close to both parallel plates.[8] Particles are accelerated by both the radial and axial electric fields. Detailed discussion is omitted here due to the limitation of the number of pages.

Three-Moment Fluid Model

The three-moment fluid model for electrons is as follows:

$$\frac{\partial n_e}{\partial t} = -\nabla \bullet (n_e u_e) + \nu n_e \tag{1}$$

$$\frac{\partial (m n_e u_e)}{\partial t} = -\nabla \bullet (m n_e u_e u_e) - n_e q E - \nabla P_e - \frac{m n_e u_e}{\tau_m} \tag{2}$$

$$\frac{\partial (n_e w)}{\partial t} = -\nabla \bullet (n_e w u_e) - n_e q u_e E - \nabla \bullet (u_e P_e) - \frac{n_e w}{\tau_w} \tag{3}$$

Eq. (1) is the particle conservation equation, where the electron density is n_e, t is the time, u_e is the electron average velocity, q is the magnitude of electronic charge and ν is the ionization frequency. Eq. (2) and (3) represent the conservation of momentum and energy, respectively. The electron mean energy is w,; E, the electric field; m, the electron mass; P_e, the electron pressure;

$$P_e = \frac{2}{3} n_e \left(w - \frac{1}{2} m u_e^2 \right) \tag{4}$$

since $P_e = n_e k_b T_e$ and $3/2\ k_b\ T_e = w - 1/2\ m\ u_e^2$, where k_b is the Boltzmann constant and T_e represents the electron temperature.

The last terms in the right-hand side of Eqs. (1)-(3) are the source terms due to the electron collisions. In Eq. (1), $n_e \nu$ is the approximate form of the zeroth moment of the collision term. In Eqs. (2) and (3), the collision terms are approximated by relaxation terms containing the effective momentum and energy relaxation times.[1] The ion fluid model is a single-moment model with inertia effect.

$$\frac{\partial n_i}{\partial t} = -\nabla \bullet (n_i u_i) + \nu n_e \qquad (5)$$

$$n_i u_i = -\mu_i n_i E^* - \nabla(n_i D_i) \qquad (6)$$

$$\frac{\partial E^*}{\partial t} = \frac{q}{\mu_i m_i}(E - E^*) \qquad (7)$$

In these equations, n_i, u_i, μ_i, and D_i are the ion density, average velocity, mobility coefficient, and diffusion coefficient, respectively. E^* is the effective ionic electric field, and the ion mass is m_i. In the right-hand side of Eq. (6), the first term represents the ion flux caused by the drift velocity and the second term represents the ion flux caused by the diffusion velocity. It is assumed that the ion temperature is constant, uniform, and equal to the neutral gas temperature.

The Poisson's equation shown below is coupled with the fluid model for a self-consistent solution.

$$\nabla \bullet E = \frac{q}{\varepsilon}(n_i - n_e) \qquad (8)$$

The zero potential reference point along the radial direction is at infinity. Because we do not know how to set up the potential on a floating dielectric surface, we integrate from the zero potential point to avoid this problem. The two-dimensional Poisson solver is based on a direct solver.[15] The Poisson's equation solver uses nonuniform cells with exponentially increasing size along the radial direction for 20 cells (outside the reactor) as the zero potential reference position which is close to infinity in comparison to the radius of the reactor. We use a 64 (z direction) by 60 cells (r direction, 40 uniformly distributed inside the reactor and 20 nonuniform grids outside the reactor).

The results will be compared with kinetic models and experiments later to assess the validity of the fluid model.

KINETIC MODELS FOR CAPACITIVELY-COUPLED DISCHARGES

Simulation Geometry

Simulations of parallel-plate rf discharges are performed in a two-dimensional cylindrically symmetrical parallel-plate geometry (in a configuration space) for a gap distance of 6.7 cm and a radius of 5.08 cm. The radial wall of the discharge chamber is assumed to be an insulator which is not grounded. The reference point of zero potential is at infinity. It is also assumed that the radius of the parallel plate is much bigger than that of the radial wall, such that a uniform applied field exists inside the discharge volume when there is no space charge. The discharge volume is divided by using 64 grids along the z direction, and 40 grids along the r direction. The velocity space for electrons in the Boltzmann equation is assumed to be independent of f in spherical coordinates (v, μ, φ) and is divided by using 16 grids in the v direction and 8 grids in the μ direction. The grid along the μ (or mu in the legend of figures) direction is numbered from 1 to 8, corresponding to angles from $0°$ to $315°$ with a grid size of $45°$. The applied rf source is connected to the left electrode (V=V_0 sin(2πft), V_0=150 V and f=13.56MHz) and the right electrode is grounded.

Initial and Boundary Conditions

The background pressure of He gas is 0.3 and 0.1 torr, respectively. The initial condition is obtained from converged fluid simulation results from a model based on a three-moment fluid model (1)-(8) to reduce the simulation turnaround time. The initial electron energy distribution function is assumed to be a Maxwell-Boltzmann distribution with local mean energy from the fluid model.

The boundary conditions are established by setting the second derivative of the distribution function and ion density function to zero, at the electrodes and radial wall surfaces. The flux is set to zero, if the direction of velocity is toward the discharge (from the electrodes or from the reactor wall). We assume that there are no secondary electrons emitted or charged particles scattered from electrodes and the radial wall.

Direct Solution of the Boltzmann Equation

The Boltzmann equation describes the transport of electrons in the phase space,

$$\frac{\partial f(r,v,t)}{\partial t} + v \bullet \nabla_r f(r,v,t) + \frac{qE}{m} \bullet \nabla_v f(r,v,t) = \left(\frac{\partial f(r,v,t)}{\partial t}\right)_{collosion} \quad (9)$$

The electron transport model is based on the Boltzmann equation which is solved by the Generalized Monte Carlo Flux (GMCF) method. The ion transport model is a nonequilibrium single-moment model with ion inertia effect represented by a phase delay between the ion motion and the local electric field, (4). The self-consistent electric field is solved from the Poisson's equation, (8).

A new kinetic scheme, the GMCF method, provides the electron particle distribution function in phase space, $f(v, \mu, r, z, t)$ (v: speed, μ: velocity angle, r: radial position, z: axial position, and t: time), for solving the Boltzmann equation in modeling capacitively coupled rf discharges. For a simulation with spatial- and temporal-varying fields in rf discharges, the GMCF method handles the collision terms of the Boltzmann equation by using *one* transition matrix to compute the collision transition between velocity space cells. An anti-diffusion flux transport scheme is developed to overcome the numerical diffusion in the velocity and configuration spaces. The major advantages of the GMCF method are the increase in resolution in the tail of distribution functions and the decrease of computation time.

Results

Overview

Simulations are carried out in a cylindrically symmetrical volume with specified boundary conditions. The amplitude of the applied sinusoidal voltage, V_0=150 V, and frequency, f, of the rf source is 13.56 MHz, and the background pressures, P, for simulations, are 0.3 and 0.1 Torr. Sampling time is defined as T_n = (n-1)(0.125T) where T is the duration of one rf cycle and n = 1, 2, ..., 8.

Comparison of electron angular velocity distribution functions in the axial sheath and the bulk

There are axial sheaths (where E_z is large and E_r is negligible) close to both electrodes and there is almost no field inside the bulk of the plasma; hence, electrons are driven significantly by the E_z close to the electrode. Electrons are driven towards the left electrode because of the positive potential applied to the left electrode at the first half cycle. It is clear that the dominant angular velocities of electrons are μ (mu) from 4 to 6, corresponding to angles from 135° to 225°. Certainly the most significant angular velocity is 180° due to a large E_z and negligible E_r inside the axial sheath and around the axial axis. The high-energy tails in the electron angular velocity distribution functions increase from t= 0 T to 0.375 T, as show in Fig. 1 (at t = 0.375 T only), due to the E_z, pushing electrons toward the left electrode and forming a sheath close to the right electrode in the first half cycle. Sheath field close to the right electrode does not decrease from t =

0 to t = 0.375 T (no collapsing of sheath yet due to a lag effect observed from our sampled E field at t = 0, 0.125 T and 0.375 T), thus, the high energy tail of EEDF keeps increasing in this period. Even no secondary electron emission (which causes low electron density in the sheath area), electron's energy can be increased by stochastic heating from the oscillating sheath. Since electrons are always heated from an initial energy (not a cold plasma) in a stochastic manner, electrons can have energy over 150V (the V_0) in a low probability (3 orders of magnitude less). Certainly, the Anti-diffusion Flux Transport (AFT) technique may not get rid of all the numeric diffusion causing high tail of EEDF. In contrast, the electron angular velocity distribution function at the center of the discharge (r=0 cm and z=3.35 cm) for P=0.1 torr and t=0 T is shown in Fig. 2 to illustrate that electrons are not as energetic as those inside the axial sheath, and there is no clear dominant angular velocity in the low-energy region. Since there is almost no electric field at the center of the discharge, the electron distribution function is close to an equilibrium distribution function. Based on the slopes of the distribution functions in Fig. 2, there are two slopes in the energy range 0-45 eV. These two slopes correspond to two thermal energies in terms of Maxwellian-Boltzmann distribution. The second slope in the range 8-45 eV, corresponding to a higher thermal energy of Maxwellian-Boltzmann distribution, is mainly due to high-energy electrons shooting from the sheath regions.

The advantage of high-resolution high-energy tails in the distribution function provides a high contrast of comparison of electron angular distribution functions inside the bulk and those inside the axial sheath. The axial sheath electrons are in a nonequilibrium state with high-energy tail in comparison to bulk electrons. Furthermore, the high-resolution angular distribution functions clearly show the electron motion which cannot be illustrated through electron energy distribution.

The effect of the radial field to electron angular velocity distribution functions inside the radial sheath

Obviously, comparing the magnitudes of E_r and E_z, the radial field cannot be neglected where r is close to the radial wall. The radial sheath is defined as the region where E_r is significant in comparison to E_z. It is important to quantitatively study the electron motion because of the presence of the radial field; hence, the electron angular velocity distribution function at r=4.33 cm, z=5.7 cm (inside the radial sheath) and t=0.375 T is shown for P=0.1 torr in Fig. 3. The distribution of electrons with angular velocity μ =5 (180°) is roughly the same as

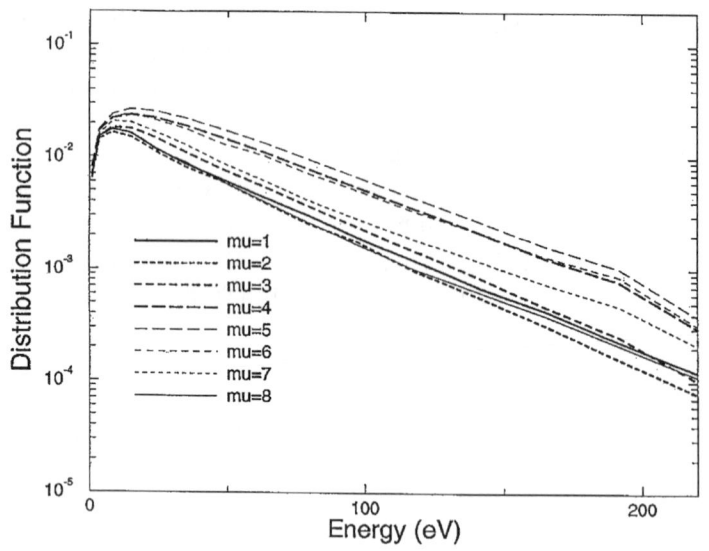

FIGURE 1. The electron velocity distribution functions in the axial sheath at t = 0.375 T, r = 0 cm, z = 6.3 cm, and P = 0.1 torr.

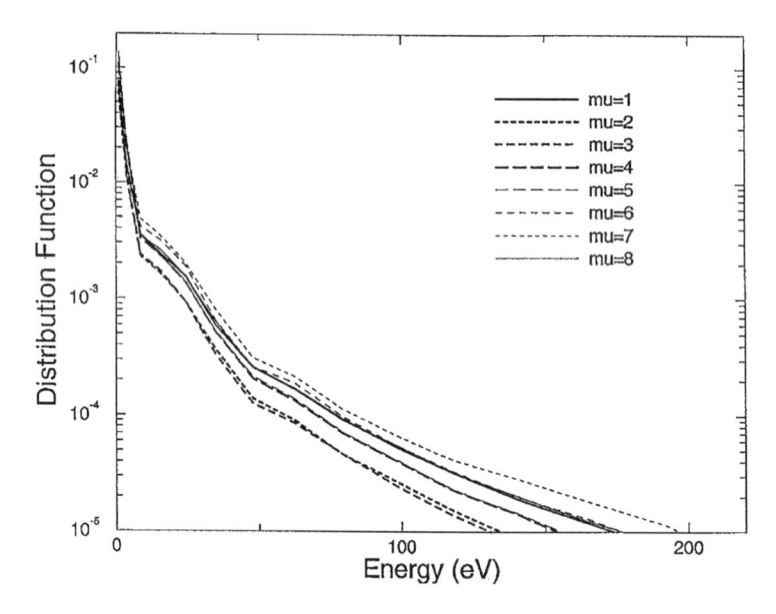

FIGURE 2. The electron velocity distribution functions in the center of the bulk at t = 0.1 torr, r = 0 cm, and z = 3.35 cm.

those with $\mu = 6$ (225°) (which is roughly parallel to the electric field direction) in the energy range from 0 to 45 eV. The radial electric field drives electrons to have such an angular velocity distribution since the radial field contributes approximately equivalent to the axial field during that period of time in an rf cycle. The axial field presents longer in an rf cycle than the radial field, which is oscillating in an rf cycle, and thus, causes the high-energy tail electrons (>50 eV) to have a higher probability along velocity direction $\mu = 5$ (corresponding to 180°).

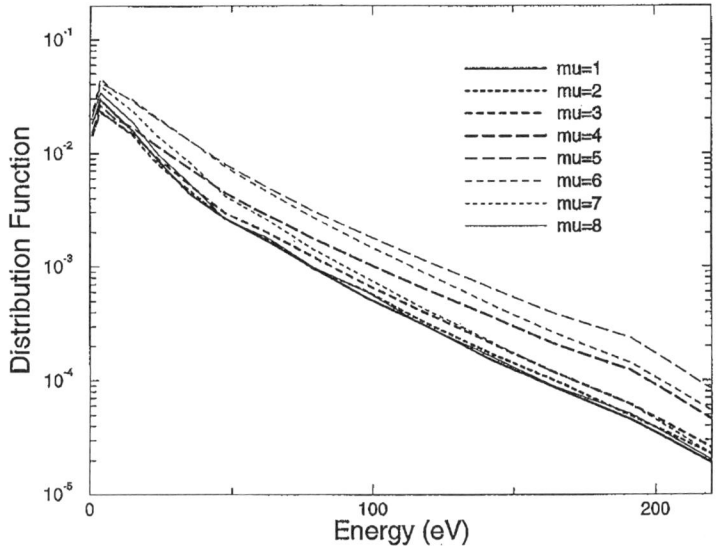

FIGURE 3.. The electron velocity distribution functions in the radial sheath at t = 0.375 T, r = 4.33 cm, z = 5.7 cm and P = 0.1 torr.

Particle Simulations

Simulations using a Monte Carlo technique (the Self-consistent Electron-Monte-Carlo and Ion-Fluid model, SEMCIF)[10] to replace the GMCF method in the 2D kinetic model is carried out. The results are compared with the other models and experiments below.

COMPARISON BETWEEN CAPACITIVELY-COUPLED DISCHARGES MODELS AND EXPERIMENTS

The macroscopic results, such as bulk ion density, peak ionization rate, peak E_z and peak E_r, of the simulation were verified with simulations using Monte Carlo technique (the SEMCIF) to replace the GMCF method as well as results from 2D fluid models, 1D PIC method and 1D convective scheme and experimental results. The 1D PIC and convective scheme results and experimental results were collected by Surendra for the Gaseous Electronics Conference (GEC) reference cell benchmark study. The difference in bulk ion density is negligible between the GMCF and SEMCIF, 7.7 % between GMCF and 2D fluid, 20 % between GMCF and 1D PIC from IBM, 57% between the GMCF and 1D convective scheme, and 48 % between the GMCF and the experiment. For the peak ionization rate, the difference is 21 % between the GMCF and SEMCIF, 10 % between GMCF and 1D convective scheme, and 22 % between GMCF and 2D fluid. The comparison of ionization rate between GMCF and convection scheme shows that the only minor difference is the rate in the bulk area and the GMCF is smaller than that of the convection scheme; this difference is due to the complicated 2D motion and induced radial field. The differences of E_z and E_r between the GMCF, SEMCIF and 2D fluid model are also in acceptable range. The comparison of the electron energy distribution function between the GMCF and experimental results shows a similar two-slope distribution function in the range 0-25 eV. The result from the convective scheme also clearly illustrates the similar two-energy distribution function as the GMCF results, whereas the PIC simulation does not show.

The computation time for the GMCF is roughly 40% less than that of the SEMCIF for one rf cycle and the convergence rate of the GMCF, 10^{-2} %, is better than that of the SEMCIF, 10^{-1} % (primarily because of statistical fluctuation in the Monte Carlo technique).

ANALYTIC MODELS FOR INDUCTIVELY-COUPLED DISCHARGES

Electromagnetic Model

We developed a 2-D *(r, z)* model to investigate the EM fields in the discharge chamber. The main assumptions used in the EM model include: (a) we approximate the spiral-like coil as separate current loops; this is reasonable when the coil contains many turns; (b) since the current loops have azimuthally symmetric configuration, the EM fields excited by the coil possess the same symmetry, namely, all the EM fields, space charges and space currents have no azimuthal dependence, so that the problem degenerates into a 2-D *(r, z)* model; (c) since the *RF* source is a purely sinusoidal source with frequency ω, all the induced terms possess the same time-dependence of $e^{-i\omega t}$; and (d) the *RF* wave traveling along chamber axis *z* impresses all the induced terms with a *z*-dependence of $e^{i\beta_n z}$. From Maxwell equations, the wave equations of vector (*A*) and scalar potentials (*Φ*) can be deduced

$$\begin{cases} \nabla^2 \mathbf{A} + \varepsilon\mu\omega^2 \mathbf{A} = -\mu \mathbf{J} \\ \nabla^2 \phi + \varepsilon\mu\omega^2 \phi = -\rho/\varepsilon \end{cases} \quad (10)$$

where e and m are dielectric constant and permeability of the studied region, respectively. **J** contains all the current sources and ρ involves all the charge sources in the considered region.

Since the azimuthally symmetric configuration of the *RF* antenna and the reactor, the dominant part of the E-field in an ICPS is its azimuthal component, which is excited by azimuthal currents flowing along the coil antenna, in the plasma region and on the metal surface of the vessel. In terms of the Lorentz Gauge, the azimuthal E-field can be expressed by its vector potential, which is the superposition of the coil field A_θ^c, plasma field A_θ^p and induced field A_θ^{in}, namely,

$$E_\theta(r,z) = i\omega \sum_n A_{\theta n}(r,z), \quad (11)$$

with

$$\begin{cases} A_{\theta,n} = A_\theta^c(r,z) + A_\theta^p(r,z), \ n = 0 \\ A_{\theta,n} = A_\theta^{in}(r,z) = \sum_{n=-\infty}^{\infty} a_n I_0(k_n r) e^{i\beta_n z}, \ n \neq 0 \end{cases} \quad (12)$$

where $I_0(k_n r)$ is the modified Bessel functions of zero-order and the mode expansion coefficients a_n can be deduced by matching boundary conditions on the vessel surface, i.e., $A_\theta(a,z) = 0$. The axial transmission factor β_n of the *n*th eigenmode is determined by the axial boundary conditions, which states a standing-wave form for each mode between the coil surface (*z*=0) and the substrate surface (*z*=*h*), because the dense coil geometry behaves roughly like a conducting surface. Thus, we have

$$\beta_n = \pm\frac{n\pi}{h}, \quad (n = 1, 2, 3, \cdots), \quad (13)$$

where the positive sign represents the forward wave transmitted by the coil antenna and the negative sign represents the backward wave reflected by the substrate. β_n determines axial phase velocity of the *n*th *RF* induced mode by ω/β_n. The radial wave-number k_n of the *n*th induced mode reads

$$k_n = \sqrt{\beta_n^2 - \varepsilon\mu\omega^2}, \quad (14)$$

where ω is the frequency of the *RF* source; ε and μ are dielectric constant and permeability of the studied region.

Kinetic Model

The collisionless electron heating is a warm plasma effect. It means that the electrons are assumed to be in thermal motion initially. In this connection, it is supposed that a uniform plasma with a Maxwellian velocity distribution $f_0(v)$ exists in the initial state. The 3-D Maxwellian distribution function of electrons is:

$$f_0(v_r, v_\theta, v_z) = \frac{n_0}{\left(\sqrt{\pi} v_{th}\right)^3} e^{-(v_r^2 + v_\theta^2 + v_z^2)/v_{th}^2}. \tag{15}$$

where n_0 is electron density and v_{th} is the thermovelocity of electrons, which are the input parameters of the kinetic model and can be obtained from experimental data. In the first order, the external EM field has been added and can be treated as the perturbation of the initial plasma motion. The distribution function now becomes

$$f(\mathbf{r}, \mathbf{v}, t) = f_0(\mathbf{v}) + f_1(\mathbf{r}, \mathbf{v}, t). \tag{16}$$

Taking the differential of (16) over time and considering the particle-preserving Krook collision term, one obtains the Boltzmann-Vlasov equation as follows:

$$\frac{\partial f_1}{\partial t} + \mathbf{v} \cdot \nabla f_1 - \frac{e}{m} \mathbf{E} \cdot \frac{\partial f_0}{\partial \mathbf{v}} = \left(\frac{n_1}{n_0} f_0 - f_1\right) \nu, \tag{17}$$

where ν is the collision frequency, and n_1 ($\ll n_0$) is the change of plasma density in the perturbation of external field, which can be found out by

$$n_1 = \int\!\!\int\!\!\int_{-\infty}^{+\infty} f_1(v_r, v_\theta, v_z) dv_r dv_\theta dv_z \tag{18}$$

where the first order perturbation $f_1(\mathbf{r}, \mathbf{v}, t) \propto E_\theta(\mathbf{r}, t)$. In the ICPS, the dominate E-field is the azimuthal component excited by coil current. Neglecting the radial component E_r, one may rewrite (17) as

$$f_1(v_\theta, v_z) = i \frac{\dfrac{e}{m} E_\theta \dfrac{\partial f_0}{\partial v_\theta} + \dfrac{e}{m} E_z \dfrac{\partial f_0}{\partial v_z} + \nu \dfrac{n_1}{n_0} f_0(v_\theta, v_z)}{\omega + i\nu + i v_z \dfrac{\partial E_\theta}{E_\theta \partial z}} \tag{19}$$

The last term of the numerator on the right side of (19) is contributed by collisions, and the first two terms result from the plasma-wave interactions by collisionless mechanisms, in which the second term expresses the normal 1-D Landau Damping, yet the first term demonstrates a 2-D coupled damping effect. Since the plasma moves with v_z along z-direction and the dominant exciting E-field is along azimuthal direction, the wave damping in ICPS may be expected to possess a 2-D coupled damping effect, in which electrons travel with v_z along the axis and meanwhile gyrate azimuthally due to the E_θ. Substituting (19) into (18) and neglecting collision terms gives

$$n_1 = \frac{-2in_0 eE_z}{m\sqrt{\pi}v_{th}^3} \int_{-\infty}^{\infty} v_z \left(\omega + iv_z \frac{\partial E_\theta}{E_\theta \partial z}\right)^{-1} e^{-v_z^2/v_{th}^2} dv_z \qquad (20)$$

where the integration concerning the first term in (19) vanishes due to the antisymmetric property of $\partial f_0/\partial v_\theta$. Because of the azimuthally symmetric structure of ICPS, there is no gradient of the E-field along θ-direction, which results in the plasma density in (20) to be determined only by the axial E-field. Since $E_z \ll E_\theta$ in ICPS, n_1 can be neglected in calculating f_1. The distribution function in (19) may thus be rewritten as

$$f_1(v_\theta, v_z) = i\frac{e}{m} E_\theta \frac{\partial f_0}{\partial v_\theta} \left[\omega + iv + iv_z \frac{\partial E_\theta}{E_\theta \partial z}\right]^{-1}. \qquad (21)$$

Considering the relation between energy and velocity coordinates, equation (21) can thus be rewritten as

$$f_1(E_e) = -i\frac{2}{v_{th}} \frac{e}{m_e} f_0(E_e) \sum_{V_z} \left[\sqrt{\frac{E_e}{E_{th}} - V_z^2} \cdot \sum_n^{c,p,in} \frac{E_{\theta,n}}{\omega + iv + iv_{th}V_z \frac{\partial E_{\theta,n}}{E_{\theta,n} \partial z}}\right] \Delta V_z \qquad (22)$$

Having obtained the perturbation term in (22), we integrate the total distribution function in the velocity domain to obtain the azimuthal current. Since the Maxwellian distribution function is isotropic, the azimuthal current j_θ can be deduced by summing the effects of the coil field, plasma field, and the induced eigenmodes as

$$j_\theta = \frac{-e^2 n_0 \omega}{m\sqrt{\pi}v_{th}} \sum_n^{c,p,in} \frac{A_{\theta,n}}{\beta_n} \int_{-\infty}^{\infty} \frac{e^{-V_z^2}}{\left(V_{\varphi,n} + i[\alpha_n + \eta_n(r,z)V_z]\right)} dV_z, \qquad (23)$$

with the definitions of

$$\alpha_n \equiv \frac{v}{\beta_n v_{th}}, \quad V_{\varphi,n} \equiv \frac{\omega}{\beta_n v_{th}}, \quad \eta_n(r,z) \equiv \frac{\partial A_{\theta,n}(r,z)}{\beta_n A_{\theta,n} \partial z}, \qquad (24)$$

where $\alpha_n = \frac{v}{\beta_n v_{th}}$ means the normalized collision velocity corresponding to the nth eigenmodes ($n=0$ corresponds to the coil and space-current fields), β_n is the transmission factor of the nth RF eigenmode given in (2) (for $n=0$, $\beta_0 = \frac{\omega}{c}$) and $V_{\varphi,n} = \frac{\omega}{\beta_n v_{th}}$ presents the normalized phase velocity of the RF field ($n=0$ corresponds to the coil and space-current fields; $n \neq 0$ correspond to the induced eigenmodes); the η_n is the normalized gradient of the nth azimuthal E-field, which is the function of coordinates. Separating the complex integral in (23) into the two real integrals, we can present the azimuthal current as

$$j_\theta = \frac{ie^2 n_0 \omega}{m_e \sqrt{\pi} v_{th}} \sum_n^{c,p,in} \left[\eta_n F_n + (iV_{\varphi,n} + \alpha_n)G_n\right]\frac{A_{\theta,n}(r,z)}{\beta_n}, \qquad (25)$$

and two real integrals are given by

$$\begin{cases} F_n(r,z) = \int_{-\infty}^{\infty} \frac{V_z e^{-V_z^2} dV_z}{(\alpha_n + \eta_n(r,z)V_z)^2 + V_{\varphi,n}^2} \\ G_n(r,z) = \int_{-\infty}^{\infty} \frac{e^{-V_z^2} dV_z}{(\alpha_n + \eta_n(r,z)V_z)^2 + V_{\varphi,n}^2} \end{cases} \qquad (26)$$

Equation (26) gives a unified form for three different parts of the azimuthal E-field considered, i.e., the induced higher-order field, the fields excited by coil antenna and the plasma current, which are considered as the 0-order of the mode expansion. The space current excited by the existing E-field describes the influence of the *RF* field on the plasma motion in the plasma-wave interaction. In order to analyze the damping mechanism that governs the interaction procedure, (25) needs to be rewritten further as

$$j_\theta = \frac{-e^2 n_0 \omega}{m_e \sqrt{\pi} v_{th}} \sum_n^{c,p,in} (V_{\varphi,n} G_n - i\eta_n F_n)\left[1 - i\frac{v}{\omega}(1+D_n)\right]\frac{A_{\theta,n}(r,z)}{\beta_n}, \qquad (27)$$

with the damping coefficient defined as

$$D_n = \frac{1}{\frac{G_n V_{\varphi,n}}{iF_n \eta_n} - 1}. \qquad (28)$$

It is obvious that the first term of the imaginary part in (27), $i\frac{v}{\omega}$, describes the wave damping by collisions, while the second term, $i\frac{v}{\omega}D_n$, demonstrates a collisionless dissipation mechanism, which is concerned with a 2-D coupled interaction, where it has demonstrated that the damping coefficients are nonvanishing only when V_φ is close to, or smaller than, 1. It means that the *RF* wave cannot be dampened by collisionless interaction when its phase velocity, v_φ, is much larger than plasma thermovelocity, v_{th}. The coupled interaction takes place only as the *RF* phase velocity is slowed down to the dimension of the plasma thermovelocity. The damping coefficient D_n of the nth eigenmode gives the criterion to describe the magnitude of the coupled damping of the nth mode with plasmas relative to the collisional damping. As D_n is much larger than 1, the coupled damping demonstrates the dominant effect in the plasma-wave interactions in comparison to the collisions among electrons and background gases.

Distribution Functions

According to (23), the poles of the EVDF (as shown in Fig. 4) are determined by the axial velocity V_z of electrons. The peak of the instability appears, as $V_z = V_{\varphi,n}$, namely, $v_z = \dfrac{\omega}{\beta_n} = \dfrac{\omega h}{n\pi}$, $(n = 1,2,3,\cdots)$. It means the RF wave in the reactor is mainly dampened by a collisionless mechanism as the RF phase velocity is slowed down to the dimension of the plasma thermovelocity. Since the RF wave is represented in its eigenmodes, which have different phase velocities, the instability is expected to appear at different V_z. For the first mode (n=1), the disturbed peak is observed at about $V_z=2$, while the peak for the second mode appears at smaller value of $V_z=1$. The instabilities for even higher eigenmodes are not as obvious as that of the first two eigenmodes. Therefore, the EVDF at the low axial velocity region is dominated by the Maxwellian distribution, namely, the collisional damping mechanism.

Fig. 5 shows the comparison of the EEDFs for four different heights. It is clear that changing the height can influence the position of the perturbed peak of EEDFs. Since the accurate relationship between the reactor height and the EEDF local peak is solvable based upon the self-consistent kinetic analysis presented in this paper, adjusting the reactor height provides an easy way to control the collisional processes by exciting the useful electrons with the required energy. With the larger height, the peak shifts towards the high energy region, but the magnitude of the peak also decreases due to the exponential dependence of the Maxwelian distribution.

The analytical results obtained in this paper agree well in many aspects with experimental results[16] except for the location of perturbed peaks due to the simplified collision mechanism.

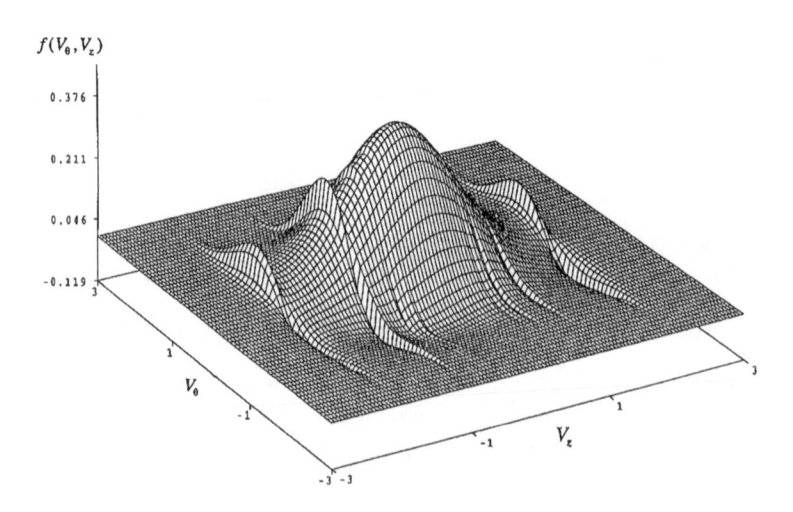

FIGURE 4. The electron velocity distribution function at r = 4 cm and z = 3 cm for a reactor with radius = 10 cm and height = 7.5 cm and 10 current loops carrying 6 A. The E_{th} = 3 eV, $\nu = 10^6$ /s and $n_0 = 10^{12}$ cm^{-3}.

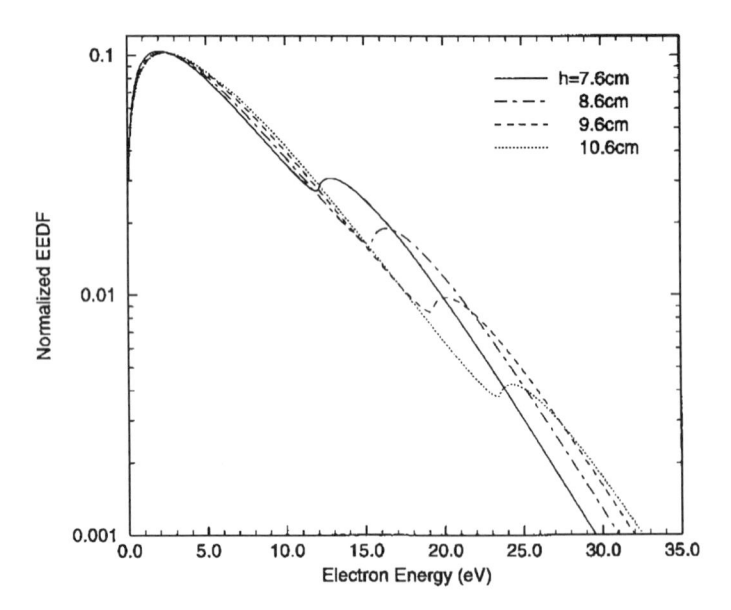

FIGURE 5. The electron energy distribution functions for four different heights of reactors at r = 0.1 cm and z = 3 cm for a reactor with radius = 10.15 cm and height = 7.5 cm and 10 current loops carrying 6 A. The E_{th} = 3.4 eV, $\nu = 10^5$ /s and n_0 = 2x10^{12} cm^{-3}.

CONCLUDING REMARKS

A review of numerical and analytical models for capacitively- and inductively-coupled rf discharges is described in this paper. This work provides a better understanding of the charged particle transport phenomenon in 2D configuration space as well as in velocity space. It enables ways to achieve a more uniform plasma for material processing and to selectively excite the desired species for chemical reactions.

ACKNOWLEDGMENTS

The authors want to thank Prof. Kunhardt from Stevens Institute of Technology for his long-term encouragement and helpful guidance. Useful communications from many research groups in University of Illinois, U.C. Berkeley, University of Houston, Scientific Research Associates, Osram Sylvania, U.C. L.A. and many friends are greatly appreciated.

REFERENCES

1. Kunhardt, E. E., Wu, C., and Penetrante, B., *Physical Review A*, 37, 1654-1662 (1988).
2. Young, F. F., and Wu, C., *Applied Physics Letters*, Vol. 62, No. 5, 473-475 (1993).
3. Young, F. F., and Wu, C., *The International Journal of Supercomputer Applications*, Vol. 7, No. 1, 50-63 (1993).
4. Young, F. F., and Wu, C., *IEEE Transactions on Plasma Science*, Vol. 21, No. 3, 312-321 (1993).
5. Li, C., and Wu, C., *Computers in Physics*, Vol. 7, No. 3, 363-375 (1993).
6. Young, F. F., and Wu, C., *Journal of Applied Physics*, Vol. 74 No. 2, 839-847 (1993).
7. Young, F. F., and Wu, C., *Journal of Physics D: Applied Physics*, Vol. 26, 782-792 (1993).
8. Tsai, J.-H. and Wu, C., *Physical Review A*, 41, 5626-5644 (1990).
9. Li, C., and Wu, C., *IEEE Transactions on Plasma Science*, Vol. 20, No. 6, 1000-1014 (1992).
10. Tsai, J.-H. and Wu, C., *Journal of Physics D: Applied Physics*, Vol. 26, 496-499 (1993).
11. Wu, C., Li, C., Tsai, J. H. and Young, F.F., *IEEE Trans. on Plasma Science*, to be published in Vol. 23, No. 4 (1995).
12. Dai, F. and Wu, C., *IEEE Trans. on Plasma Science*, Vol. 23, No. 1, 65-73 (1995).
13. Wu, C. and Dai, F., *IEEE Trans. on Plasma Science*, Vol. 23, No. 1, 74-82 (1995).
14. Dai, F. and Wu, C., *IEEE Trans. on Plasma Science*, to be published in Vol. 23, No. 4 (1995).
15. Wu, C., and Kunhardt, E. E., *Journal of Computational Physics*, 84, 247-254 (1989).
16. Barnes, M.S., Forster, J. C., and Keller, J. H., *Appl. Phys. Lett.*, Vol. 62, No. 21, 2622-2624 (1993).

Physics of high-power electron beam-plasma interaction

V.S.Koidan

SSC RF Budker Institute of Nuclear Physics, 630090 Novosibirsk, Russia

Abstract. Results related to physics of collective interaction of a high-power relativistic electron beam with a magnetized plasma are presented. The main regularities of this process are given. The observed peculiarities of hot plasma producing by E-beam heating on the GOL-3 device are discussed. Some results on study of strong Langmuir turbulence by laser scattering method on GOL-M device are presented.

I. INTRODUCTION

The invention of the first generators of high-power relativistic electron beams (REB) in late 60s has given a new pulse to the development of the theoretical and experimental studies on collective relaxation of electron beams in a plasma. In addition to the fundamental scientific interest, these studies are also important for various fields of physics and applications, namely: a possibility to use REB for the fast plasma heating in thermonuclear devices (especially in open systems); simulation of astrophysical phenomena; a study of various phenomena in physics of gas discharge; generation of powerful microwaves; interaction of a electron beam with gaseous medium of powerful lasers, etc. As the Coulomb mean free path of relativistic electrons is too large even in a very dense plasma, a noticeable relaxation of REB in a plasma is possible on a reasonable length only as a result of the collective interaction. One of the main mechanisms of such an interaction is a two-stream instability leading to excitation of a small-scale Langmuir turbulence in a plasma. As a result of energy dissipation of this turbulence an efficient plasma heating is possible.

The first theoretical and experimental studies of REB collisionless relaxation in a plasma, which in particular are related to the plasma heating in mirrors, have been started in late 60s - early 70s in many groups. These studies have been especially extensive during the 70s-80s and they are still carried out up to now. The results of these studies are presented in numerous original papers as well as in a number of reviews and summary papers (see [1-18] and references therein).

Experimental studies of the REB-plasma interactions at the Novosibirsk Institute of Nuclear Physics have been carried out from the early 70s. The major part of physical data has been obtained on the installations INAR, INAR-2, GOL-1, GOL-M (in operation), where the electron beams of a nanosecond duration have been used, as well as recently, on a new operational large facility GOL-3 with the use of a powerful REB of a microsecond duration [16]. At present, the beam-plasma interaction studies are carried out too on the Chekhian REBEX facility [19].

This paper is devoted to a short presentation of the most important (to our point of view) results obtained on this problem in recent years.

II. MAIN REGULARITIES OF REB-PLASMA INTERACTION

Experimental studies carried out on the devices with nanosecond beams to the middle of 80s enabled to reveal the main regularities of collective interaction of the high-power electron beam with plasma. Remind, that the principal scheme of such experiments is quite simple (see, e.g., [10]). In a long (up to 7 m) either glass or metal vacuum chamber the initial plasma is produced in some or other way. Plasma column is placed in a quite strong longitudinal magnetic field B_z which is substantially stronger than that of a beam B_φ. Quite often, the field has the end mirrors. As a rule, the plasma column is limited by thin metal foils at the chamber ends. An electron beam from the pulse accelerator is injected into a plasma through the input foil

Variation range of experimental parameters was quite broad: (see, for example, [9,18]):

Electron beam:
Energy   0.3-3 MeV
Current   2-100 kA
Duration   30-200 ns
Particle density 10^{10}-10^{14} cm^{-3}

Plasma:
Density 10^{11}-10^{17} cm^{-3}
Length   0.2-7.5 m
Magnetic field 0.2-9 T

In some experiments, quite a broad set of diagnostics including new one (see [12,16]) was developed and used.

Let us give a brief summary of the experimental results obtained and understanding achieved in the REB-plasma collective interaction process.

If an angular spread of injected beam is quite large $\theta_0 > 20°$ and a plasma has high density n_p, so that parameter $n_b/n_p < 10^{-3}$, where n_b - density of the beam, collective interaction is very weak even on the length of a few meters. According to theory (see, e.g., [2]), the kinetic growth rate of the two-stream instability has the form:

$$\Gamma_b \cong \omega_{pe} \cdot \frac{n_b}{n_p} \cdot \frac{1}{\gamma \cdot \theta_0} \cdot \frac{\omega_{pe}^2}{\omega_{pe}^2 + k_\perp^2 c^2}$$

i.e., it increases with a decrease in the beam angular spread and with the growth of parameter n_b/n_p. Therefore, at large value θ_0 and low value of parameter n_b/n_p the beam instability is suppressed by binary collisions in a plasma. A substantial growth of efficiency of the beam-plasma interaction occurs at considerable increase in the beam initial angular spread and with an increase in external magnetic field. This fact was especially distinct demonstrated in [20]. It was shown that a three times decrease in an angular spread leds to an increase in an energy release in plasma nearly by an order of magnitude. In this case, a strong inhomogeneity of an energy release in a plasma was observed along the device length.

An efficiency of REB- dense plasma interaction grows also with an increase in parameter n_b/n_p. As a result, for example, at beam density $n_b = 1.5 \cdot 10^{12}$ cm^{-3} and an angular spread $\theta_0 < 10°$ the beam deceleration up to 40% was achieved at the INAR-2 device at $n_p = 10^{15}$ cm^{-3} on the device length of 0.75 m [16].

The energy released during the collective beam deceleration transforms into an energy of Langmuir turbulence that transfers mainly to plasma electrons. In direct experimental measurements [10,12] it was shown that the hot plasma electrons have the distribution function being strongly nonequilibrium in a wide energy range, as it was predicted in theory [1]. It is established that the major part of energy lost by a beam is transformed into the high energy non-Maxwellian fraction of distribution function. The "hot" electron concentration amounts a few percent of initial plasma density.

As it shown in experiments [12], at intense beam-plasma interaction, both the energy (ΔE) and angular ($\Delta \theta$) spreads of a beam are increased simultaneously and practically in the same value, so that $\Delta E/E \sim \Delta \theta$. This fact enabled the evaluation of energy density for resonance oscillations excited by the beam. The level of oscillations evaluated in this way is $W_r/n_p T_e \cong 0.3$-0.5, and this value $W_r/n_p T_e \geq (\omega_{He}/\omega_{pe})^2$, i.e., it achieves and can be higher than the level at which (according to theory given, e.g., [21]) the modulational instability should develop.

Note, that during injection of a high current beam into rare enough plasma with parameter $n_b/n_p \sim 10^{-1}$-10^{-2}, it is quite probable to get a substantial influence of the effect of return current generated by beam in plasma. In this case, an incomplete compensation for a beam current by return current is observed and some noticeable shifts of a beam are possible (see [22]). As a result of return current microinstability an anomalous resistance occurs. In this case, an energy deposition into a plasma because of the return current energy dissipation turns to be essential.

Under conditions of the beam total compensation over the current density, when the dissipation of return current energy is inessential, a number of experimental factors was obtained [5,12] which indicate the fact that the dominant mechanism of collective interaction of REB with plasma is a two-stream instability with excitation of Langmuir turbulence. These factors are the following: 1) strong influence of a beam initial angular spread; 2) considerable inhomogeneity of energy release along the plasma column length; 3) dependence of interaction effect on parameter n_b/n_p; 4) presence of microwave radiation from plasma in the frequency range $2\omega_{pe}$; 5) considerable energy and angular spreads acquired by a beam because of interaction with a plasma; 6) appearance of an anomalous large number of "non-Maxwellian" particles in the spectrum of plasma hot electrons.

All these factors are at least in a qualitative agreement with the available theoretical understanding of a nonlinear mechanism of collective interaction between the relativistic electron beam and magnetized plasma (see, e.g.,[15]).

The regularities found and understanding achieved of interaction of the high-current electron beam with plasma enabled us to begin the new experiments on this problem, namely, the study of interaction between plasma and electron beams of microsecond duration with an energy content of ~100 kJ in a pulse as well as the experiments on direct observation and study of the beam excited Langmuir turbulence in plasma by laser scattering method.

III. EXPERIMENTS WITH A MICROSECOND ELECTRON BEAM ON THE GOL-3 DEVICE

By the middle of the 80s, the physics and technology for generation of powerful REB of a microsecond duration with an energy content of ~100 kJ per pulse and higher were developed in a number of laboratories including the Novosibirsk INP [23]. This made the basis for performance of new experiments on a study of interaction of such beams with a plasma which was especially applicable to the problem of plasma heating up to subthermonuclear temperatures. To this aim, the GOL-3 device has been constructed and put into operation in 1988 [16,24] which is being operational by the present time.

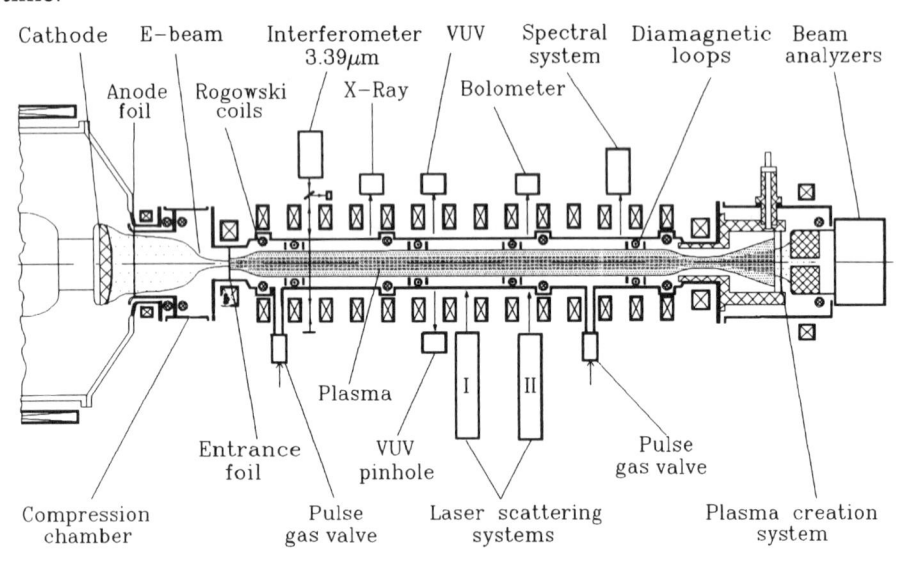

FIG. 1. Layout of the GOL-3 experiment.

Fig. 1 shows the schematic of the experiments on this device. It consists of an electron beam generator U-3, a plasma chamber inside a solenoid with 6 T field in the 7-m-long homogeneous section and 12 T in the single mirrors. It also comprises a 10 MJ capacity storage for energy supply of the solenoid, and systems of control, monitoring and diagnostics. In the experiments the electron beam (energy 0.8-0.9 MeV, maximum current density ~1 kA/cm^2, diameter 6 cm, duration 3-5 μs, total beam energy content of 20-90 kJ) was injected into a column of hydrogen plasma of 8 cm diameter and 10^{14}-10^{17}cm^{-3} density. Plasma density can be either uniform over the device length or strongly nonuniform in the two-stage heating experiments [24,25].

Diagnostics covered a wide set of techniques for both beam and plasma measurements and studies on collective plasma-beam interactions and plasma heating (see, [16,24,25]).

In experiments performed at the GOL-3 device it has been shown that the beam can release up to 25% of its energy in plasma under optimal conditions. The diamagnetic measurements have indicated, that the energy lost by the beam is transferred in plasma heating, the energy in the plasma being confined for rather long time even after the beam injection.

The special attention in the experiments was basically paid to the study of the plasma electron bulk [25,26]. The energy distribution function of the E-beam-heated plasma electrons which was experimentally studied at the previous generation facilities (see, e.g., [12]), is complicated. Together with the bulk Maxwellian electrons there exist suprathermal electrons, which may contain a major part of the energy lost by the beam in the plasma. At large beam durations (up to 5 μs in GOL-3 experiments) fast electrons may leave the plasma during the beam injection time. The experiments showed that after the beginning of the beam injection the plasma electron temperature increased from 1-3 eV up to 0.5-1 keV. The plasma density in these experiments was 10^{15} cm^{-3}. Fig. 2 shows the typical time evolution of the effective plasma temperature, obtained from the diamagnetic measurements. The temperature obtained from laser measurements (Z=270 cm) is shown to be 0.6±0.2 keV in the heating maximum, and the values of the "laser" and the "diamagnetic" temperatures coincide within the measurement accuracy (20%). In this case, in the vicinity of the device entrance the plasma electron temperature in the heating maximum should be 0.8-1 keV at the given above density.

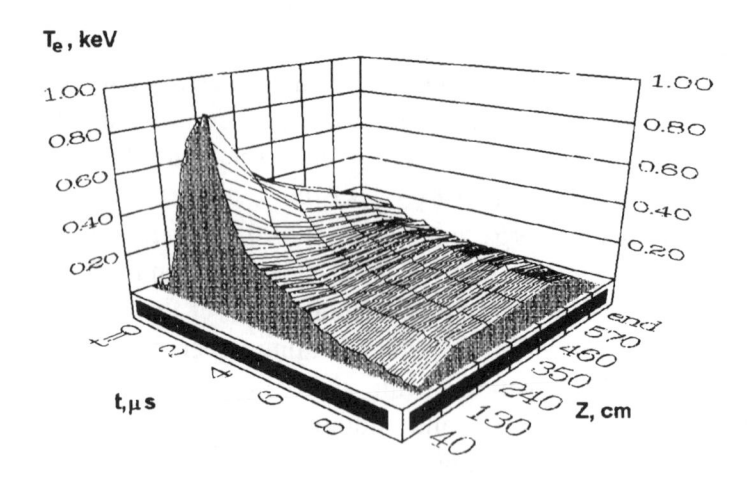

FIG. 2. Time evolution of the electron temperature along the device.

The data on suprathermal electrons was obtained mainly by X-ray measurements. It was determined that the high-energy "tail" of electron distribution function has the mean energy of at least 10 keV and density of these electrons is several percent of plasma one to the end of heating pulse. The instantaneous concentration of suprathermal electrons in the plasma is much less than total amount of such electrons generated during the heating pulse due to non-classical nature of their scattering (see later) and short transit time. It means that hot electrons can receive a major part of the energy, lost by the beam.

The very nonordinary fact is that the axial distribution of the plasma temperature (Fig.2) remains strongly nonuniform during the beam duration. The calculations of heat transport in the plasma showed [26] that the experimentally observed temperature distribution cannot be explained by classical heat conductivity (Fig.3). High temperature gradients in the relatively long-lived sub-keV plasma can be maintained if scattering rate of plasma electrons is sufficiently higher than classical one. The heat conductivity suppression in the plasma region near the device entrance should reach a factor of 100-1000 to the beam pulse end (details see in [26]). This fact demonstrates brightly the existence of the strong developed Langmuir turbulence in a plasma during a collective relaxation of the E-beam.

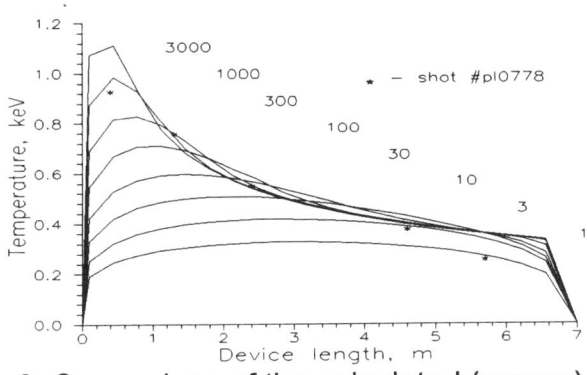

FIG. 3. Comparison of the calculated (curves) and experimental (points) axial distribution of the plasma temperature. The values of the heat conductivity suppression in comparison with classical one are indicated.

After the end of heating pulse the level of plasma turbulence and additional electron scattering decreases rapidly. Thus plasma heat conductivity becomes classical after short time since the beam end. This result was compared too with the plasma temperature decay calculations based on the classical heat conductivity [26]. In the vicinity of the plasma central cross section the temperature time evolution follows the classical dependence:

$$T = T_{max} / (1 + \alpha t)^{2/5}$$

with $\alpha = f(n, Z_{eff})$ calculated for the given point of the device. The measured decay rate of the plasma temperature allows to estimate Z_{eff}, which should be 1.2-2 to fit the experimental data.

Thus, as a result of the intense microsecond E-beam collective relaxation on the GOL-3 device the hot plasma of 10^{15} cm^{-3} density and 1 keV temperature was obtained.

IV. DIRECT OBSERVATION OF LANGMUIR TURBULENCE ON THE GOL-M DEVICE

As was established, Langmuir waves have to be excite during beam-plasma interaction. In the last years the experiments on the direct observation and study of the Langmuir turbulence exited by a E-beam in a plasma are performed on the GOL-M device [27-29]. It is used the laser scattering method, detecting the scattering light at small angles from the direction of the incident light.

The schematic of these experiments is shown in Fig.4. The electron beam (700 keV, 2-3 kA, 2 cm diam.,250 ns) is injected into initial hydrogen plasma with the density n_e~10^{15} cm^{-3} and the electron temperature T_e~1 eV. The length of the plasma column is 250 cm and its diameter is 8 cm, longitudinal magnetic field is 2.5 T, in mirrors 4.5 T.

It is used the CO_2-laser collective scattering technique for measurements of frequency (ω) and wave vector (k) spectra of Langmuir turbulence [28]. The spatial and temporal scales of turbulence are determined by the electron plasma frequency (300 GHz at $n_p=10^{15}$ cm^{-3}) and by characteristic wave number of Langmuir waves coupled directly with relativistic electrons (k_r=60 cm^{-1}).

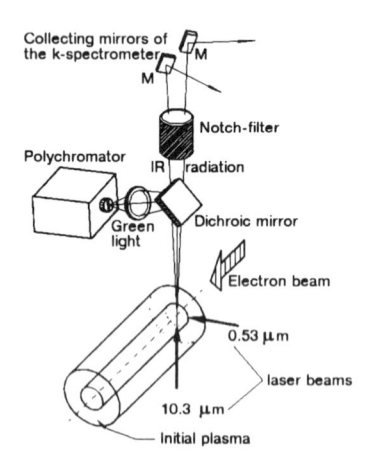

FIG.4 Layout of the GOL-M experiment.

The CO_2-laser scattering signal is observed only during E-beam injection. Its duration (τ_s=200 ns) is much shorter than that the typical laser pulse duration ($\tau\approx1.5$ μs). Frequency spectra are observed at the scattering angles θ=0.5-2°. They were measured the spectra at different plasma densities. The location of the frequency spectrum peak is consistent with the $\omega_{pe}/2\pi$ positions at each plasma density, which was determined independently by 90° Thomson scattering of ruby or 2nd harmonic Nd-laser in the same shot.

The k-spectra of turbulent waves were studied at the electron plasma density around $1.5 \cdot 10^{15}$ cm^{-3} in several experiments using different detecting systems and two configurations of the laser probing beam incidence: 90° and 30° to the E-beam direction. Fig.5 demonstrates the spectrum of resonant Langmuir oscillations over thermal level which is given as a function of transverse component of wave vector. On the whole, these experiments showed the following results: [28,29]

- the frequency and k-spectra of strong developed Langmuir turbulence driven by an electron beam are measured in a broad spectral region;
- the spectral density of the turbulence energy exceeds the thermal level by 5-11 orders of magnitude;
- the strong Langmuir turbulence has a broad spectrum with the power law of the drop to the short wave region;
- the threshold of modulational instability [21] is exceeded by an order of magnitude for perturbations transverse to a magnetic field.

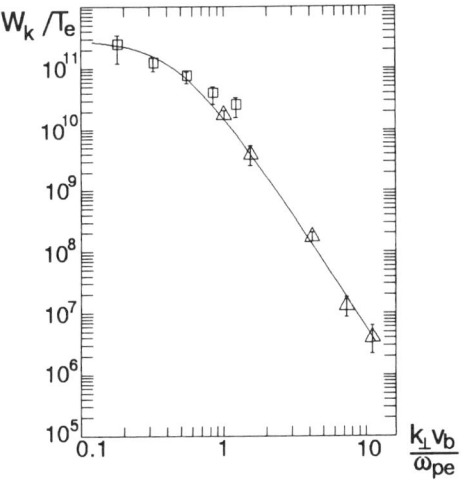

FIG.5 Spectral density of resonant Langmuir oscillations (over thermal level) as a function of transverse wave vector.

These experiments confirmed in particular some conclusions on Langmuir turbulence made in papers [10,12] on the basis of the received that time experimental results.

Recently at the GOL-M device it is also observed the ion acoustic oscillations in non-isothermal plasma with developed Langmuir turbulence (see [30]).

V. CONCLUSION

1. Main regularities of the physics of interaction of high-power electron beams with magnetized plasmas is found at present.
2. Strong Langmuir turbulence driven by an electron beam is observed and studied by direct laser scattering measurements on the GOL-M device.
3. Hot plasma of 1 keV temperature and $10^{15} cm^{-3}$ density is obtained with use of high-power microsecond E-beam on the GOL-3 facility.

ACKNOWLEDGMENTS

Author is very grateful to A.V.Burdakov, K.V.Tsigutkin and L.N.Vyacheslavov for useful discussions and help in preparation of this paper.

This work was supported in part by Soros Foundation under Grant № NQF000.

REFERENCES

1. Sudan, R. N., *Proc. 6th Europ. Conf. on Control. Fusion and Plasma Physics, Moscow*, 1973, v. 2, p.184
2. Brejzman, B.N. and Ryutov, D. D., *Nucl. Fusion* 14, 873 (1974)
3. Thode, L. E., *Phys. Fluids* 19, 831 (1976)
4. Arzhannikov, A. V., Burdakov, A. V., Breizman, B. N. et al. *Plasma Physics and Contr. Nucl. Fusion Res. (Proc. 7th Int. Conf., Insbruck, 1978). IAEA, Vienna*, 1979, v.2, p. 623.
5. Arzhannikov, A.V., Burdakov, A.V., Burmasov, V.S., et al. *Proc. 3rd Int. Conf. on High Power Electron and Ion Beam Res. and Tech., Novosibirsk*, 1979, v.1, p.29.
6. Koidan, V. S., *Proc. Int. Symp. on Physics in Open Ended Fusion Systems, Japan, Tsukuba*, 1980, p.365.
7. Koidan, V.S., Kruglyakov, E.P., Ryutov, D.D. *Proc. 4th Int. Conf. on High Power Electron and Ion Beam Res. and Tech., Palaiseau*, 1981, v.2, p.531.
8. Breizman, B.N., Kruglyakov, E.P. *Proc. 15th Int. Conf. on Phenomena in Ionized Gases, Minsk, USSR*, 1981, v.2, p.93.

9. Sunka, P., *Proc. Xth Europ. Conf. on Control. Fusion and Plasma Physics, Invited Papers, Moscow, USSR*, 1981, v.2 p.96.
10. Arzhannikov, A.V., Burdakov, A.V., Koidan, V.S., Vyacheslavov, L,.N. *Physica Scripta* **T2/2**, 303 (1982).
11. Sunka, P., *Proc. 16th Int. Conf. on Phenomena in Ionized Gases, Dusseldorf, Invited Papers*, 1983, p.232.
12. Arzhannikov, A.V., Burdakov, A.V., Koidan, V.S. et al. *Proc. 1984 Int. Conf. on Plasma Physics, Lausanne, Invited Papers*, 1984, v.1, p.285.
13. Kruglyakov, E.P. *Proc. of the 6th Intern Conf. on High Power Particle Beams, Kobe, Japan*, 1986, p.49.
14. Arzhannikov, A.V., Astrelin, V.T., Avrorov, A.P. et al. *Proc. of the 11th Intern. Conf. on Plasma Physics and Contr. Nucl. Fusion Res., Kyoto. IAEA, Vienna*, 1986, v.2, p.323.
15. Breizman, B.N., *in Review of Plasma Physics, edited by Kadomtsev, B.B. (Consultants Bureau, New York)* **15** (1987).
16. Arzhannikov, A.V., Burdakov, A.V., Kapitonov, V.A. et al. *Plasma Physics and Contr. Fusion (Proc. 15th Europ. Conf., Dubrovnik)*, 1988, v.30, p.1571.
17. Ryutov, D.D. *Proc. 7th Int. Conf. on High Power Particle Beams, Karlsruhe*, 1988, v.1, p.208.
18. Koidan, V.S., *Proc. IV Latin-American Workshop on Plasma Physics, Invited Papers, Buenos Aires, Argentina*, 1990, p.257.
19. Piffl, V., Bohacek, V., Clupek, M. et. al. *Proc. 9th Intern Conf. on High-Power Particle Beams, Washington*, 1992, v.2, p.1221
20. Arzhannikov, A.V., Burdakov, A.V., Koidan, V.S. et. al. *Pisma Zh. Exp. Teor. Fiz.*, **27**, 173 (1978), (*JETF Lett.*, **27**, 161 (1978))
21. Pozzoli, R., Ryutov, D.D., *Phys. Fluids*, **22**, 1782 (1979).
22. Abrashitov, Yu.I., Koidan, V.S., Konyukhov, V.V. et. al. *Zh. Exp. Teor. Fiz.*, **66**, 1324, (1974).
23. Voropaev, S. G., Knyazev, B.A., Koidan, V.S. et al. *Pisma Zh. Tech. Fiz.*, **7**, 431 (1987) - in Russian; *Proc. 18th Intern. Conf. on Phenomena in Ionized Gases, Swansea, U.K.*, 1987, v.1, p.204.
24. Arzhannikov, A.V., Burdakov, A.V., Chikunov, V.V. et. al. *Proc. 8th Intern Conf. on High-Power Particle Beams, Novosibirsk*, 1990, v.1, p.14.
25. Arzhannikov, A.V., Burdakov, A.V., Chikunov, V.V. et. al. *Proc. 9th Intern Conf. on High-Power Particle Beams, Washington*, 1992, v.1, p.127
26. Burdakov, A.V., Postupaev, V.V. *Preprint INP 92-9, Novosibirsk*, 1992 - in Russian.
27. Vyacheslavov, L.N., Kandaurov, E.V., Kruglyakov, E.P. et al, *Pisma Zh. Exp. Teor. Fiz.*, **50**, 379 (1989). (*JETF Lett.*, **50**, 410 (1989).
28. Vyacheslavov, L.N., Kruglyakov, E.P., Losev, M.V., Sanin, A.L. *Rev. Sci. Instrum* **64**, 1393 (1993).
29. Vyacheslavov, L.N., Burmasov, V.S., Kandaurov, I.V. et al, *Phys. Plasmas* **2**, 2224 (1995).
30. Burmasov, V.S., Kandaurov, I.V., Kruglyakov, E.P. et al, *ICPIG-22 Conf., 1995, contributed papers.*

Progress in the Understanding of non-1D Glow Discharges: Experiment and Theory

Jacques Derouard[*] and Leanne Pitchford[+]

[*]*Laboratoire de Spectrométrie Physique (UA CNRS 08)*
Université Joseph Fourier (Grenoble-I), BP 87, 38402 Saint Martin d'Hères, France
[+]*Centre de Physique des Plasmas et Applications de Toulouse (UA CNRS 277)*
Université Paul Sabatier, 31062 Toulouse, France

Abstract. This contribution is a joint presentation on the structure of non unidimensional glow discharges experimentally observed using in particular optical diagnostics and predicted using self consistent numerical models. Results are presented concerning the effect of radial losses and electrode edges on the structure of DC and RF glow discharges, and the inititation phase of the breakdown in planar hollow cathode discharges. Most of the observed features of the discharges can be well reproduced by the models, which thus can be used to predict the effect of geometry on plasma devices and guide the optimization of their design.

INTRODUCTION

In many cases glow discharges appear like a plasma column, showing the well known axial structure as described in every textbook. The fact that this structure is mostly axial tends to give credit to the fact that glow discharges can be considered as one-dimensional systems. This is often true, and impressive agreement between experiment and 1-D numerical models have been now obtained in the case of DC, RF and transient glow discharges, including the description of both physical and chemical processes occuring in the gas phase and at the surface.

However there are many cases where the radial structure of the discharge cannot be ignored.

In this paper we shall both present experimental evidence of these non 1-D effects and show how numerical simulations can be applied to their description and can help to the understand the phenomena.

NUMERICAL MODELS

Self consistent models must solve the Poisson equation together with the continuity equation and the transport of particles, their kinetics of creation and losses, the kinetics of energy exchange (1). One distinguishes usually between "fluid models", where particles are characterized by some mean values of their distribution function in space and momentum, and "particle models", where the whole distribution function is sampled by following the dynamics of individual particles subjected to both external forces (electric field) and internal forces (collisions, usually treated by a Monte Carlo procedure).

Fluid models (2,3) are valid for pressures such that the velocity distribution can be well characterized by their mean value, typically above 100 mT at room temperature. In addition, some assumptions must be done about the ionization processes in order to close the system of differential equations, and to take into account the fact that these processes often involve the non Maxwellian, high energy, tail of the electron energy distribution function. These models do not require large computation times, even in multidimensional geometries.

Particule models (4) are valid in principle at any pressure, although they are more suited (and more useful) for the description of low pressure discharges. The description of ionisation processes is as realistic as it can be. However the required computation time is *much* larger than for fluid models and their application to the modelling of 2-D discharges is beyond the capability of most existing computers.

In *hybrid models* (5-7) the ionization source term is calculated using a microscopic, Monte Carlo simulation, while the transport of particules coupled to the Poisson equation is self consistently solved using a fluid computation; The ionisation source term is calculated again from time to time as the electric field configuration evolves with time. This combines a realistic description of ionization processes for any geometry with the efficiency of fluid models, making these models still tractable in multidimensional geometries. Of course their validity is still limited to the pressure range where fluid models are valid. Most of the numerical results shown here have been obtained using this kind of model.

In the following we shall consider axysymmetric systems, which can be considered as 2-D systems, the two relevant coordinates being of course the axial and radial coordinates.

DC DISCHARGES

As a first example let us consider the case of DC discharges. Many

1-D models have been developped and have successfully reproduced the cathodic sheath and negative glow regions of parallel plate DC discharges. An interesting prediction is the existence of a field inversion in the negative glow, which has been experimentally pointed out by Gottscho et al (8). However, these 1-D models are not able to predict the existence of well known features of these discharges such as the positive column (unless some ad hoc hypothesis concerning loss processes like volume recombination is introduced), or the "normal glow regime". Recently, 2-D hybrid model simulations have been able to well reproduce the whole anatomy of DC glow discharges, from the cathodic sheath to the anode fall regions (6). Fig. 1 shows the structure of a DC discharge which is predicted by this 2-D hybrid self consistent numerical model (6,7). The radial constriction of the glow which does not cover the whole cathode, corresponding to the "normal glow regime" is clearly seen. The ion current density distribution shows the ion motion in the sheath and in the presheath, the reversal of the velocity in the negative glow associated with the field inversion mentionned above (and which is nothing else than ambipolar diffusion), how the ions come to the wall where they recombine.

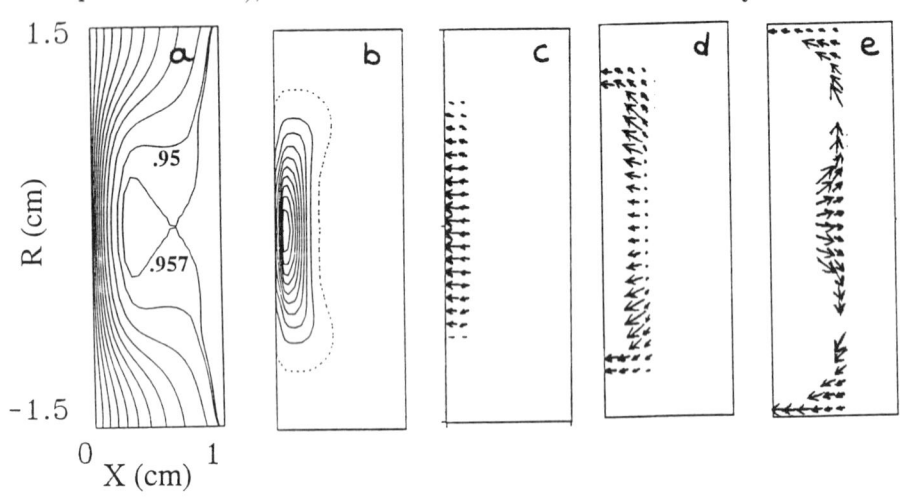

FIGURE 1. Contours of constant potential (a), and constant ionization source term (b), and 2-D plot of ion current density with scales in the ratio 120 (c), 10 (d), 0.5 (e) calculated using a 2-D hybrid numerical model (6). Argon discharge, 3 Torr, 130 V, 175μA/cm^2 on axis.

To continue with DC discharges let us consider now the edge effects at the electrodes. More specifically, Fig. 2 shows the results of the 2-D hybrid numerical model for discharges with cylindrical, rod shaped electrodes (5,6). The important point is that for low currents the discharge is like a plane parallel electrode system, it is confined in the volume facing the parallel faces of the electrodes and is not sensitive to the side faces of the rods. However, when the current grows and the sheath contracts the discharge eventually tends to develop at the side faces, while it can become strongly inhomogeneous on the front face of the electrode. We note also the enhancement of the space charge field at the edge.

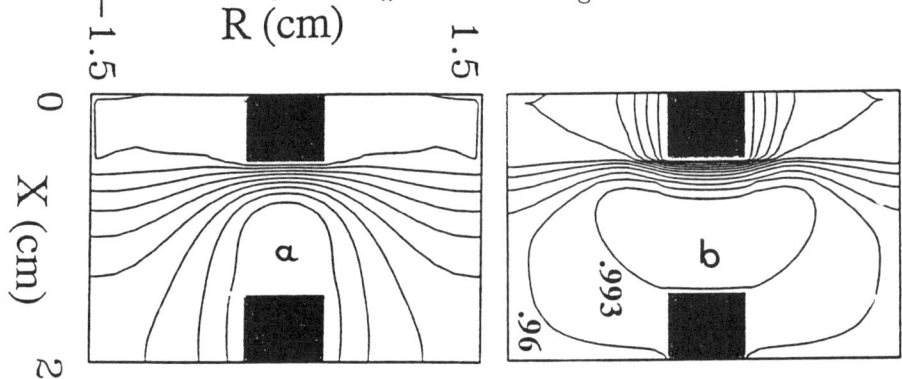

FIGURE 2. Equipotential contours calculated using our 2–D hybrid numerical model for discharges in Argon with rod shaped electrodes. Discharge conditions are 1 Torr, 142 V, 163μA/cm^2 (a) and 149 V, 1900μA/cm^2 (b) (on axis).

This is in good agreement with experimental observations of the electric field vector distributions measured in DC discharges in Argon-Potassium mixtures using Laser Stark spectroscopy of NaK molecules (9) (Fig. 3).

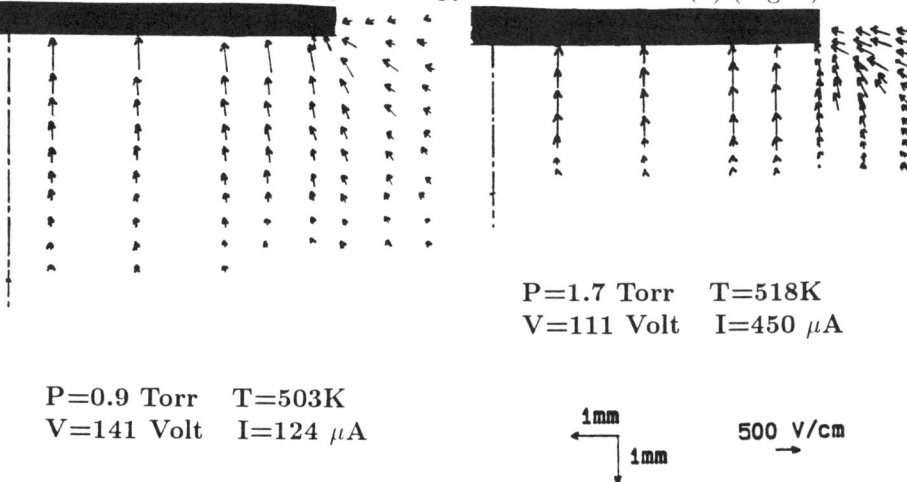

P=0.9 Torr   T=503K
V=141 Volt   I=124 μA

P=1.7 Torr   T=518K
V=111 Volt   I=450 μA

FIGURE 3. Cathodic sheath electric field vector distribution in discharges in Ar/K mixtures measured using Laser Stark spectroscopy of NaK molecules.

This shows two discharges in conditions which are relatively similar with respect to the reduced current density j/p^2, corresponding to "abnormal" discharge conditions. However they operate at two different pressures, hence the different sheath lengths. We actually see that when the sheath is short, the space charge field at the cathode edge is especially intense, and this can be associated with the spreading of the glow to the sides of the electrode.

RF DISCHARGES

Because RF discharges are so widely used in microelectronics manufacturing it is especially important to control the radial non uniformities in the plasma and to understand their origin.

The 2-D electric field distribution and its time evolution has been measured by Alberta et al (10) in a RF discharge using the same laser Stark spectroscopy as above. It has been found that although the axial field profile and the sheath length strongly vary with time, surprisingly the shape of the field lines does not change significantly during the RF cycle. This indicates that the field lines distribution (and thus the shape of the potential contours) is more determined by the geometry of the reactor than by the kinetics of the discharge. The potential contours shown on Fig. 4 have been determined experimentally from the orthogonal lines to the field lines, measured at a time when the applied RF voltage is close to its negative maximum. From these, one can see the extent of the region where the discharge can be considered as uniform over the electrode. Like in DC discharges, we observe some enhancement of the electric field intensity at the electrode edge.

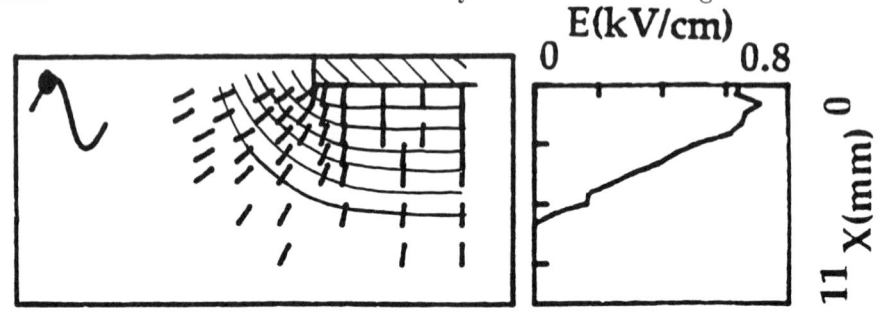

FIGURE 4. Axial electric field profile, electric field lines and equipotential lines measured in a RF discharge in Ar/K mixture at a time where the applied voltage is close to its maximum. 255 V, 4 MHz, 10^{16} atoms/cm^3.

Radial non uniformities in RF discharges have been also pointed out in the GEC reference cell by Overzet and Hopkins (11), Greenberg and Hebner(12), and more recently by McMillan and Zacharias (13) who used planar Laser induced fluorescence to measure the Ar metastable density 2-D distribution. This density distribution can be more or less correlated with the excitation rate of the plasma, which is itself strongly correlated in this case (low RF voltage) with the electron density. Fig. 5a shows the results of 2-D fluid model calculations of Boeuf and Pitchford (3) performed in the GEC reference cell geometry in discharge conditions similar to the experiment by McMillan and Zacharias (13). The ressemblance with the experimental Argon metastable Argon spatial distribution (See Fig. 5 of Ref. (13)) is striking. The comparison with Fig. 5b demonstrates that the existence of a maximum of electron density off axis is due to the guard ring around the powered electrode. Further experimental evidences for these effects have been given by plasma induced optical emission measurements and laser light scattering of dust trapped in electrostatic potential wells (14). This clearly points out the potential of 2-D numerical simulations to help design and optimize the shape of plasma reactors.

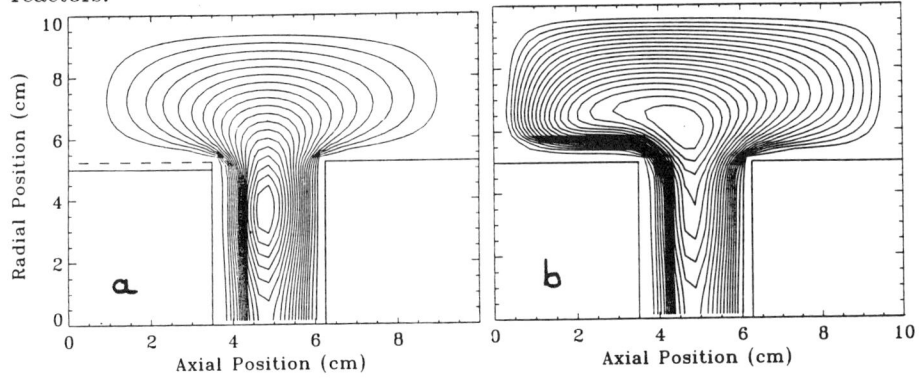

FIGURE 5. Contours of constant averaged electron number density calculated in the GEC reference cell geometry with (a) and without (b) a grounded guard ring around the powered electrode, using the 2-D fluid numerical model of Boeuf and Pitchford (3). V_{RF}=100 V, 13.6 MHz, 0.1 Torr Argon.

HOLLOW CATHODE BREAKDOWN

Finally we will consider the case of discharges in planar hollow cathode cells such as shown in Fig. 6. This type of geometry is similar to those of "pseudospark devices" (15), which are known in particular for their interesting ability to switch very high currents (10's of kA) in very short times (ns). The initiation phase of the switch involves the breakdown in the hollow cathode volume between the two parallel grounded plates facing the anode. The discharge current begins to grow very slowly after the voltage has been applied, then rises up very fast after a given delay time (16,17) (Fig. 7).

The question which we would like to answer is: "How does the development of the discharge inside the hollow cathode volume proceed, and how can it be triggered most efficiently ?"

Because of the hole on the axis in the cathode plate in front of the anode, the discharge configuration is obviously two dimensional. The evolution of the electric field predicted by the 2-D hybrid numerical model (16) following the application of the applied voltage can be seen on Fig. 8, and the corresponding current wave form is shown on Fig. 7b. The initial peak in the total current corresponds to the penetration of the space charge field in the hollow cathode volume, and is actually a displacement current at the cathode (the ions dont have time to move at a time scale of 100 ns). Once the space charge field has penetrated inside the hollow cathode volume, ionisation processes occur efficiently and the current and plasma density grow faster and faster.

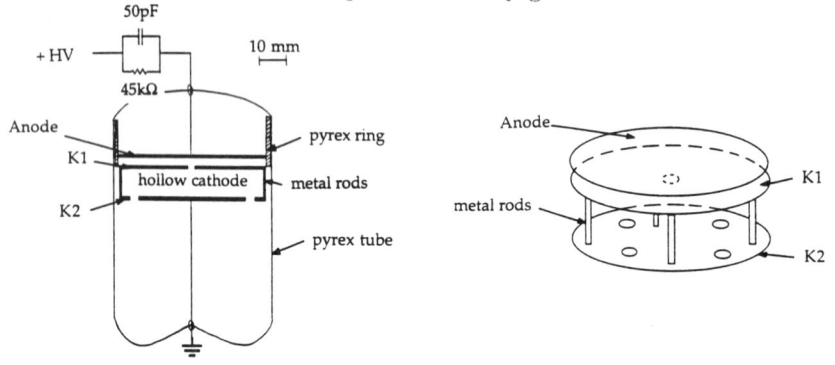

FIGURE 6. Planar hollow cathode cell with transparent walls used for the experiments.

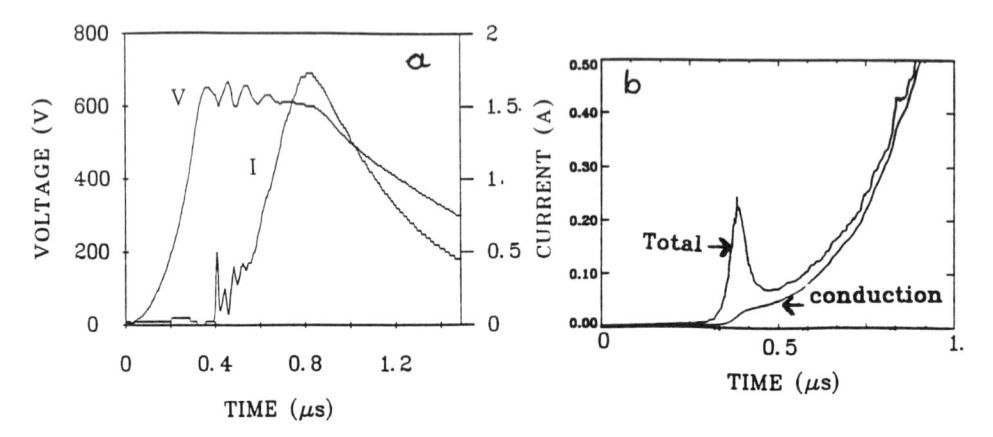

FIGURE 7. Typical experimental current and voltage waveform observed with 0.5 Torr pure Argon (a), and calculated current waveform (b) using a 2-D hybrid numerical model in similar conditions (0.65 Torr, 500 V).

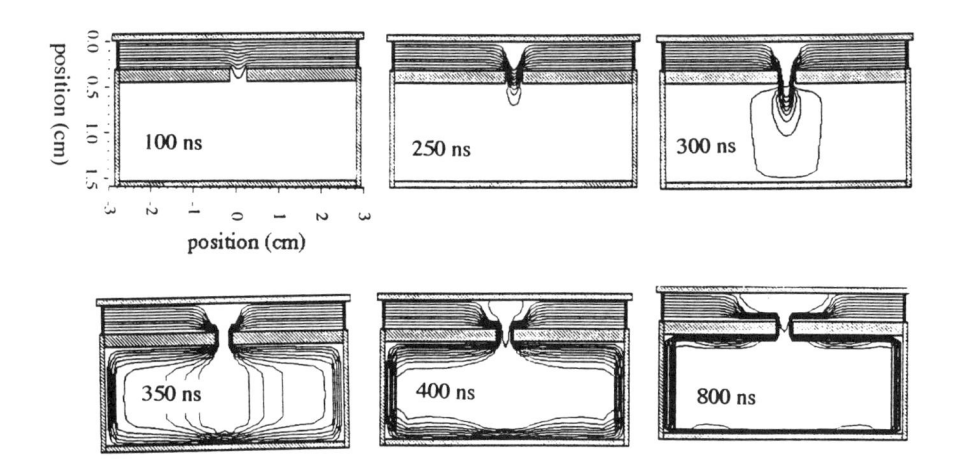

FIGURE 8. Equipotential contours during the discharge initiation calculated using a 2-D hybrid numerical model in 0.65 Torr Argon, 500 V constant voltage applied from t=0.

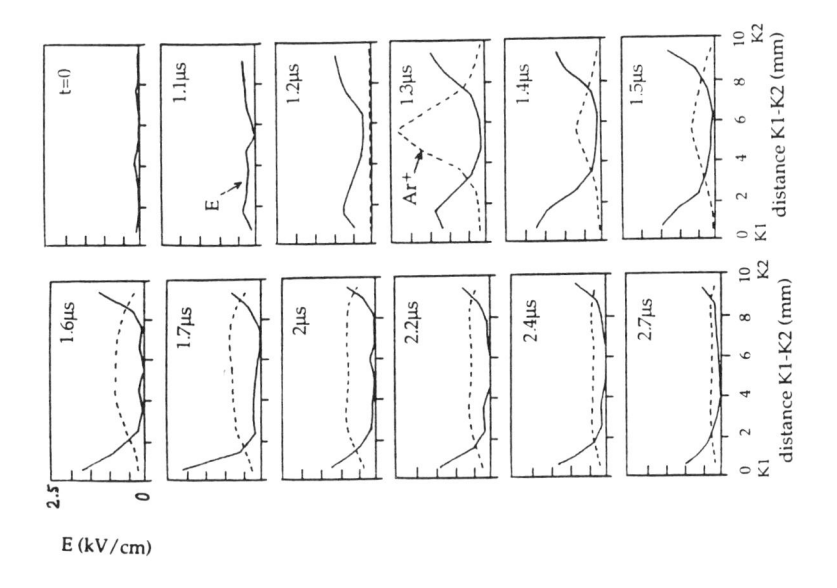

FIGURE 9. Space and time variation of the electric field intensity (continuous line) and Ar$^+$ emission at 457.9 nm (dashed line) measured in the hollow cathode as a function of the axial position for a position 8mm off axis. The initial applied voltage applied at $t=0$ is 400 V. The gas is a Ar/K mixture at a density of 1.8×10^{16} cm^{-3}.

The comparison with the measurements shown on Figs. 7a and 9 indicates that the phenomenon is well understood and can be accurately described by the numerical model. The flash of plasma optical emission which is observed at the time when the space charge field penetrates into the hollow cathode volume is also well predicted by the numerical model (16).

The mechanism is the following: the breakdown will occur when a sufficient number of positive ions have accumulated in the main gap in front of the central hole to distort the geometrical field inside the hollow cathode, and this results in a very large increase of the ionization source term due to the onset of the hollow cathode effect.

Initially, charges are introduced in the hollow cathode volume (experimentally it might be using for example an auxilliary discharge, laser impact on the electrode surface, or field emission from a high voltage tip). Afterwards there is a competition between the diffusion of these charges toward the inner walls of the hollow cathode where they recombine, and the accumulation of a sufficient number of ions in the main gap formed by the impact of electrons escaping the hollow cathode volume and then strongly accerelated by the geometrical field.

Surprisingly, numerical simulations indicate that this mechanism is the same for both low (some 100's Volts) and high (several kV) voltage conditions (5,17), and this is confirmed by the comparison in both cases between observations and predictions concerning the evolution of the delay time with pressure, applied voltage, and initial quantity of charges introduced to trigger the discharge (17).

The numerical model has been eventually used to predict the influence of geometrical factors (such as cathode hole diameter and width, location of the triggering charges inside the hollow cathode volume) on the time to breakdown (17).

CONCLUSION

It is clear that the interplay between experiment and numerical modelling have led to improve our understanding of glow discharges, and to the validation of numerical models. We are now to the point where numerical modelling of glow discharges can help in shaping the geometry of plasma devices, which obviously requires multidimensional computing codes. Progress in computer techniques should help to further extend the applicability of these numerical codes to low pressure conditions for which particulate models are needed. From the experimental point of view, optical imaging associated with various spectroscopic techniques is the method of choice to investigate these discharges and will therefore be developed in the future to more fully characterize the structure of these discharges.

ACKNOWLEDGMENTS

We gratefully aknowledge the contributions of our coworkers to the results presented in this paper: M.P. Alberta, H. Debontride, N. Sadeghi in Grenoble, and J.P. Boeuf, A. Fiala, N. Ouadoudi in Toulouse. This work has been supported in part by DRET and CNET.

REFERENCES

1. Kushner, M. J., and Graves, E. D. (eds), Special issue on modeling collisional low temperature plasmas, *IEEE Trans. Plasma Sci.*, **PS-19**(2), (1991).
2. Boeuf, J. P., J. Appl. Phys. **63**, 1342–1349 (1988).
3. Boeuf, J. P., and Pitchford, L. C., *Phys. Rev. E*, **51**, 1376–1390 (1995).
4. Boeuf, J. P., Pitchford, L. C., Fiala, A., and Belenguer, P., *Surf. Coat. Technol.*, **59**, 32–40 (1993).
5. Boeuf, J. P., and Pitchford, L. C., *IEEE Trans. Plasma Sci.*, **PS-19**, 286–296 (1991).
6. Fiala, A., *Thesis*, Université Paul Sabatier de Toulouse, 1995.
7. Fiala, A., Pitchford, L. C., and Boeuf, J. P., *Phys. Rev. E*, **49**, 5607–5622 (1994).
8. Gottscho, R. A., Mitchell, A., Scheller, G. R., Chen, Y. Y., and Graves, D. B., *Phys. Rev. A*, **40**, 6407–6414 (1989).
9. Alberta, M. P., and Derouard, J., *J. Phys. B*, **24**, 904–908 (1991).
10. Alberta, M. P., Debontride, H., Derouard, J., and Sadeghi, N., *J. Phys. III France*, **3**, 105–124 (1993).
11. Overzet, L. J., and Hopkins, M. B., *Appl. Phys. Lett.*, **63**, 2484–2446 (1993).
12. Greenberg, K. E., and Hebner, G. A., *J. Appl. Phys.*, **73**, 8126–8133 (1993).
13. McMillin, B. K., and Zachariah, M. R., *J. Appl. Phys.*, **77**, 5538–5544 (1995).
14. Dorier, J. L., and Hollenstein, C., *Private communication*, (1995).
15. Gundersen, M. A., (ed.), *The Physics and Applications of Pseudosparks*, New York: Plenum, 1990.
16. Alberta, M. P., Derouard, J., Pitchford, L. C., Ouadoudi, N., and Boeuf., J. P., *Phys. Rev. E*, **50**, 2239–2252 (1994).
17. Pitchford, L. C., Ouadoudi, N., Boeuf, J. P., Legentil, M., Puech, V., Thomaz, J. C., and Gundersen, M. A., *J. Appl. Phys.*, **78**, 77–89 (1995).

External Magnetic Field Influence on Properties of High-Power Laser-Produced Plasma

Wołowski J. Kasperczuk A. Pisarczyk T.

Institute of Plasma Physics and Laser Microfusion
23 Hery St., 00-908 Warsaw, Poland

Abstract. The paper presents results of formation of expanding plasma by combining laser-produced plasma with an external strong magnetic field. The plasma was generated by means of a Nd-glass laser which was focused on a solid target located on the axis of a single-turn coil providing magnetic field of up to 50T. Spatial characteristics of the dynamics of interaction of the plasma with the magnetic field were registered by means of a three-frame interferometry. For registration and analysis of interferograms, CCD cameras and a multichannel image acquisition system were used. An interesting influence of the strong magnetic field on the plasma dynamics and shape was observed. Preliminary results of numerical modelling are compared with the experimental data.

1. INTRODUCTION

Plasma produced with a high-power laser (of power densities of $\sim 10^{14} W cm^{-2}$) focused on a solid target placed in a vacuum chamber is characterized by high concentration and temperature gradients as well as fast temporal variation of all parameters. These properties of plasma are utilized in various applications (laser fusion, subnanosecond laser-plasma sources of X-rays, recombination XUV lasers, etc.). Strong external magnetic field (of the strength of some Teslas) changes the properties of a laser-produced expanding plasma considerably. In particular, spatial distributions of density and temperature as well as temporal characteristics of such plasma undergo changes. Therefore, an experimental system in which laser-produced plasma is created in external magnetic field makes possibilities to study the physics of phenomena occuring during interaction of strongly nonuniform plasma with strong magnetic field (e.g., (1-4)). Modification of laser-produced plasma properties with the use of such a field can be also applied to obtain amplified spontaneous emission in the XUV range (e.g., (5)). Unfortunately, the mechanisms and detailed hydrodynamics, especially interaction of the laser-produced plasma with a strong external magnetic field and the process of the plasma column formation, are not so well understood yet as for free expansion approaches, and without the careful diagnostics and modelling, the extrapolation to shorter wavelengths will remain a matter of conjecture. It is also proposed to apply the external magnetic field in optimization of plasma streams in the interior cavities of laser thermonuclear targets (6). The formation of such plasma streams is an important aspect of the indirect-compression targets and of targets with direct heating of the interior. In particular, the degree of conversion of the laser radiation into X-rays depends on the parameters of the plasma streams, as does the efficiency of the entry of the laser radiation into the target, since the produced plasma streams fill the input aperture and can hinder the entry of the laser radiation into the target. One possibility of controlling the parameters of these plasma streams is by using an external magnetic field. The relatively thin plasma jet, which appears when a laser plasma interacts with external magnetic field, can probably be used also to tackle the urgent problem of transport in a laser ion injector. Experiments on laser-produced plasma crossmotion for simulation of interaction between plasma clouds and space plasma are carried on with reference to better understanding of the interaction processes of ionospheric and magnetospheric barium releases with surrounding magnetized plasma (7) and interactions between expanding Supernova remnants and magnetized media in space (8). The idea of our experiments was to investigate the dynamics of a laser-produced plasma in a strong external magnetic field and the possibility to produce a plasma column such as is necessary to fulfill the X-ray lasing conditions.

2. EXPERIMENTAL SETUP

The experimental setup consists of three main parts: a magnetic field generator, a plasma chamber, and a Nd-glass laser system. The magnetic field is generated by a single-turn magnetic coil located inside the plasma chamber. The axial field-amplitude component, B_0, of the damped sinusoidal magnetic field pulse generated by this coil is $5-15T$. The Nd laser of the energy $E_l = 5J$ and the pulse duration of about $1ns$ (FWHM) is focused on a teflon $(-CF_2-)_n$ target, which is placed on the axis of the coil. To study the interaction of the laser plasma with the external magnetic field, we used a three-frame interferometry (9) and high-luminosity spectrograph [10] with spectrally and spatially resolved images. The scheme of the experimental setup and the location of the diagnostics are shown in Fig.1. Longer wavelength of the probing beam was chosen in order to increase the sensitivity of the interferometer for low electron densities ($< 10^{18} cm^{-3}$). The laser probing pulse could be delayed by $1-50ns$ with respect to the laser heating pulse and it was possible to make three interferograms of the plasma for chosen time moments of the plasma expansion. Individual frames from the three-frame interferometer were registered by means of CCD cameras of the Pulnix TM-565 type with the matrix of 512x512 pixels. The interferograms of the plasma were processed using a computer

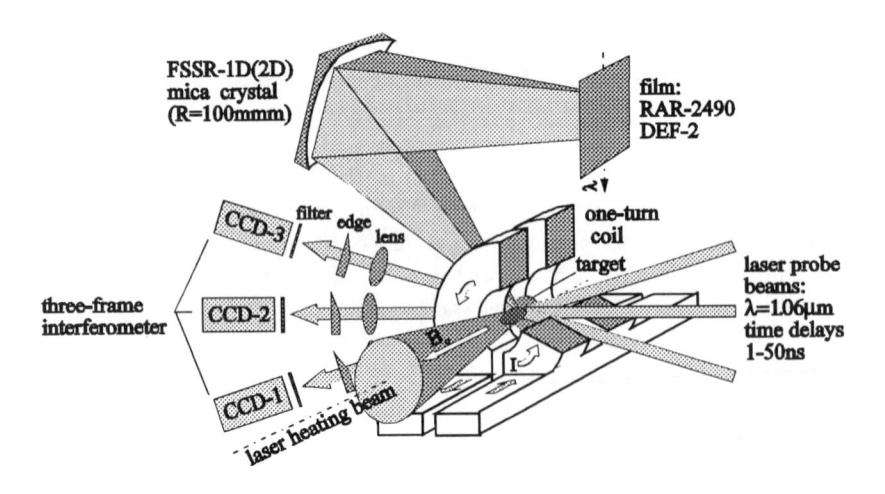

FIGURE 1. Schematic drawing of location of the diagnostics in the experiment.

digitizing system. From the shifts of the interference fringes, the phase and plasma electron density distributions have been calculated (11).

The spatial distribution of soft X-ray emission was measured with the aid of a pinhole camera equipped with a pinhole of about $200\mu m$ in diameter and covered with Be filters ($10 - 20\mu m$ thick). The spectral studies of spatial plasma-cloud parameters distribution were carried out in cooperation with the AU-Russian Scientific Research Institute of Physico-technical and Radio-Engineering Measurements (Moscow Region, Russia) and with the aid of X-ray spectroscopic apparatus developed at that institute. The X-ray emission characteristics were determined making use of two focusing spectrographs in which the spatial resolution was provided by mica crystals bent to form a spherical surface with the radius $R = 100mm$ (one-dimensional system, FSSR-1D, and two-dimensional system, FSSR-2D (10)).

A pulsed magnetic field (up to 50T) was generated by discharging a bank of low-inductance capacitors through the coil (with the outer diameter of 50mm, the inner diameter of 6mm, and the diagnostic slit of 8mm) synchronously with the laser pulse.

3. RESULTS

3.1 Free expansion of laser-produced plasma

The knowledge of the free expansion (at $B_0 = 0$) of the laser-produced plasma was necessary to determine essential differences in the case when $B_0 \neq 0$ in successive phases of the process.

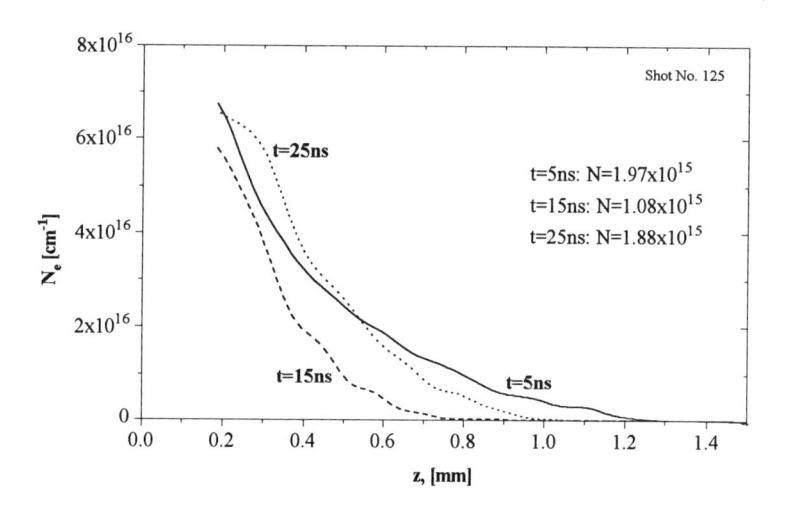

FIGURE 2. The linear electron density distributions for the free expansion of the laser-produced plasma.

On the basis of the analysis of the electron density distributions, obtained at different moments of the plasma expansion, we can distinguish two plasma components: fast and slow. The fast part of the plasma occurs in the initial phase of the plasma expansion and is identified as the thermal plasma due to the plasma ablation. This plasma component has the axial velocity of more than $10^7 cm/s$ and it disappears relatively rapidly. The slow component is generated from a crater made in the target due to such processes as the shock wave, thermal conductivity, generation of soft X-rays, etc. The axial velocity of this component is about one order of magnitude less than of the fast one. The existence of these two components is confirmed by both the linear electron density distributions and the total number of electrons as shown in Fig.2. The lifetime of the fast component is shorter than $15ns$, hence, at $t = 15ns$, the linear electron density drops rapidly along the whole plasma length and an entire plasma disappearance in the area of $z > 0.8mm$ is practically observed (the total electron number in the region of observation drops by more than 50%). At $t = 25ns$, both the linear density and the total electron number grow again. It is connected with the axial propagation of the slow plasma component.

Interferometric investigations of the plasma expansion made it possible to distinguish two essentially different kinds of the geometry of the plasma expansion, angular and axial, related to the target illumination changing from one laser shot to another (due to diffractive disturbances of the laser beam resulting from aperture limitations and the thermo-optical effects in the laser rods, as well as due to the precision of the laser beam focusing on the target (12).

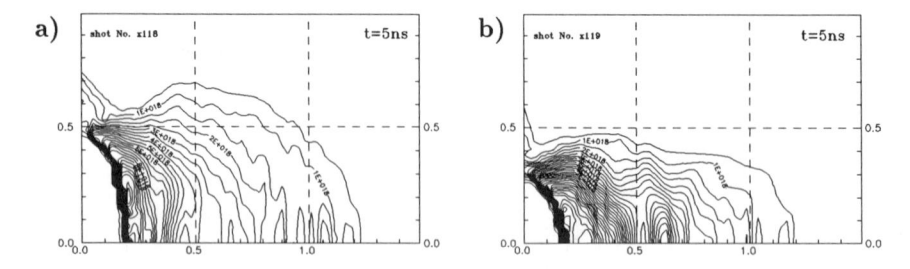

FIGURE 3. The free expansion of laser-produced plasma: a) the angular and b) the axial expansion.

The isodensitograms for both these expansion kinds are shown in Fig.3. The angular expansion (Fig.3a), in comparison with the axial expansion (Fig.3b), is characterized by clear angular plasma emission with respect to the axis and less plasma elongation (the plasma elongation is defined as the ratio of the maximum length and the maximum radius of the plasma blob measured on the electron density level $n_e = 10^{17} cm^{-3}$). At $5ns$, the plasma elongation for the angular expansion is roughly equal to 1.5 and for the axial one it is about 3. Moreover, in case of the angular expansion, the front of the slow plasma component is wide and flat or slightly concave as can be seen in Fig.4a for $t > 15ns$. This front has the axial velocity of about $3 * 10^6 cm/s$. The radial distribution of the electron density has its minimum on the axis and the maximum electron density in the whole observed region is not higher than $1.5 * 10^{19} cm^{-3}$. The lifetime of the plasma does not exceed $50ns$.

In the case of the axial expansion, the slow plasma component has the shape of the blob front similar to a rotary paraboloid with its top directed in the off-target direction. In the whole time of observation, the distribution of $n_e(r)$ has the maximum value more than twice higher compared with the former case. Due to the concentration of the plasma on the axis, the lifetime of the plasma blob in this case is longer by $20 - 30ns$.

3.2 Influence of the magnetic field on hydrodynamics.

The investigations were carried out for the magnetic field induction in the range of 5-15T and the laser pulse energy of about 5J. The essential investigation was made at the 10-T magnetic field, all other values of the magnetic field were used as auxiliary. The analysis of the plasma-magnetic field intraction were made for the both kinds of the expansion geometry. In the case of the angular plasma expansion at the 10-T magnetic field, only fundamental differences with respect to the free expansion for $t > 5ns$ are seen (see Fig.4). The flat front of the plasma blob seen at zero the field (Fig.4a, $t = 15ns$)

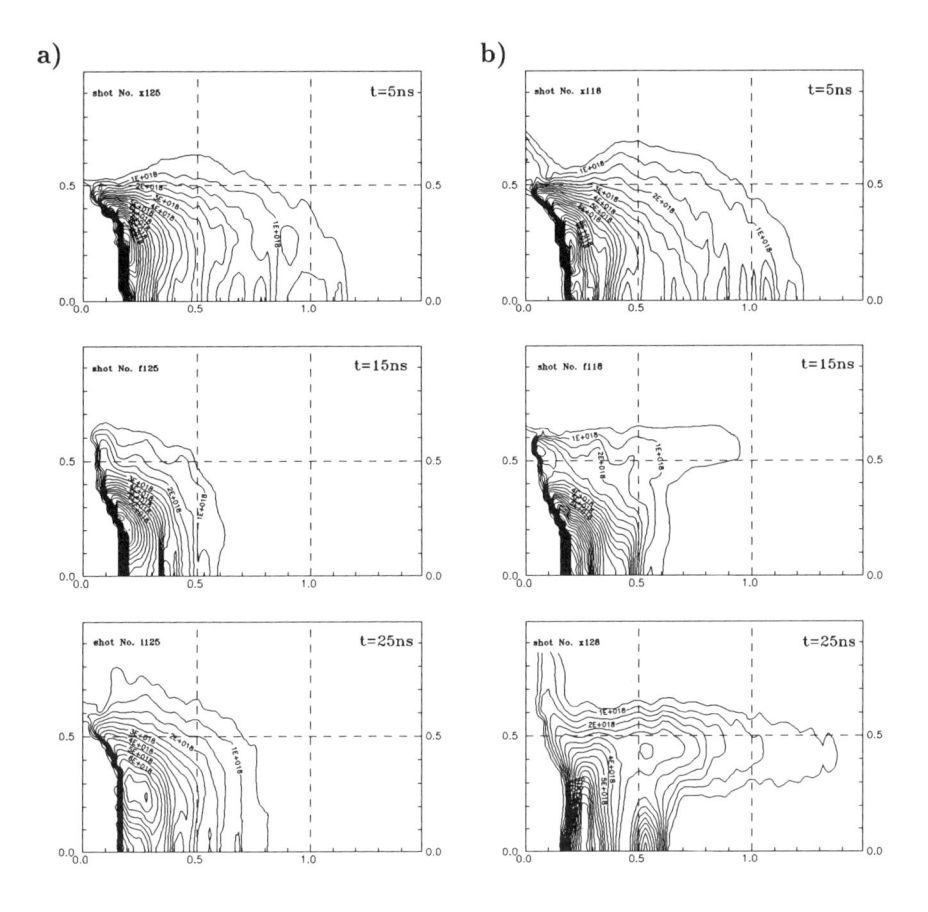

FIGURE 4. Isodensitograms of plasma electron density for the cases: a) $B_0 = 0$, b) $B_0 = 10T$

undergoes deformation and becomes concave very clearly in the 10-T magnetic field (Fig.4b, $t = 15ns$). It is connected with creation of additional wings of about 0.5 mm in length. In the beginning, the electron density in these wings does not exceed $10^{18} cm^{-3}$. The radial plasma expansion across the magnetic field is limited by magnetic pressure. After $15ns$, plasma implosion near the target and associated with this implosion strong plasma stream in the form of a sharp jump of $n_e(r)$ on the axis are observed (Figs.4b or 6a for $t = 25ns$). Next, the strong plasma stream flows along the concave plasma front to the wings. Thanks to it, the electron density in the wings grows a few times. When the maximum electron density during the implosion reaches the axis, the plasma reexpansion is observed. Due to the radial implosion, the maximum of n_e moves to the axis (Fig.4, $t = 15ns$). Later, the thick maximum is being fixed on the axis but its value drops as a result of the reexpansion (Fig.5, $t = 25ns$). Finally,

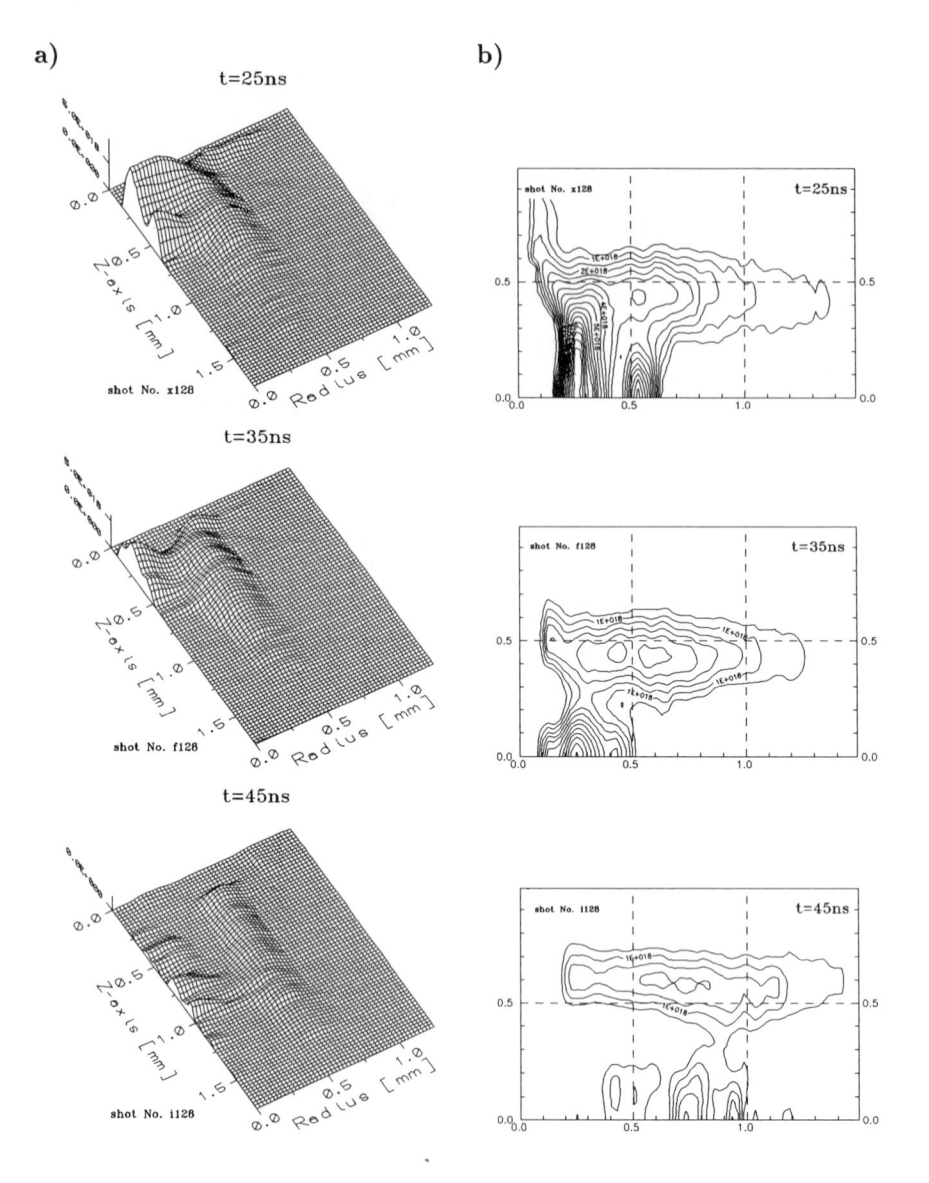

FIGURE 5. Electron density distribution of plasma in the presence of magnetic field ($B_0 = 10T$): a) radial profiles, b) isodensitograms

the above mentioned processes result in a situation that, for $t > 25ns$, most of the plasma is located outside the axis creating a plasma cylinder with the length of about $1mm$ and the inner radius of about $0.5mm$ (Fig.5, $t = 45ns$).

Similarly as in the case of angular expansion, no differences are observed in the axial plasma expansion, during the first 5 nanoseconds, with reference to the free expansion. After further 10 nanoseconds, characteristic wings of low-density plasma appear which are about twice longer than in the case of angular expansion and which decay rapidly.

These wings are supplied by partial flow of the plasma from the extreme regions of the main stream of the plasma. However, the supply is not strong and it has no influence on the behaviour of the slow component of the central dense plasma. The central plasma (as in the case of $B_0 = 0$), during the whole time of observation of the phenomenon, has the paraboloidal shape of its front and the maximum electron density about twice higher than in the case of angular expansion, always located on the axis. In case of axial expansion, the shape of magnetic-field lines undergoes only minor deformation. The fast thermal plasma deflects external magnetic-field lines off the axis along considerable length, changing slightly the central plasma which can move with no obstacles. The decay of cool low-density wings allows to suppose that the external magnetic-field lines become parallel to the axis in the later phases of the process.

4. DISCUSSION

In our opinion, the deformation of the magnetic-field lines by the fast plasma component in the beginning of the plasma blob expansion is responsible for the later behaviour of the slow plasma component. At the moment of the plasma generation on the target surface, the magnetic-field lines become frozen in the hot thermal plasma. Due to the plasma radial expansion, the magnetic-field lines are taken by the plasma and the diamagnetic cavity is being created. The radius of this cavity is given by:

$$R_B = \left(\frac{3E_k\mu_o}{2\pi B_0^2}\right)^{1/3} \quad (1)$$

where: E_k is the kinetic energy of the plasma (in our case, E_k is a function of the laser energy). If we assume that the outer radius of the wings, R_s, is approximately the cavity radius, R_s should be the same function of B_0. The measurements of R_s performed for various values of the magnetic field have proved that $R_s \sim B_0^{-2/3}$.

This conclusion allows reconstruction of the expansion process shown in Fig.6a. The process continues until the magnetic pressure becomes higher than the plasma one. Then, the implosion process is observed (Fig.6b). This process is followed by strong plasma flow directed to the plasma wings along the magnetic-field lines which results in a characteristic plasma configuration in a

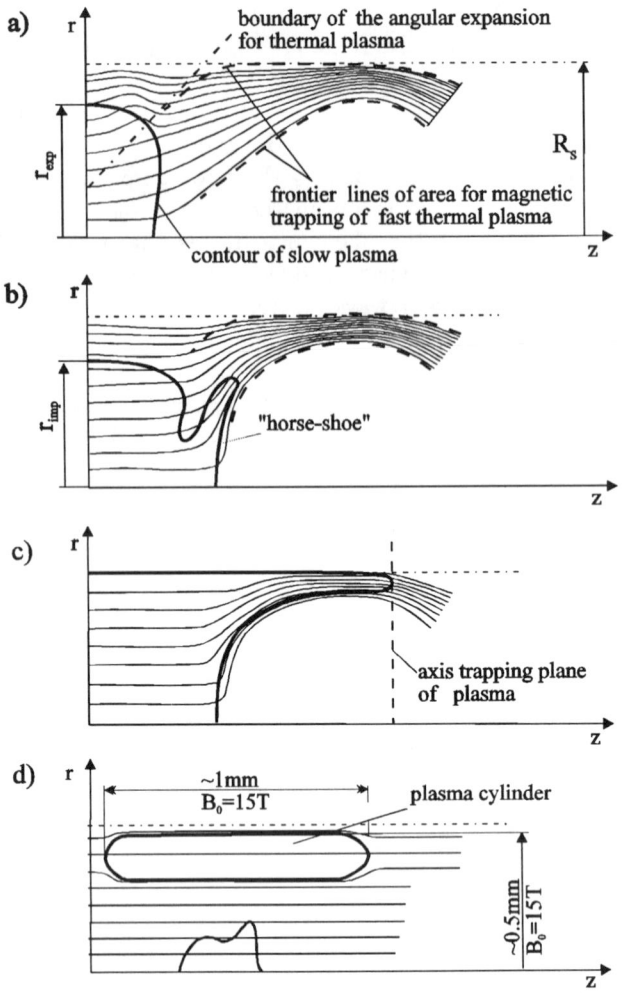

FIGURE 6. The illustration of the plasma - magnetic field interaction for the case of the angular expansion.

"horse-shoe" shape. Because of this flow and the plasma reexpansion close to the target surface (Fig.6c), most of the plasma is dislocated to the off-axis region and a cylinder is created (Fig.6d). Longer lifetime of the plasma cylinder with respect to the dense plasma located on the axis is probably related to higher temperature of this plasma. It is evident because the plasma in the cylinder consists of the hot thermal plasma and the thermalized plasma of the strong implosion stream. It is interesting that the reach of the wings and later the cylinder are unchanging and don't cross the characteristic plane (Fig.6c). This axial plasma limitation results from the plasma magnetic trapping.

5. CONCLUSIONS

The presented results of interferometric studies of laser-produced plasma expansion in an axial magnetic field of induction $B_0 = 5 - 15T$ demonstrate that it is practically impossible to generate a relatively homogeneous stream of such a plasma during the early phase of expansion, which could be the medium of the X-ray laser. However, an interesting matter in this context is creation of a plasma cylinder in the final stage of evolution. The cylinder can be a favourable configuration for propagation and amplification of coherent X-rays. This has been suggested in [13] and other papers.

ACKNOWLEDGEMENTS

The authors wish to acknowledge Dr.L.Karpiński for the arrangement of the magnetic-field generator, Mrs.E.Zielińska, M.Sc., and Mr.P.Parys, M.Sc., for their aid in processing of the measuring results, as well as Mr.J.Makowski and Mr.J.Król for preparation of the laser and technical assistance in the experiment.

REFERENCES

1. Sudo S. et al., *J.Phys. D:Appl. Phys.* **11**, 389(1978)
2. Pisarczyk T. et al., *Laser and Particle Beams* **10**, 767-776(1992)
3. T.Pisarczyk et al. *Physica Scripta* **50**, 72-81 (1994)
4. Bryunetkin B.A. et al., *Sov. J. Quant. ELectron.* **22**, 223(1992)
5. Suckewer S. et al., *J.Quantum Electron.* **19**, 72(1983)
6. Guskov S.Yu., Rozanov V.B., Pisarczyk T., *J.Sov. Laser Research* **14**, 219-222(1994)
7. Zakharov Yu.P., et al., *Proc. XIX ICPIG* **V2**, 276(1989)
8. Antonov V.M. et al., *Proc. 23rd ECLIM, Oxford*, Sep. 19-23, 1994, 167-171.
9. Kasperczuk A. et al., *Proc. "Plasma'93" Conf., Warsaw*, Sep. 29-30, 1993, 211
10. Bruynetkin B.A. et al., *Laser and Particle Beams* **10**, 849(1992)
11. T.Pisarczyk et al., *Laser and Particle Beams* **12**, 549-561(1994)
12. A.Dubik, A.Sarzyński, *Optica Applicata* **17**, 211(1987)
13. H.M.Milchberg et al., *J. Opt. Soc. Am.* **12**, 731-37(1995)

Air Ions and Aerosol Science

Hannes Tammet

Department of Environmental Physics, Tartu University, Tartu, Estonia EE2400

Abstract. Collaboration between Gas Discharge and Plasma Physics, Atmospheric Electricity, and Aerosol Science is a factor of success in the research of air ions. The concept of air ion as of any carrier of electrical current through the air is inherent to Atmospheric Electricity under which a considerable statistical information about the air ion mobility spectrum is collected. A new model of air ion size-mobility correlation has been developed proceeding from Aerosol Science and joining the methods of neighboring research fields. The predicted temperature variation of the mobility disagrees with the commonly used Langevin rule for the reduction of air ion mobilities to the standard conditions. Concurrent errors are too big to be neglected in applications. The critical diameter distinguishing cluster ions and charged aerosol particles has been estimated to be 1.4–1.8 nm.

INTRODUCTION

Air ions are a common research subject of Atmospheric Electricity, Gas Discharge and Plasma Physics, and of Aerosol Science. During the development of the science, the research fields are expanding and folding over each other as shown in Fig. 1.

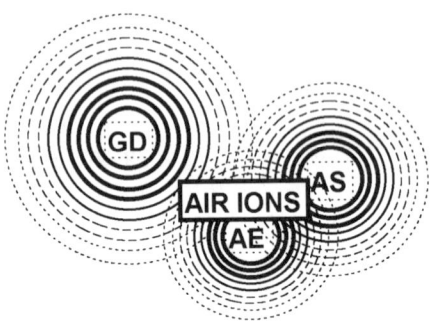

FIGURE 1. An illustration of the expanding research fields and of the position of air ions as a common research subject. GD – Gas Discharge and Plasma Physics, AE – Atmospheric Electricity, AS – Aerosol Science.

The overlapping of GD&AE and AE&AS in Fig. 1 is an old feature and the study of air ions has been a roundabout bridge between GD and AS for many years. A new direct bridge between GD&AS is created by the research of the dusty plasma. However, the bridge is built only from one side and it is almost idle by now.

Separated research fields form their own paradigms (1). Each paradigm consists of specific knowledge about the carriers of electric current in the air. Today, the study of electrical phenomena in natural air is expanding, owing to applications in air quality control and monitoring. Interaction of paradigms is a factor of success in the study of shared subjects. The aim of the present paper is to help the scientists engaged in separated research fields to join their efforts for a better understanding of air ions.

CONCEPT OF AIR ION

When J. J. Thomson started the study of the passage of electricity through gases, the nature of the carriers of electrical current was unknown and the word "ion" was denoting any carrier, the electrons excluded. This original meaning has steadily been preserved in Atmospheric Electricity. A considerable role in the formation of terminology was played by the prestige of Paul Langevin whose paper "Sur les ions de l'atmosphère" (2) is often cited as a first description of charged particles of a mobility of about $1/3000$ $cm^2V^{-1}s^{-1}$. These particles are now called the Langevin ions or large ions. They carry up to one per cent of the electrical current in the atmospheric air. The diameter of a single charged Langevin ion is about 0.09 µm and it consists of about 10^7 atoms. In Aerosol Science, the Langevin ions are called the charged aerosol particles.

Originating from Atmospheric Electricity, the term "air ion" or "atmospheric ion" has been fixed by Atmospheric Electricity and other narrow research fields are not authorized to redefine the meaning of this word combination. There are no gaps in the air ion mobility spectrum in a range of 0.0001–2.5 $cm^2V^{-1}s^{-1}$. The term "air ion" denoting all these particles could be considered an inseparable phrase independent of the term "ion" used in general physical context. The terms "small ion", "intermediate ion" and "large ion" used in Atmospheric Electricity have been derived from "air ion", not from the term "ion" denoting ionized atoms or molecules.

In the general physical context, the small air ions could be called the cluster ions, and the large air ions – the aerosol ions (3). The term "cluster ion" is commonly acknowledged, in spite of the controversy with the usage of the word "ion" in a general physical context. The principle of discrimination between cluster ions and aerosol ions is explained hereinafter.

The lowest size of a particle considered a subject of Aerosol Science has been decreased during recent years. In the last volumes of *Journal of Aerosol Science* one can find the titles "A new electromobility spectrometer for the measurement of aerosol size distributions in the size range from 1 to 1000 nm" (4), "Electric mobility measurements of small ions ..." (5), etc. Today, all kinds of air ions are accepted as research subjects of Aerosol Science.

CHARACTERISTICS OF AIR IONS

The frequently used characteristics of air ions are: mass and electrical mobility in Gas Discharge and Plasma Physics, electrical mobility in Atmospheric Electricity, size and mass in Aerosol Science. Mass, electrical mobility, and size of air ions are closely correlated.

Direct measurement of the mass of a microscopic particle is possible when the particle has been transmitted into the vacuum. The processes of composition and decomposition of clusters, and condensing and evaporating of compounds from particles during the rapid expansion of the air are disturbing the measuring of mass of an air ion as it is in the normal air. Nevertheless, important results in small air ion research are achieved using mass spectrometers that enable an excellent resolving power and the best capacity for identification of chemical compounds. There is a huge amount of clusters of the same mass assembled from various compounds. Reliable information about chemical composition of small air ions in the ground layer of the atmosphere was first obtained by Eisele who introduced a sophisticated technique in case of which the clusters are disassembled in the first section of the instrument (6). After this the masses of molecular fragments are measured and identified.

The electrical mobility is an in situ measurable parameter of an air ion; most of the information available about air ions is expressed in terms of mobility. Measurement of air ion mobilities in the natural air is complicated when compared with the mobility measurements in laboratory experiments. The reason is the low concentration of air ions combined with the low mobility that follows in extremely low values of electrical current of the collected air ions. The instrumental broadening of the mobility lines, on the occasion of tropospheric measurements, is typically exceeding ten per cent. Better resolution is achieved, when the ions artificially created in the natural air are measured to get the information about trace gases. This method is known as plasma chromatography (7).

A simple equation of empirical regression between mass and mobility presented in the handbook (8) is based on the classic data set by Kilpatrick (9, 10) where the measured mobilities and masses of 36 ions are presented over the mass range of 35–2122 u. The original data set (9) is slightly deformed because Kilpatrick has reduced the mobilities to standard conditions using an incorrect procedure. While correcting the data according a new model (11), the numerical values of regression coefficients presented in (8) should be changed, and the improved equation for standard conditions would be:

$$K \approx \left(\sqrt[3]{\frac{1210u}{m}} - 0.21 \right) \text{cm}^2 \text{V}^{-1}\text{s}^{-1}. \qquad (1)$$

While restoring the mobilities measured by Kilpatrick, the mean-square and maximum relative errors of the equation above are 3.3% and 7.8%, respectively. If the smallest ion and biggest ion in the Kilpatrick data set are neglected, the mean square and maximum errors for other 34 ions will be 2.8% and 5.3%.

A real particle has no exactly determined geometric surface; and it can be non-spherical. Size is a parameter for a model of air ion. The concepts of collision size, mobility size, and mass size are discussed in the paper (11). The mass size is preferred as a simple and natural extension to the macroscopic concept of size. The mass diameter of a particle is defined as

$$d_m = \sqrt[3]{\frac{6m}{\pi \rho}}, \qquad (2)$$

where ρ is the density of the particle matter.

The size of large ions can be measured directly using an electron microscope. Usually, the size is a parameter to be determined by an indirect way, e.g. measuring the mobility, and calculating the value with the help of the size-mobility relation.

MOBILITY SPECTRUM OF AIR IONS

A typical air ion is encountering about 10^{12} neutral molecules during its existence in the atmosphere as a cluster ion. Therefore, the trace admixtures of relative concentration down to 10^{-12} can participate in forming the composition and determining the mobility of a cluster. The result is a big variety of cluster ions in the air, and the mobility spectrum consisting of many neighboring lines not resolvable in measurements. The mobility spectrum of cluster ions is ultrasensitive to some trace gases; the mobility spectrometry is accepted as a promising technique for environmental analysis (12).

About 3% of cluster ions in the ground level atmospheric air is neutralized encountering another cluster ion of opposite polarity. Most of the cluster ions are attaching to initially neutral or charged aerosol particles. Some of cluster ions are collecting an unlimited number of ligand molecules, and become aerosol particles. The latter process is called the ion induced nucleation (13), its role in the nature is a hot research problem today.

Mobility spectrum of air ions is formed in the process of all possible ion-molecular reactions, ion-to-particle attachment, ion induced nucleation, particle coagulation and sintering. The elementary processes are well known in the limits of low and high Knudsen numbers (the ratio of the molecule mean free path to the particle size), and satisfactorily known in the domain of medium Knudsen numbers. The models of Aerosol Science for high Knudsen numbers consist of the knowledge that could be useful in the research of dusty plasma.

The processes involving all components of the real air are too complicated to be covered by a universal model. Even the narrow partial models are technically complicated. A model by Luts and Salm (14) has been presented as a huge system of differential equations, it contains 1518 kinds of ion-molecular reactions. The model is satisfactorily describing the early stages of the evolution of cluster ions. A possible application of the model in Gas Discharge Physics is calculation of the current-voltage characteristics of corona discharge, where the evolution of mobility in the discharge gap is a reason of the inconsistency of the simplified models and experimental results. The models simulating the evolution of aerosol systems can be used in order to solve other partial problems. A new model by Kerminen (15) includes the Van der Waals forces and can be applied to the particles with a size down to 1 nm. Unfortunately, the model by Kerminen does not include the effect of free charges.

Today, the most reliable information about the mobility spectrum of air ions can be obtained by measurements. Long term measurements of air ion mobility spectrum in full range of mobilities are running in Tahkuse, Estonia. The published results (3, 16) are describing the statistical distributions, averages, and some exceptional processes. The long-term average mobility distribution of aerosol ions is consistent with the average aerosol particle size spectrum, and the theoretical model of ion-particle attachment (3). The spectrum over high mobility subrange is shown in Fig. 2. The resolving power of the instrument is not high and the data about the fraction of 2.5–3.2 $cm^2V^{-1}s^{-1}$ can be a result of instrumental broadening of the spectrum lines.

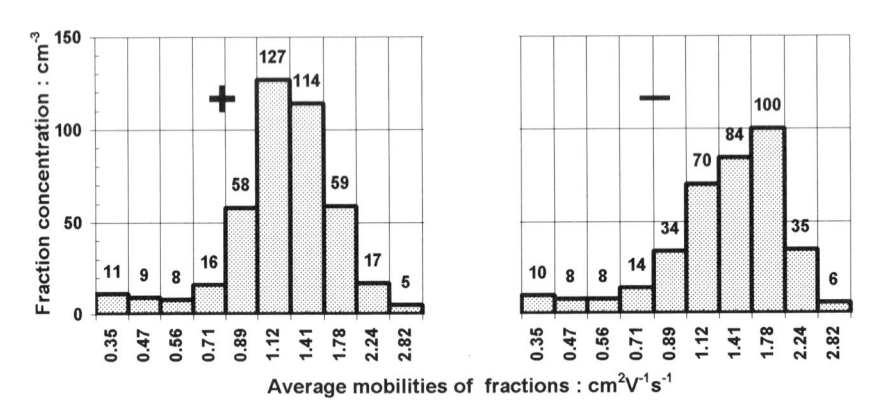

FIGURE 2. The average spectrum of positive and negative air ions with the mobility of 0.32–3.2 $cm^2V^{-1}s^{-1}$ in a rural observatory Tahkuse, Estonia during June–September 1985.

The spectrum has a steady minimum near the mobility of 0.5 cm^{2}V^{-1}s^{-1} for both polarities. Two fraction concentrations are positively correlated in time if they both are below or above the border of 0.5 cm^{2}V^{-1}s^{-1}. Concentrations of fractions separated by this border are statistically independent in the measurement series. It could be concluded, that the air ions with the mobility below 0.5 cm^{2}V^{-1}s^{-1} and above 0.5 cm^{2}V^{-1}s^{-1} have different nature. This observation will be essential when the discrimination between cluster ions and charged aerosol particles is discussed.

A phenomenon of special interest is the sporadic enhancement of air ion concentration in the mobility interval of 0.32–0.5 cm^{2}V^{-1}s^{-1} observed in Tahkuse. In most occasions the concentration of both positive and negative ions was enhanced but in some occasions the concentration was enhanced only for one polarity that can be positive or negative. An example is given in Fig. 3.

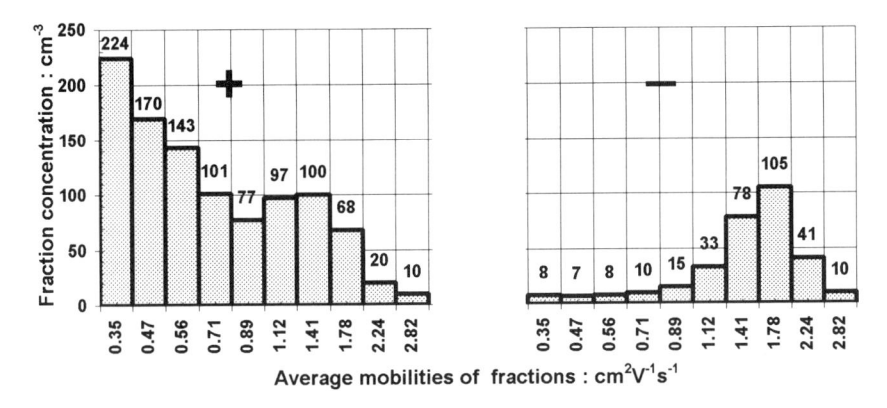

FIGURE 3. Hourly average air ion mobility spectrum in Tahkuse 13 June 1985 at 3 a.m.

The phenomenon above cannot be explained using the standard model of creating the ions with the mobility of 0.32–0.5 cm^{2}V^{-1}s^{-1} by diffusion attachment of cluster ions to initially neutral nanometer particles. A hypothesis was advanced that the ions of this mobility range are intermediate products of ion induced nucleation (17). According to the hypothesis, the enhanced concentration of air ions with the mobility of 0.32–0.5 cm^{2}V^{-1}s^{-1} is indicating the appearance of some trace gases in the air, which are able to condense on positive or negative cluster ions, or on cluster ions of both polarities. Fig. 3 is demonstrating the situation where a trace gas is condensing on positive cluster ions. The origin and chemical composition of such gases in the Tahkuse Observatory has remained unknown.

SIZE-MOBILITY CORRELATION

In the free molecule regime, the velocities of ambient gas molecules are independent of the air ion velocity; the equation of size-mobility correlation is given in the Chapman-Enskog kinetic theory. A complication occurs on the occasion of big clusters. In this case the inelastic interactions between air ions and gas molecules become essential. The drag on macroscopic particles at low Knudsen numbers is given by the Stokes-Cunningham-Knudsen-Weber-Millikan equation called below the Millikan equation for the sake of brevity. The Millikan equation considers the inelastic effect of collisions independent of the particle size and it is not valid in the domain of cluster ions. Ramamurthi and Hopke (18) proposed the first empirical equation fitting the kinetic theory in the free molecule limit, and Millikan equation in macroscopic limit. An advanced model is developed in the paper (11).

While discussing the size of a particle, the model of particle-molecule interaction should include a parameter which could be interpreted as the size. A continuous potential model like the Lennard-Jones model or the Tang-Toennies model does not include a proper parameter. Therefore, some modification of the model of rigid spheres should be used in the discussion.

The basic features of the size-mobility correlation model proposed in (11) are:

◊ The interaction between an air ion and ambient gas molecule is described by the ($\infty - 4$) potential, where the collision distance is written as a sum of three addends: the collision radius of the gas molecule, the mass radius of the air ion and an extra distance that is completing the mass radius to fit the collision radius of the air ion.

◊ The collision radius of the gas molecule is considered to depend on the temperature as in the Chapman-Hainsworth model, and on the energy of the polarization interaction.

◊ The extra distance is regarded as an empirical parameter that should be estimated fitting the model to the experimental data.

◊ The transition from the elastic collisions specific of molecules to the inelastic collisions specific of macroscopic particles is described using the Einstein factor of "melting" of the particle internal energy levels.

◊ The model is written as Millikan equation completed by additional factors describing the transition to the Chapman-Enskog equation in the microscopic limit.

Three parameters should be determined while fitting the model to the experimental data: the density of the ionic matter ρ, the extra distance h and the critical radius of transition from elastic to inelastic collisions r_{cr}. The fitting of the model to the data by Kilpatrick (9) yields following estimates:

$\rho = 2.07$ g cm^{-3}, $h = 0.115$ nm, $r_{cr} = 1.24$ nm.

The model cannot be presented by an explicit function and it is presented for applications by a computational algorithm (11). The size-mobility curve covering the transition region for air ions in standard conditions is presented in Fig. 4.

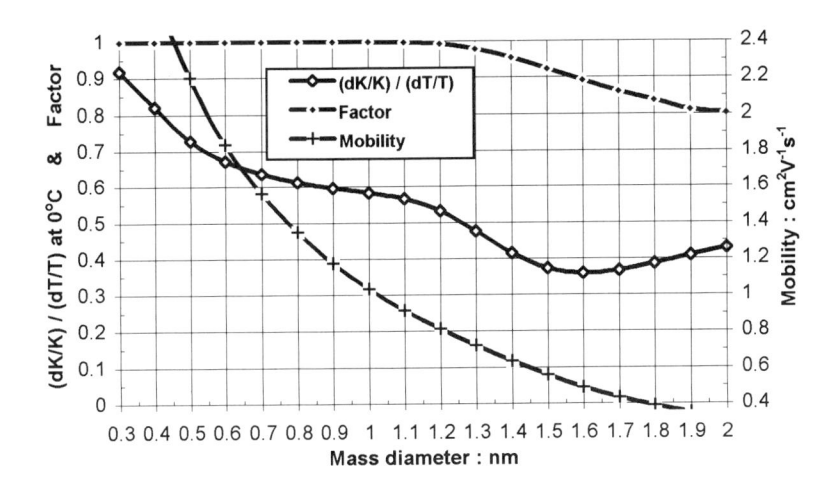

FIGURE 4. Size-mobility correlation, the mobility factor of inelastic collisions, and relative temperature coefficient of the mobility according to the model (11) for air ions in the standard conditions (101325 Pa and 0°C). The mobility factor of inelastic collisions is approaching a value of 0.754 in the macroscopic limit.

REDUCTION OF MOBILITY TO STANDARD CONDITIONS

Traditionally, the mobilities of molecular and cluster ions measured in various experiments are numerically reduced to the standard conditions according to the Langevin rule

$$K_{reduced} = K_{measured} \frac{273.15K}{T} \frac{p}{101325 Pa}, \qquad (3)$$

and the reduced values are presented in the publications. It is well known that the Langevin rule is correct only in the polarization limit and not exact when applied in the case of cluster ions. However, no alternative has been available and the error made while using the Langevin rule has been unknown. Therefore, most of the published data about small air ions with a mobility down to 0.5 $cm^2V^{-1}s^{-1}$ is calculated according to Eq. 3 without any special notice.

The mobility is reduced to the standard conditions according to the model (11) in two steps: at first, the size of an air ion is calculated in the measurement conditions, and then, the standard mobility is calculated according to the size. A measure of the error made by using the Langevin rule is the relative temperature coefficient of the mobility $\frac{dK}{K} \Big/ \frac{dT}{T} = \frac{dK}{dT} \Big/ \frac{T}{K}$. If the Langevin rule were correct, the coefficient would equal one. The size variation of the relative temperature coefficient according to the model (11) is shown in Fig. 4. As it should be, the coefficient is approaching one in the zero size limit. The values of the coefficient for the real air ions are considerably less than one and the error of the Langevin rule is essential even on the occasion of the air ions of the highest mobility. The size variation of the coefficient is peculiar in the size range of transition from elastic to inelastic collisions between air ions and molecules that is between 1.2 and 2 nm.

Langevin rule for the reduction of air ion mobilities can cause substantial errors as demonstrated by the following example. The mass-mobility data by Kilpatrick have been measured at a temperature of 200°C and the reduced values of the mobility have been published in the paper (9). Fitting of these data yields the regression equation similar to Eq. 1 but with numerical values of the coefficients of 850 and 0.3 (8). In the present research the original 200°C data were restored and the mobilities at 0°C were recalculated according to the model (11). The results are essentially different from the published data (9) and the best fit of the mass-mobility regression was achieved at numerical values of the coefficients of 1210 and 0.21. The ratio of air ion masses estimated according to the measured mobilities is 1210/850 which could not to be neglected in applications.

The quantitative results about the size and temperature variation of the cluster ion mobility above are depending on the values of the empirical parameters of the model (11) that are based on the old measurement data by Kilpatrick. The revision of numerical results will be required when improved experimental data become available.

DISTINCTION BETWEEN CLUSTERS AND MACROSCOPIC PARTICLES

According to the long term measurements (3) a border of 0.5 $cm^2V^{-1}s^{-1}$ appears as critical in statistical behavior of air ion fraction concentrations in atmospheric air. The same border is critical in the transition from elastic to inelastic collisions if the particle size is increasing. The distinction between clusters and macroscopic particles in physics is based on fitting of the models of the particle internal electron structure (19). If the orbital electron structure model is suitable, the particle is called the cluster. If the zone model is fitting, the particle is called the macroscopic particle. The transfer from elastic to inelastic collisions is closely related to the internal electron structure of the particle. The transfer curve is presented in Fig. 4. A conclusion is that the air ions of a size less than 1.4 nm or of a mobility greater than 0.6 $cm^2V^{-1}s^{-1}$ could be called the cluster ions, and the air ions of a size greater than 1.8 nm or of a mobility less than 0.4 $cm^2V^{-1}s^{-1}$ could be called the charged aerosol particles or aerosol ions.

ACKNOWLEDGMENTS

This study was supported in part by International Science Foundation Grants LGE000 and LKE100, and by Estonian Science Foundation Grants no. 622 and 1226. The author is grateful to Jaan Salm, Hilja Iher and Urmas Hõrrak for co-operation in air ion measurements, and for discussions.

REFERENCES

1. Kuhn, T. S., *The Structure of Scientific Revolutions*, Chicago, The University of Chicago Press, 1970, 210 pp.
2. Langevin, P., *C. R. Acad. Sci.* **140**, 232–234 (1905).
3. Hõrrak, U., Iher, H., Luts, A., Salm, J., and Tammet, H., *J. Geophys.Res. (D)* **99**, 10697–10700 (1994).
4. Winklmayr, W., Reischl, G. P., Lindner, A. O., Berner, A., *J. Aerosol Sci.* **22**, 289–296 (1991).
5. Thuillard. M., *J. Aerosol Sci.* **26**, 219–225 (1995).
6. Eisele, F. L., *J. Geophys. Res. (D)* **93**, 716–724 (1988).
7. Carr, T. W. (Editor), *Plasma chromatography*, New York, Plenum Press, 1984, 259 pp.
8. *CRC Handbook of Physics and Chemistry, 74th Edition*, Boca Raton, CRC Press, 1993, Section 14, pp. 25–26.
9. Kilpatrick, W. D., "An experimental mass–mobility relation for ions at atmospheric pressure", in *Proc. 19th Annu. Conf. Mass Spectrosc.*, 1971, pp. 320–325.
10. Böhringer, H., Fahey, D. W., Lindinger, W., Hovorka, F., Fehsenfeld, F.C., Albritton, D.L., *Int. J. Mass Spectrom. Ion Processes* **81**, 45–65 (1987).
11. Tammet, H., *J. Aerosol Sci.* **26**, 459–475 (1995).
12. Brokenshire, J. and Pay, N., *International Laboratory* Oct. 1989, 38–41.
13. Mäkelä, J., *Acta Polytechnica Scandinavica, Appl. Phys.* **182**, 115 pp. (1992).
14. Luts, A. and Salm, J. *J. Geophys. Res. (D)* **99**, 10781–10785 (1994).
15. Kerminen, V.-M., *Aerosol Sci. and Tehcnology* **20**, 207–214 (1994).
16. Tammet, H., Iher, H., and Salm, J., *Acta et comm. univ. Tartu.* **947**, 35–49 (1992).
17. Tammet, H., Salm, J., and Iher, H., *Lecture Notes in Physics* **309**, 239–240 (1988).
18. Ramamurthi, M. and Hopke, P. K., *Health Physics* **56**, 189–194 (1989).
19. Petrov, Yu. I., *Clusters and Fine Particles* (in Russian), Moscow, Nauka, 1986, 367 pp.

Molecular Beam Studies of Collisional Autoionization Processes

Brunetto Brunetti,* Stefano Falcinelli,* and Franco Vecchiocattivi[†]

*Dipartimento di Chimica and [†]Istituto per le Tecnologie Chimiche,
Università di Perugia, 06100 Perugia, Italy

Abstract. The general features of collisional autoionization reactions involving metastable rare gas atoms studied in recent years are outlined. Atom-molecule autoionization processes can be seen as the result of two "half-collision" events: the autoionization of the neutral collision complex and the following dynamics of the ionic products. The Ne*+N_2 system is discussed in terms of the vibrational predissociation which occurs in the Ne-N_2^+ product ion. The results for the collisional autoionization of H_2, D_2, and HD by Ne* are presented and discussed from the point of view of the "post-ionization" ion-molecule reaction. Finally, involvements of these processes in some plasma applications are discussed.

INTRODUCTION

Autoionization of collisional complexes are possible in slow molecular collisions because the autoionization time (~10^{-15} s) is shorter than the characteristic molecular collision time at thermal energies (~10^{-12} s). The basic requirement is that the two partners should have enough internal energy to produce an autoionizing collision complex.

A collisional autoionization process can be schematically written as follows:

$$X + Y \rightarrow [X \cdots Y]^*$$

where X and Y are atoms or molecules and $[X \cdots Y]^*$ the collision complex in an autoionizing state, that is degenerate with the ionization continuum,

$$[X \cdots Y]^* \rightarrow [X \cdots Y]^+ + e^-.$$

After an $[X \cdots Y]^+$ ionic complex is formed, the collision continues towards the final ionic products

$$[X \cdots Y]^+ \rightarrow \text{ion products}$$

In the literature this collisional autoionization process is often called "Penning" ionization after the early observation in 1927 by F.M.Penning (1). However other names, such as "chemi-ionization", are also used. This process has long attracted the attention of the scientific community, as shown by the large number of papers and review articles (2,3) on this topic. Some specific features make these processes very interesting from a fundamental point of view. However many applications of collisional autoionization to important fields like radiation chemistry, plasma physics and chemistry, and the development of laser sources, are also possible.

Several experimental techniques are used to study the microscopic dynamics of these collisional autoionization processes. It is well established that the most valuable information about the dynamics of a collisional process is provided by molecular beam scattering experiments. In fact, in these cases it is possible to study single collision events and also to define the translational and internal energy of the two collision partners, thus avoiding statistical averaging. In some systems, molecular orientation or orbital polarization during the collision can also be achieved by using appropriate external fields or laser light. The most detailed studies on collisional autoionization are those where the molecular beam technique is coupled with an appropriate detection method to study product particles such as electrons, ions, neutral atoms or molecules and, in some cases, photons.

The rare gas atoms, excited to their first levels, are very suitable for these experimental studies because of their high energy content and relatively long lifetime which allow them to survive along beam paths in typical molecular beam apparatuses. The metastable rare gas energy values are large enough for an autoionizing complex to be formed in most cases (2). Metastable helium atoms have enough energy to ionize all known atomic and molecular species, except ground state helium and neon atoms, while in the case of metastable neon atoms the exceptions also include fluorine atoms.

In the present paper we briefly report recent experimental studies on the ionization of molecules by collision with metastable rare gas atoms with a particular emphasis to the results from our laboratory in Perugia. Finally we include some considerations regarding as yet unsolved problems and discuss future directions for development in this field.

EXPERIMENTAL APPARATUS

The crossed beam apparatus, used in Perugia for collisional autoionization studies, has been already described (4) and it is schematized in Fig.1. It consists of a vacuum chamber system where two beams, one of metastable rare gas atoms and the other of target molecules, cross each other in a well defined collision volume. The product ions are then mass analyzed and detected.

The rare gas atom beam can be produced by two different sources which can be used alternatively. The first one is a standard effusive beam source at room temperature, the second one is a supersonic source which can be heated at different temperatures. The rare gas beam is passing through an electron bombardment exciter where metastable atoms are produced. The metastable atom velocity can be analyzed by a time-of-flight technique.

The molecule target beam is produced by a room temperature glass multicapillary array, and crosses the metastable beam at right angle. An electric field extracts the ions formed in the beam crossing volume in a direction perpendicular to the plane of the two beams. These ions are then focused, by an Einzel lens system, into a quadrupole filter, mass analyzed and detected by a channel electron multiplier.

Photoionization measurements can be performed by replacing the primary beam source with a microwave discharge in pure rare gas. Such a discharge produces

Rg(I) photons but also metastable atoms. In this case the time-of-flight system is used to separate photoionization from collisional autoionization. In fact, since the signal for the ions produced by photoionization and collisional autoionization are well separated in the time spectra, the discharge can be simultaneously used as a windowless lamp and as a metastable beam source to study photoionization and collisional autoionization in the same experimental conditions.

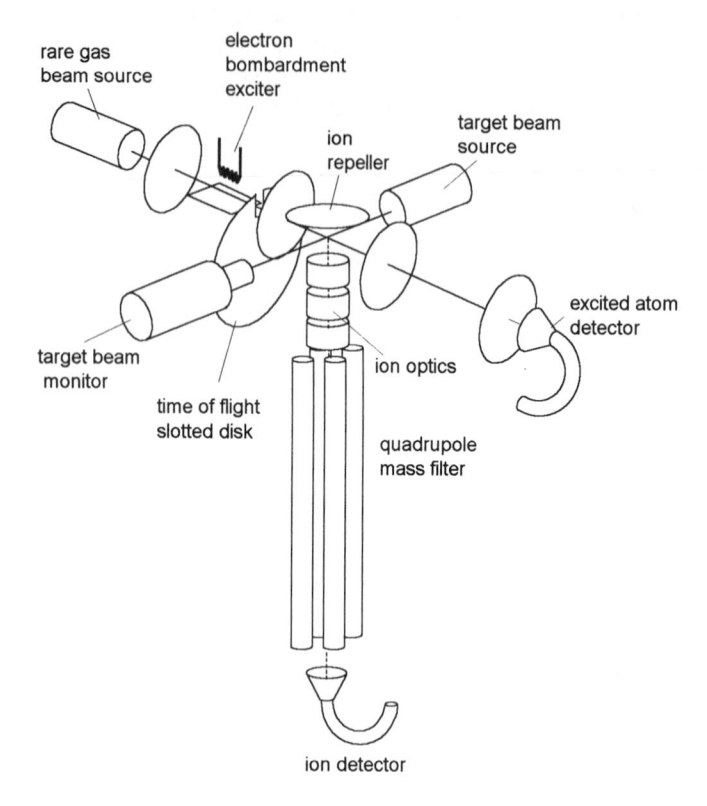

FIGURE 1. A scheme of the molecular beam apparatus used in Perugia for collisional autoionization studies.

COLLISIONAL AUTOIONIZATION DYNAMICS

When the two collision partners are atoms two ionization channels are possible:
$$Rg^* + A \rightarrow Rg + A^+ + e^- \text{ (Penning ionization)}$$
and
$$Rg^* + A \rightarrow RgA^+ + e^- \text{ (associative ionization)}.$$
However, when one collision partner is a molecule, in addition to the two above channels, either
$$Rg^* + AB \rightarrow RgA^+ + B + e^- \text{ (rearrangement ionization)}$$
or
$$Rg^* + AB \rightarrow Rg + A^+ + B + e^- \text{ (dissociative ionization)}$$

are also possible.

The simplest atom-molecule system is obviously the one where only associative and Penning ionization occur. This is the case when a weak interaction is established between the partners before and after the ionization event. Such a situation is schematically illustrated in case (a) of Fig.2, where other possible cases for a hypothetical Rg^*-AB system are also represented. The potential energy surface of the Rg^*-AB system is embedded in the ionization continuum of $RgAB^+ + e^-$. Therefore the collision dynamics evolve on the upper surface until the instant of the autoionization, when, because of electron ejection, the collision continues on the lower surface. In case (a) of the figure, both surfaces are of a non-reactive type, that is, the A-B and $[A-B]^+$ bonds are much stronger than Rg^*-AB and Rg-AB^+. The other two cases in the figure represent two other limiting situations: the ones where the surfaces for the neutral Rg^*-AB or for the ionic Rg-AB^+ systems are reactive. In these limiting situations a chemical reaction is possible, before or after the ionization event respectively.

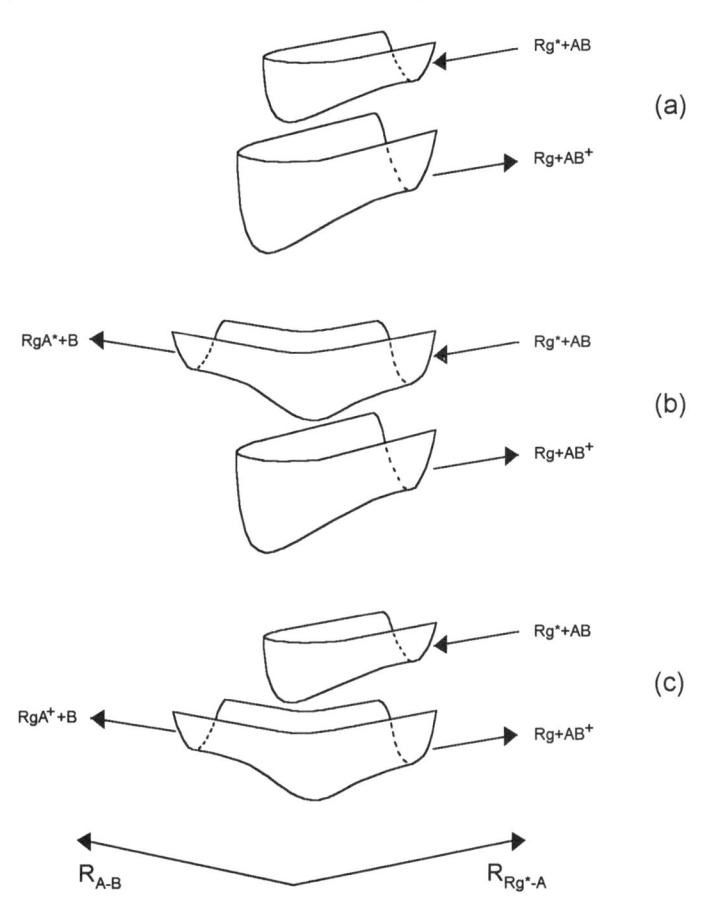

FIGURE 2. Sketch of the potential energy surfaces involved in some possible cases of Rg*-AB ionization systems.

Several studies have been done on the dynamics of the atom-atom collisional autoionization processes and for a recent updating on this topic we refer to a review paper by B.Brunetti and F.Vecchiocattivi (2).

In the interpretation of experimental results it is customary to describe the ionization reaction in two steps: the dynamics of the neutral reactant collision in the presence of a coupling with the ionization continuum and then, the dynamics of the ionic system which continues the collision after the ionization event. For the first step, the optical potential model is usually adopted (2,5,6).

In the following sections, two cases of autoionization by atom-molecules collisions are discussed: the first is the case of Ne*-N$_2$ system, where weak interactions are effective between the collision partners, before and after the ionization event (case (a) in Fig.2); second is the case of Ne*-H$_2$, where a chemical interaction is present between the neutral partners after the electron ejection (case(c) in Fig.2).

THE Ne*-N$_2$ COLLISIONAL AUTO-IONIZATION CASE

In the case of weak interaction systems, characterized by potential energy surfaces such as those schematically represented in Fig.2(a), only Penning and associative ionization are possible. The collisional autoionization of a nitrogen molecule by a metastable rare gas atom is a good example of such a system. Because of its ionization potential, N$_2$ can be ionized by both He* and Ne*. However, while the neon metastable atom can only lead to the ground N$_2^+$(X$^2\Sigma_g$) ion, the helium metastable atom has enough energy to produce N$_2^+$ in the two excited A$^2\Pi_u$ and B$^2\Sigma_u$ states too. The collisional autoionization of nitrogen molecules by metastable helium atoms has long been studied using several experimental techniques (8,9,10).

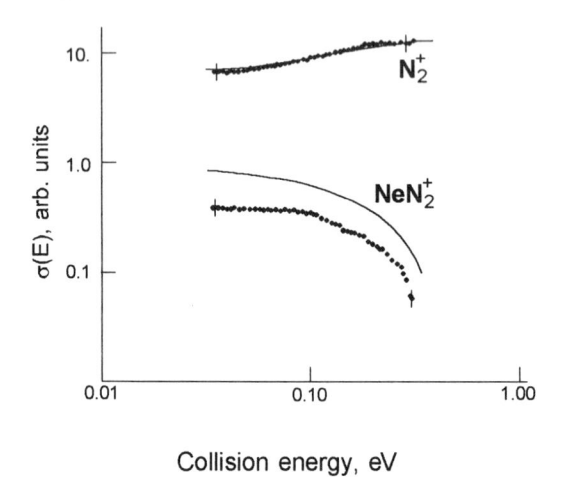

FIGURE 3. Penning and associative ionization cross sections for Ne*-N$_2$ as a function of collision energy.

The ionization of N_2 by collision with metastable neon atoms produces N_2^+ in the ground state only. Total ionization cross sections have recently been measured, as a function of the collision energy, with and without j-selection of Ne^* atoms (11,12,13,14). The Penning and associative ionization cross sections, as a function of the collision energy, was measured by Appolloni et al. (13) who found that the relative yield for the associative NeN_2^+ ion production was much lower than that previously observed for $NeAr^+$ in Ne^*-Ar collisions (15). In Fig.3 the Ne^*-N_2 ionization cross sections of Ref.13 are compared with a theoretical calculation that appears to overestimate the experimental associative ionization. As we discuss below, this is an evidence that a predissociation phenomenon in the associative ionization channel is operative.

Isotropic optical potentials for Ne^*-Ar and Ne^*-N_2 were recently obtained by Baudon et al. (16) through a multiproperty analysis. The real part of the potential for Ne^*-N_2 describes the spherical average of the potential energy surface, while the imaginary part, which is obtained mainly from the analysis of the total ionization cross sections, is an effective function able to simulate the "opacity" of the system. From this work it appears that the differences between the spherical components of the Ne^*-N_2 and Ne^*-Ar interaction potentials are not large enough to justify the different behaviors of these systems in the associative ionization. This difference must be attributed to the higher dimensionality of the atom-molecule system. In fact, the interaction between Ne and N_2^+ is expected to be of a non-chemical character and should be weaker than all the excited vibrational energies of $N_2^+(v \geq 1)$. The PIES spectrum measured by Hotop and coworkers (17,18) shows that the nascent N_2^+ ions are formed in vibrational states up to v=4. Therefore when associative ions NeN_2^+ are formed, those having the internal vibrational N_2^+ state with $v \geq 1$ have enough energy to pre-dissociate in a time typically in the range of 10^{-10} s. The mass spectrometric study by Appolloni et al. (13) detected the ions about 10^{-5} s after the ionization event and therefore only after their pre-dissociation. In other words, this process can be schematically written as follows: a primary ionization event,

$$Ne^*(^3P) + N_2 \rightarrow [Ne \cdots N_2]^+ + e^-,$$

is followed by Penning ionization,

$$[Ne \cdots N_2]^+ \rightarrow Ne + N_2^+,$$

or associative ionization,

$$[Ne \cdots N_2]^+ \rightarrow NeN_2^+.$$

However when the internal N-N vibration in NeN_2^+ has $v \geq 1$ the ion predissociates before its detection

$$NeN_2^+(v \geq 1) \rightarrow Ne + N_2^+.$$

This predissociation can also explain the general observation that associative ionization is a very minor channel in most atom-molecule collisional autoionization systems (13,19).

Sonnenfroh and Leone (20) measured the rovibrational states of $N_2^+(X^2\Sigma_g)$ produced in an Ne^*-N_2 crossed beam experiment, by using the laser-induced-fluorescence technique. The rotational states of N_2^+ observed by these authors showed a bimodal distribution, which could indicate that more than one mechanism is responsible for the N_2^+ ion production in Ne^*+N_2 thermal energy collisions.

This information is consistent with and supports the previous observations, giving deeper insight into the details of the ionization dynamics in this system.

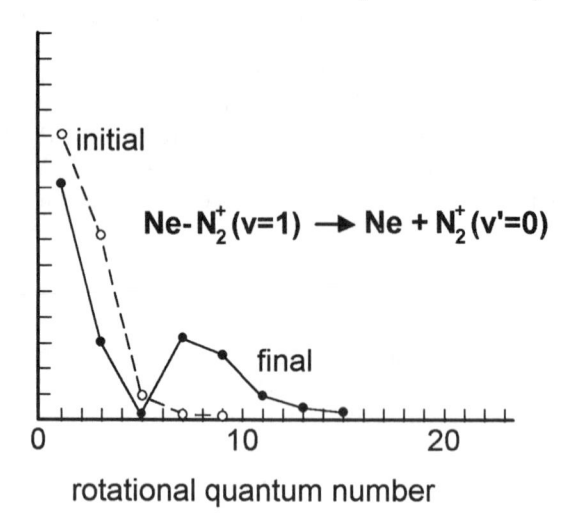

FIGURE 4. Rotational distribution of N_2^+ ion as obtained in a theoretical calculation on the Ne-N_2^+ ion vibrational predissociation.

Recently we have started a theoretical study, in collaboration with G.Delgado-Barrio and P.Villareal, of the vibrational predissociation of the Ne-N_2^+ ion. Preliminary results clearly indicate that this process proceeds producing N_2^+ ions characterized by a bimodal rotational distribution, as indicated in the Fig.4.

THE Ne*-H$_2$ COLLISIONAL AUTOIONIZATION CASE

The ionic complex formed by collisional autoionization of a molecule can give rise to a number of chemical changes which depend on the complexity of the molecule and also on the electronic energy of the metastable atom. In the case of the ionization of a diatomic molecule AB, one has the primary ionization step,

$$Rg^* + AB \rightarrow [Rg \cdots AB]^+ + e^-,$$

which can be followed by simple Penning and associative processes,

$$[Rg \cdots AB]^+ \rightarrow Rg + AB^+$$
$$\rightarrow RgAB^+,$$

by ion decomposition,

$$[Rg \cdots AB]^+ \rightarrow Rg + A + B^+,$$

or rearrangement,

$$[Rg \cdots AB]^+ \rightarrow RgA^+ + B.$$

The primary ionization step has the dynamical role of establishing a distribution of translational energy, angular momenta, and internal states of the $[Rg \cdots AB]^+$ ionic complex, which are crucial for the dynamical evolution of the collision complex after the electron ejection.

He*-H$_2$ is the simplest case of a system belonging to case (c) schematized in Fig.2. This system is of great interest for the prospect it holds out for a direct comparison between experiment and theory: new experimental studies and more detailed theoretical schemes are continuously being proposed (21,22,23).

The collisional autoionization of H$_2$ by metastable neon atoms is expected to present many similarities with that by He*. However this system has not been studied as intensively especially because comparison with theoretical results appears to be more difficult. The possible ion products are H$_2^+$, NeH$^+$, and NeH$_2^+$, which have been determined to be in a 78:20:2 ratio at a collision energy around 0.045 eV (19). A PIES and mass spectrometric study was performed by Hotop and coworkers (24). In this study the authors determined the relative vibrational population of the nascent H$_2^+$ ion for j selected metastable Ne*(^{3}P$_j$) atoms. They also compared such spectra with the one they obtained with the laser excited Ne*(^{3}D$_3$) atoms, finding a substantially different behavior, which can be attributed to the different interaction between the two partners in the incoming channel. Differential elastic and rotationally inelastic cross sections were measured by Baudon and coworkers (25) in a crossed beam experiment. They succeeded in distinguishing elastically and inelastically scattered particles by using the time-of-flight technique. Tuffin et al. (26) measured the kinetic energy distribution of the product ions in Ne*-H$_2$ collisions.

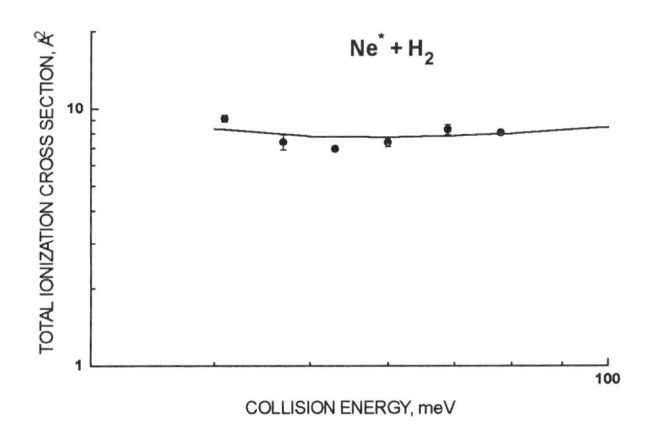

FIGURE 5. Total ionization cross section in Ne*-H$_2$ system, as a function of the collision energy.

Recently in our laboratory we have done a mass spectrometric study of collisional autoionization for Ne*-H$_2$, D$_2$ and HD systems. The Figg.5 and 6 show the total ionization cross section and the branching ratios respectively for Ne*-H$_2$, D$_2$ and HD cases. In these measurements we covered the collision energy range (0.04-0.10eV) by using a supersonic neon beam source at different temperatures. Looking at Ne*-H$_2$ and D$_2$ ionic branching ratios, the most interesting feature to note is the difference in the production of NeH$^+$ and NeD$^+$ rearranged ions for the two investigated systems.

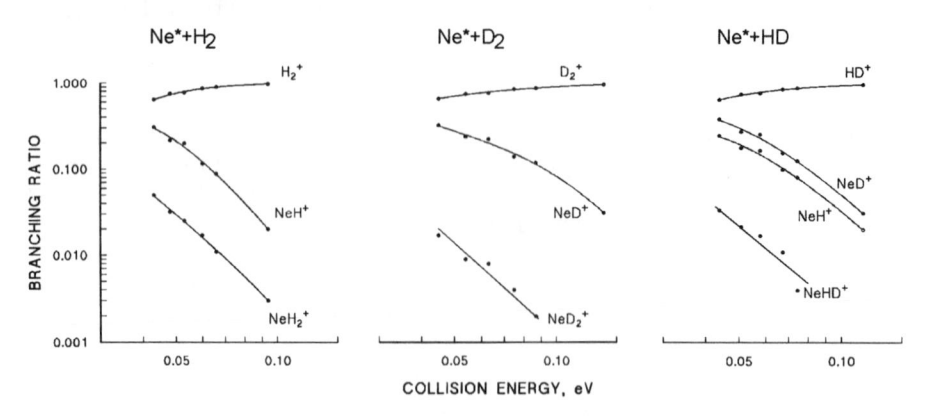

FIGURE 6. Branching ratios for the possible product ions in the Ne*-H2, D2, and HD collisional autoionization.

Trying to evidence a possible isotopic effect we measured the rearranged ions branching ratio by following the autoionization dynamics of the Ne*-HD system. In fact, this system allows to study the isotopic effect on the rearrangement channel, toward NeH^+ and NeD^+ ions, starting from the same initial dynamical conditions. The NeH^+/NeD^+ branching ratio, as obtained from our experiment, clearly shows a different behavior on the dynamical evolution of $[Ne\cdots HD]^+$ and $[Ne\cdots DH]^+$ intermediate collision complexes (Fig.6).

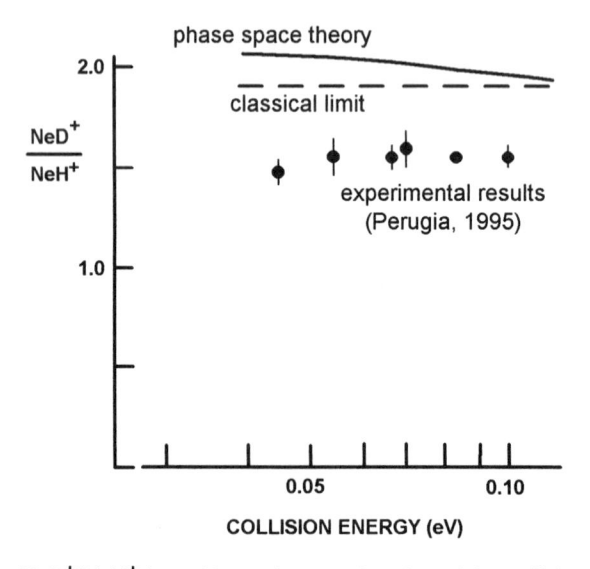

FIGURE 7. The NeD^+/NeH^+ branching ratio, as a function of the collision energy, in the Ne*-HD collisional autoionization. The experimental results are compared with Phase-Space-Theory and classical statistics expectations.

In Fig.7 the NeD$^+$/NeH$^+$ branching ratio in the Ne*-HD collisional autoionization, as a function of the collision energy, are compared with the results obtained from Phase-Space-Theory and classical statistics estimates. The experimental results appear rather lower than the statistical expectation, indicating a non statistical behavior of the [NeH$_2$]$^+$ dynamics in the thermal energy range.

APPLICATIONS TO PLASMA SYSTEMS AND CONCLUDING REMARKS

In the last decade the studies of collisional autoionization by metastable rare gas atoms have increased considerably both qualitatively and quantitatively. New experiments are giving further insight into the microscopic dynamics of these processes, and allow a systematic study of a large number of different systems where the collision partner is an atom and where it is a molecule.

For atom-atom systems the studies with j-selected Rg* atoms have clarified several aspects of the microscopic ionization mechanism revealing the role played during the reaction by the symmetry of the atomic orbitals involved. This has stimulated theoretical effort (27) in the direction of clarifying the potential energy functions describing the interactions between the collision partners as well as in the direction of a rigorous quantitative description of the ionization dynamics.

In spite of this effort, several open problems still exist. For example, in the Ne*-heavier rare gas atom systems the associative ionization cross sections, in the limit of low collision energies, appear to be in disagreement with theoretical predictions (15,28,29). This discrepancy could be attributed to a failure of the optical model theory commonly used to compute these cross sections or to the inadequacy of the potential functions used in the calculation, especially those for the final ions.

The study of atom-molecule collisional autoionization systems appears to be more difficult, but the last few years have seen considerable advances due to the application of powerful experimental techniques: ion-electron coincidence experiments have allowed different energy states produced during the ionization to be correlated with the final ionic products, while state resolved ionization cross sections and PIES spectra as a function of the collision energy have allowed the role of molecular orbital symmetry in ionization probability to be understood.

For collisional autoionization in atom-molecule systems, the theory is currently unable to provide a unified dynamical treatment as found in the atom-atom case. However recent progress in chemical reaction dynamics theory can be applied here with appropriate adjustment for the ionization occurring during the collision. The "Surface Trajectory Leaking" treatment used for the He*-H$_2$ system (23,30,31) could be a promising development for the studies of the dynamics of collisional autoionization involving molecules. The statistical treatment used by Siska and coworkers (32) for the He-H$_2^+$ dynamics, which follows the He*-H$_2$ collisional autoionization, also provides an interesting example of a dynamical study of these post-ionization reactions, as well as simpler dynamical theoretical models which have also been used for He*-H$_2$ (33) and, more recently, for Ne*-HCl (34).

An other important research field where collisional autoionization processes are relevant is the plasma physics and chemistry. Studying the microscopic dynamical evolution of these processes we are able to clarify the kinetic codes of the many reactions that happen in plasmas for a number of applications (35). For example amorphous carbon and diamond thin films suitable for mechanical and electronic applications can be prepared by plasma-enhanced chemical vapor deposition using hydrocarbon gases (36). Although different methods are presently used, most make use of low pressure discharges which involve CH_4 and H_2 as starting materials.

Since the discovery of this technique, many researchers have speculated on the nature of the growing mechanism, the common goal of such studies being the determination of optimal operating conditions to enhance both the deposition rate and the structural properties of the diamond films. For example, the dilution effect of rare gases has been investigated by using different experimental techniques (37,38). It was found that the introduction of rare gases changes the morphology of the diamond film and leads to fewer crystalline defects. The average size of the diamond crystals increases with increasing concentration of noble gases (4). The dilution effect of rare gases is believed to be related to their chemistry, i.e. rare gases induce additional ion-molecule and excited atom-molecule reactions (2), which can significantly affect the plasma composition. Although many reactions involving $Ar^+(^2P_J)$, $Ar^*(^3P_{2,0})$, $Ne^+(^2P_J)$ and $Ne^*(^3P_{2,0})$ with methane have been previously studied, some processes still remain insufficiently investigated, as in the case of collisional autoionization $Ne^*(^3P_{2,0})$-CH_4. In particular, branching ratios for the possible CH_4^+, CH_3^+, CH_2^+ product ions have never been determined as a function of the collision energy.

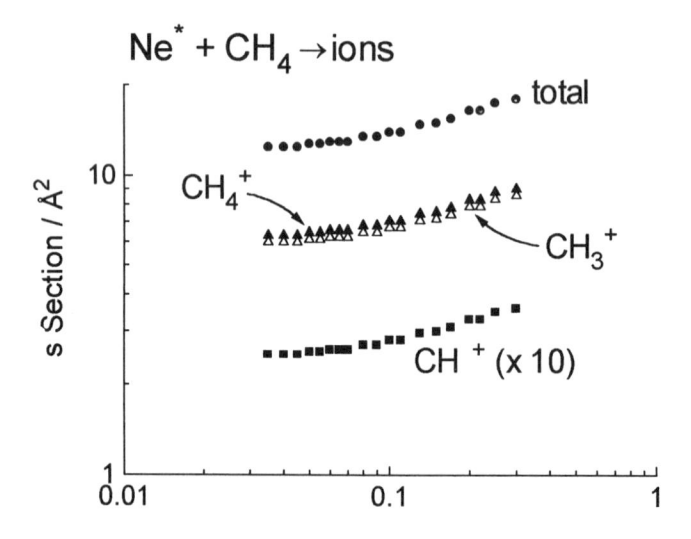

FIGURE 8. The branching ratios of product ions in $Ne^*(^3P_{2,0})$-CH_4 collisional autoionization

In a recent work performed in Perugia and Trento laboratories (4) branching ratios, cross sections and rate coefficients for Ar^+-CH_4 and Ne^+-CH_4 charge-exchange, and Ne^*-CH_4 collisional autoionization, have been measured as a function of collision energy. The branching ratios in Ne^*-CH_4 collisional autoionization are reported in Fig.8. The rate coefficients as a function of the temperature which can be derived from such data may be useful as input data for the further development of computational models, which could contribute to the clarification of the very complicated mechanism responsible for the growth of diamond.

It must be considered that the primary function of the plasma is to dissociate methane, thus providing CH_x radicals which are the natural precursors of carbon films. Actually, molecular dissociation can be achieved by different processes as electron impact or reactions involving ionic species, neutral radicals and metastable atoms produced in the plasma. Further studies are therefore necessary to clarify the relative contribution of different processes for the production of relevant CH_x radicals. In particular, computational code can be powerful tools to investigate complex kinetics (39), providing that accurate branching ratios and rate coefficients are used as input parameters.

ACKNOWLEDGMENTS

The authors wish to thank P.Candori for help given in the experiments. This work has been partially supported by EC through the SRMI European Network (HCMP).

REFERENCES

1. Penning, F.M., *Naturwissenschaften*, **15**, 818 (1927); *Z. Phys.*, **46**, 335 (1928); *Z. Phys.* **57**, 723 (1929); *Proc. Roy. Acad. Sci.* Amsterdam, **32**, 341 (1929); *Z. Phys.* **72**, 338 (1931).
2. Brunetti, B., and Vecchiocattivi, F., in *Cluster Ions,* Ng, C.Y., ed., Wiley&Sons (1993), p.359-445 and references therein.
3. Siska, P.E., *Rev. Mod. Phys.*, **65**, 337 (1993) and references therein.
4. Brunetti, B., Vecchiocattivi, F., Bassi, D., and Tosi, P., *Int. J. Mass Spectr. Ion Proc.* (1995) in press.
5. Nakamura, H., *J. Chem. Phys. Japan* **26**, 1973 (1969); **31**, 574 (1971).
6. Miller, W.H., *J. Chem. Phys.* **52**, 3563 (1970).
7. Bethe, H.A., *Phys. Rev.* **57**, 1125 (1940).
8. Tuffin, F., Le Nadan, A., and Peresse, J., *J. Phys. (Paris)* **46**, 181 (1985).
9. Ohno, K., Takami, T., and Mitsuke, M., *J. Chem. Phys.* **94**, 2675 (1990).
10. Dunlavy, D.C., Martin, D.W., and Siska, P.E., *J. Chem. Phys.* **93**, 5347 (1990); Dunlavy, D.C., and Siska, P.E., private communication.
11. Verheijen, M.J., and Beijerinck, H.C.W., *Chem. Phys.* **102**, 255 (1986).
12. van den Berg, F.T.N., Schonenberg, J.H.M., and Beijerinck, H.C.W., *Chem. Phys.* **115**, 359 (1987).
13. Appolloni, L., Brunetti, B., Vecchiocattivi, F., and Volpi, G.G., *J. Phys. Chem.* **92**, 918 (1988).
14. Aguilar, A., Brunetti, B., Gonzalez, M., and Vecchiocattivi, F., *Chem. Phys.* **145**, 211 (1990).

15. Aguilar, A., Brunetti, B., Rosi, S., Vecchiocattivi, F., and Volpi, G.G., *J. Chem. Phys.* **82**, 773 (1985).
16. Baudon, J., Feron, P., Miniatura, C., Perales, F., Reinhardt, J., Robert, J., Haberland, H., Brunetti, B., and Vecchiocattivi, F., *J. Chem. Phys.* **95**, 1801 (1991).
17. Hotop, H., private communication; Lorenzen, J., Zastrow, A., Ruf, M.W., and Hotop, H., to be published.
18. West, W.P., Cook, T.B., Dunning, F.B., Rundel, R.D., and Stebbings, R.F., *J. Chem. Phys.* **63**, 1237 (1975).
19. Hotop, H., in *Electronic and Atomic Collisions*, Oda, N., and Takayanagi, K., eds., North Holland, Amsterdam 1980, p.217.
20. Sonnenfroh, D.M., and Leone, S.R., *Int. J. Mass Spectr. Ion Processes* **80**, 63 (1987).
21. Martin, D.W., and Siska, P.E., *J. Chem. Phys.* **82**, 2630 (1985); **89**, 240 (1988).
22. Isaacson, A.D., Hickman, A.P., and Miller, W.H., *J. Chem. Phys.* **66**, 370 (1977).
23. Preston, R.K., and Cohen, J.S., *J. Chem. Phys.* **65**, 1589 (1976).
24. Bussert, W., Ganz, J., Hotop, H., Ruf, M.W., Siegel, A., Waibel, H., Botschwina, P., and Lorenzen, J., *Chem. Phys. Lett.* **95**, 277 (1983).
25. Bocvarski, V., Robert, J., Colomb de Daunant, I., Reinhardt, J., and Baudon, J., *J. Phys. (Paris) Lett.* **46**, L13 (1985).
26. Tuffin, F., Le Coz, G., and Le Nadan, A., *J. Phys. (Paris)* **48**, 1291 (1987).
27. Morgner, H., *J. Phys. B* **18**, 251 (1985).
28. Weiser, C., Siska, P.E., *J. Chem. Phys.* **85**, 4746 (1986).
29. Brunetti, B., Vecchiocattivi, F., and Volpi, G.G., *J. Chem. Phys.* **84**, 536 (1986).
30. Vojtik, J., and Paidarova, I., in *The Dynamics of Systems with Chemical Reaction*, Popieleski, J., ed., World Scientific, Singapore (1989), p.241.
31. Vojtik, J., and Paidarova, I., *Chem. Phys.* **157**, 67 (1991).
32. Martin, D.W., Weiser, C., Sperlein, R.F., Bernfeld, D.L., and Siska, P.E., *J. Chem. Phys.* **90**, 1564 (1989).
33. Martin, D.W., Bernfeld, D.L., and Siska, P.E., *Chem. Phys. Lett.* **110**, 298 (1984).
34. Aguilar, A., Brunetti, B., Falcinelli, S., Gonzalez, M., and Vecchiocattivi, F., *J. Chem. Phys.* **96**, 433 (1992).
35. Aquilanti, V., Brunetti, B., Vecchiocattivi, F., Letardi, T., Fang, F., and Fu, S., *Chem. Phys. Lett.* **205**, 229 (1993).
36. Angus, J.C., and Hayman, C.C., *Science* **241**, 913 (1988).
37. Yoder, M.N., *Diamond Relat. Mater.* **2**, 59 (1993).
38. Zhu, W., Inspector, A., Badzian, A.R., McKenna, T., and Messier, R., *J. Appl. Phys.* **68**, 1489 (1990).
39. Sekiya, H., Hirayama, T., and Nishimura, Y., *Chem. Phys. Lett.* **138**, 597 (1987).

THE PROPAGATION OF POSITIVE STREAMERS IN AIR AND AT AIR/INSULATOR SURFACES

N.L. Allen

Department of Electronic and Electrical Engineering

The University of Leeds

LEEDS, U.K.

ABSTRACT

The propagation of streamers in air, and the precision with which their properties can be measured, are discussed. The effects of an insulating surface on the growth of avalanches is reviewed; it is evident that electron emission by photon bombardment of the surface is a significant mechanism. The development of corona on surfaces is discussed. Recent experiments on streamer growth, in the absence of significant space charge, are described. It is concluded, from all the evidence, that each material has an individual, characteristic, effect on the growth of avalanches and streamers.

1. INTRODUCTION

1.1. Streamers in air

It is well-known that breakdown in air at atmospheric pressure requires an electric field of the order 3×10^6 Vm^{-1} when the field between the electrodes is uniform. However when, as is more often the case, breakdown occurs in a highly non-uniform field, it can do so when the average electric field, crudely taken as the ratio of voltage to electrode spacing, is of the order 5×10^5 Vm^{-1} or less. Indeed, in an electrode geometry such as the positive point-plane, the applied field near the plane may be less than 2×10^5 Vm^{-1}. Both of these values are far less than the field required for ionization to occur, which is of the order 2.5×10^6 Vm^{-1}.

In these cases, a discharge can propagate across a non-uniform field gap as the result of the formation of a "positive" streamer when the more highly stressed electrode is positive. The streamer is able to propagate itself in a relatively low field, since the formative avalanches develop in the localised high field of a space charge "head", which has originated from the growth to a critical size of an avalanche near the highly stressed electrode. Continuation of the process, by repeated formation of critical avalanches in the positive space charge set up by preceding ones, requires only that the applied field shall supply sufficient energy to maintain the ionization process in the avalanches. It does not require that the applied field shall be sufficient also to produce impact ionization. A model of this process has been given by Gallimberti [1].

The applied field needed for streamer propagation which, as will be shown, is less than 5×10^5 Vm^{-1}, has assumed some technical importance in recent years, since it

© 1996 American Institute of Physics

determines the sparkover voltage of non-uniform field gaps encountered in practice under positive lightning (1.2/50 μs) impulse and direct voltages. Under both conditions, streamers cross to the lightly stressed cathode whereupon an intensification of ionization caused by interaction with the electrode establishes a highly conducting channel leading to breakdown. It has proved useful, in engineering practice, to adopt the "streamer gradient" (at a round figure of 5×10^5 Vm^{-1}) as a reference against which the sparkover behaviour of electrode gaps of various geometries can be measured [2]. The variation of streamer gradient with atmospheric density and humidity has become well established[3] and it is now adopted by the International Electrotechnics Commission for the normalisation of sparkover voltages to those which would be obtained at a standard atmospheric condition, namely a pressure of 1013 millibar, temperature of 293K and absolute humidity of 11 g of moisture per m^3 of air [4].

Recent work has shown that streamers in air have properties which can be measured quite precisely, particularly when they are studied in a uniform electric field [5] [6] [7]. Thus, the minimum field that is required for propagation can be measured to an uncertainty of ± 1 per cent and their speed of propagation can be measured to ± 2 per cent. Both of these properties depend upon the initial energy of the streamer, determined by the field and voltage applied at the point of initiation. The properties that have been measured are generally consistent with the model that the streamer develops from an avalanche head containing about 10^8 ions, a value that changes only slowly when conditions such as air temperature and density are changed.

The knowledge of basic properties of streamers in air makes them a useful tool for the study of the interaction of discharges with insulating surfaces, an attribute that is of practical importance, since the weakest feature of most insulating systems is the propensity to breakdown over insulator surfaces. Moreover measurement of properties such as propagation fields and speeds permits an assessment of a material to be made as a prospective insulator.

In what follows, the discussion will be confined to positive streamers, directed from a highly stressed anode towards a cathode. Most published work has been carried out on positive streamers, since they require a lower electric gradient for propagation than negative streamers; they are therefore associated with lower breakdown voltages than is the negative case. For this reason, they are technologically more "dangerous" than negative streamers, which have been very little studied.

1.2 Avalanches on insulating surfaces

Streamer propagation depends upon the successful development of avalanches and the proximity of a surface may affect ionization growth in several ways. Thus, photo-electron emission from the material, photo-detachment of electrons from negative ions on the surface, and detachment of negative ions followed by collisional detachment of electrons are all plausible processes which could enhance the ionization process in air. Competing with these is possible attachment of electrons and ions to the surface as the result of diffusion from the avalanche. All of these processes can be expected to depend upon the nature of the material and will be discussed after a description of relevant published work.

Jaksts and Cross [8] measured avalanche currents in pure nitrogen in a plane parallel electrode gap and then repeated the experiments with polyethylene and P.T.F.E. spacers. Careful measurements were made of the current generated by the "primary" avalanche initiated by a narrow UV laser pulse, followed by a sequence of subsequent "secondary" current pulses due to avalanches initiated by photons originating in the gas near the anode and liberating photo-electrons at the cathode. These were clearly defined with nitrogen alone, but an additional "subsidiary" avalanche was detected shortly after the primary one when the insulator was inserted (Figure 1). This was ascribed to photo-electron emission from the insulator surface, initiated by photons from the primary avalanche, generated by an excited state having a lifetime of the order of 20 µs; this was of the same order of delay as that between the primary and subsidiary avalanches.

There was evidence presented that a subsidiary avalanche also occurred in pure nitrogen; this was plausible since the same excited states could be involved, but the increase in the subsidiary avalanche current due to the spacer was not large. Results with polyethylene and P.T.F.E. were similar, suggesting a common excitation process in the gas and also similar photoemission coefficients under UV in the two materials. Unfortunately, values of reduced electric field E/p (where p = pressure of neutral gas) were not given, so that full comparison with other work is difficult. However, the rise of current in the primary avalanche was not significantly affected by the presence of the insulating surfaces, which suggests that emission by photons had no significant effect at this stage of avalanche growth.

In an interesting contribution, Verhaart et al [9] also made measurements of the temporal growth of avalanches along a P.T.F.E. surface which had been either positively or negatively charged by means of an auxiliary corona. In this case, a laser beam was directed at the insulator surface itself, near the cathode, so releasing a burst of electrons by photo-emission. An estimate of the contribution to ionization due to photo-electron emission from the surface during subsequent avalanche development was made by introducing a coefficient δ, denoting the number of photo-electrons produced per centimetre length from the molecules excited by one collision. It could thus be taken into account with the Townsend coefficient α and attachment coefficient η to find an avalanche number:

$$N = \exp(\alpha - \eta + \delta) x \quad (1)$$

Computations of the growth of current in an avalanche, adapting equation (1) and using published values of α and η, were compared with observed currents, enabling estimates of δ to be made.

With carbon dioxide as the ambient gas, the best fit between measurement and computed currents was obtained with $\delta = 0$ for the case of both polarities of surface charge. When repeated in nitrogen, a value of $\delta = 0.25$ cm^{-1} (E/p = 41.9 Vcm^{-1} torr^{-1}) was obtained with negative surface charge and $\delta = 0$ with the positively charged surface. A further experiment in SF_6 with an uncharged surface showed values of δ around 0.7 (E/p = ~ 120 V cm^{-1} torr^{-1}) for good agreement between computation and experiment. A test with a negatively charged surface in SF_6 gave a value of δ of 4.0 cm^{-1} for the same value of E/p.

In contrast with the work of reference [8], it was not possible to compare the current measurements with those in the absence of the insulator, since the initial electrons were released at the insulator surface rather than the cathode. The computed current for $\delta \doteq 0$ may be considered equivalent to the "gas only" case. Thus, in certain cases, there was no apparent photo-emission from the surface, while in other cases there was. Also, in the work of Verhaart et al, it was implicit that photo-emission would occur with no significant delay during avalanche development, whereas in that of Jaksts and Cross, evidence of a delay of the order of the lifetime of the excited nitrogen atom was present. Verhaart et al found no evidence of secondary avalanches.

While accepting these reservations, it is possible to consider that the evidence obtained with uncharged P.T.F.E. in SF_6 points to a photo-electron emission effect. Likewise, the evidence of an absence of energetic photons of wavelength less than 190 nm. in the emission spectrum of CO_2 also explains the absence of photo-emission in this case. In the case of nitrogen, it must be assumed that positive charging of the surface would result in combination with photo-electrons as they are produced and that the negative charging, the density of which was not recorded, was insufficient to retard the emission of photoelectrons.

The alternative of photo-detachment of electrons from negative ions on the surface appears improbable. Thus, the cross-section for this process is of the order $10^{-21} m^2$ [10] and, assuming the number of photons produced in an avalanche to be of the same order as the number of ions [11], a maximum value of 10^8 is taken, so that the number of detachments induced by a single avalanche cannot exceed $10^{-21} \times 10^8 n_s$, where n_s is the surface density of ions. To replicate an avalanche, one photoelectron would be required, so that the minimum value of n_s would be $10^{13} m^{-2}$. This very high density is clearly unrealistic since it would produce an associated normal electric field of the order $10^5 Vm^{-1}$. Thus, the likely density os surface negative ions is too low to contribute to a photo detachment mechanism.

It appears likely, therefore, that the large value of δ obtained with negatively charged P.T.F.E. may be comprised of true photo-emission of electrons aided by other effects such as detachment of the electrons from the surface ions by ion bombardment from the avalanche and the neutralisation by the negative ions of any electric field resulting from positive charging of the material after emission of photo-electrons. The physical processes expressed by the coefficient δ may thus be complex. The inability to deduce any significant value of δ where the ambient gas was CO_2 suggests, however, that ion bombardment had little effect in this case.

2. STREAMERS ON INSULATING SURFACES

The more complex case of a streamer corona, propagated under impulse voltage over a plane insulator surface has been studied by Gallimberti et al [12]. Here, current pulses and charge injected into a rod-plane electrode gap spanned by the plane insulator surface were studied. Insulators used were PVC and glass in atmospheric air, and in this case, comparative measurements were made in air alone.

The electric field, between rod and plane, was highly non-uniform, but measurements of current against time showed that the peak current was higher, with propagation along the PVC surface, than in air (Figure 2) and was very much higher than either, with glass. The width of the current peak was, however less, with PVC, than with air and very much less with glass. On balance, the total charge injected with the insulators was much reduced compared with that in air.

Interpretation of the results was, as in the previous cases, made in terms of the effects of the surface on avalanche growth, in particular, of photo-electron emission and electron and ion attachment. Here again, it was assumed that photons reach the surface from the avalanche with no significant delay time.

The structure of the surface corona was found to be more complex than in air; this was attributed to photo-emission providing more electrons for subsequent ionization, so that the degree of branching of the corona was greater, leading to the observed higher currents. A competing process was electron attachment at the surface, so reducing the decay time of the corona current compared with that in air.

The authors note that the discharge propagated more rapidly, and farther, on PVC than in air. It is necessary to ask, however, whether these facts, accompanied by higher currents are the result of more efficient ionization over the surface, due to photo-emission (in spite of more rapid attachment) or whether the higher current is the result partly of a greater speed of propagation and larger displacement current. No quantitative data on the speed of propagation was presented, but calculations from a theoretical model indicated that both the effective ionization and attachment coefficients were increased by the presence of the PVC.

It is clear, from the published work, that there is significant evidence to suggest that electron emission from the surface by photons from the avalanches in the gas tends to increase the efficiency of ionization close to an insulator. The effect of attachment of electrons and ions to the surface may be apparent only when streamers are studied, and may be too slow a process to be significant in modifying avalanche growth.

3. SURFACE STREAMER CHARACTERISTICS

3.1. Minimum field for streamer propagation

While the foregoing measurements on avalanche and streamer corona growth have provided valuable information on the effects of photo-emission and possible attachment, there is a need for work on the intrinsic properties of streamers as influenced by an insulating surface. For example corona growth, where extensive branching occurs, results in the development of a space charge field which cannot easily be quantified, so obscuring properties that are required to be known for possible modelling.

In an effort to address this problem, measurements have been made in a uniform field arrangement similar to that used in the study of streamer propagation in air [13]. Two quantities in particular have been studied, namely the minimum field required for streamer propagation and the speed of propagation.

The apparatus used is shown in Figure 3. Streamers are propagated from a point set in a shaped receptacle in a ground plane; a positive voltage pulse 200 ns in duration is applied to the point. They traverse the gap towards an upper plane at a steady negative potential. Their progress is monitored by photomultipliers having narrow vertical fields of view directed at the point of origin and at the cathode.

In order to reduce space-charge-effects to the minimum practicable level, the propagation field was measured when the voltage at the point was reduced to the lowest value at which streamer initiation could be detected by the lower photomultiplier; this voltage was about 2.5 kV. The electric field in the gap was then reduced to a value at which streamer traverse did not occur and then raised again in small steps, using 20 trials at each voltage level, until traverse occurred with a probability of 1. A distribution was obtained and, in Figure 4, it is seen that this differed between different materials; the result for air is regarded as "reference". Thus, the insulator material affects the minimum conditions required for streamer propagation.

3.2. Speed of propagation

For the measurement of streamer propagation speed, it is inconvenient to work at this "minimum" condition, since the fact that it is a threshold leads to difficulties in working. Therefore, a pulsed voltage of 4.25 kV has been used. Here, the minimum propagation field is 400 kVm^{-1} ± 5 kVm^{-1} and this is used as a basis for comparison. It was shown previously [7] that with this pulsed voltage, space-charge effects due to branching are negligible and that the reduction of propagating field from 430 kVm^{-1} resulting from the increased pulse voltage, was due to the increased energy of the streamers initiated.

Examples are shown, in Figure 5, of oscillograms obtained in this apparatus with streamers in air and when propagated along the surface of P.T.F.E. In air, the speed of propagation is determined by measurement of the time delay between the start of the rise of detected light at the lower photomultiplier and the corresponding point on the upper trace. With P.T.F.E., two peaks of light are detected at the cathode. The delay to the first peak, for a given applied field, is shorter than in the case of air, indicating a higher propagation speed. However, tests have shown that the second peak corresponds approximately to the propagation speed in air. Thus, the two peaks appear to be due to "surface" and "air" components of the streamer system, which has divided shortly after its initiation. Figure shows the propagation speeds associated with the first and second peaks as a function of the electric field between the planes, calculated from the ratio of voltage to separation; a comparison with the speed in air alone is given. In all cases, the speed is a linear function of electric field over the range studied.

3.3. Variations with material

Measurements of the minimum propagation field show significant dependence upon the material. Figure shows results obtained with several materials, where a comparison is made with the result for air.

Direct photographs, taken with a camera sensitive in the ultra-violet, show evidence of the two components of the streamer system with P.T.F.E., Figure . However, where glazed porcelain is substituted, light is recorded only for an air component; there is no evidence of a component along the surface after the first 2 to 3 centimetres of traverse. This effect is confirmed by the photomultiplier oscillograms which show only one peak of light at the cathode with a time delay corresponding to a propagation speed equal to that in air alone. A similar result is obtained with an unglazed porcelain.

A survey of effects with several materials has shown that silicone rubber, epoxy resin with aluminium filler, glass reinforced plastic (fibreglass), and PVC all give clear evidence of "surface" and "air" components, though the relative amplitudes of the light peaks at the cathode vary with material. The speed of the surface component also varies with material, as shown in Figure 7.

The increase in propagation speed of the surface streamers, compared with that in air alone, is further evidence of the importance of electron photo-emission from the surface. Such emission effectively increases the ionization coefficient, so reducing the distance x (equation (1)) required to achieve the critical avalanche size required for streamer initiation. As a result, the time required for avalanche growth is reduced so that streamer development occurs more rapidly.

4. CONCLUSIONS

When avalanches and streamers are propagated close to an insulating surface, evidence has been obtained that:

(i) the properties of the discharges depend on the nature of the material;

(ii) photon stimulated emission of electrons occurs from insulator surfaces, with a yield depending upon the material and the ambient gas;

(iii) the decay of corona current may be associated with differing rates of attachment of electrons or ions to the material surface;

(iv) the speed of propagation of streamers over a surface depends on the material; it is ascribed to differences in the rate at which avalanches grow.

In order to pursue further the effects of surfaces on streamers, systematic measurements of current and charge need to be correlated with measurements of surface charge deposition and examination of surface properties.

5. ACKNOWLEDGEMENT

The work on streamer propagation has been supported by Grants from the Science and Engineering Research Council of the U.K. and the National Grid Company. The author acknowledges the contribution of his colleague, Dr. A. Ghaffar, in this work.

6. REFERENCES

1. GALLIMBERTI I, J. Physics D: Appl. Phys. 5, 2179-2189 (1972).

2. ALLEN N.L, FONSECA J, GELDENHUYS H.J, and ZHENG J.C. Electra No. 134, p.63-90 (1991).

3. ALLEN N.L. and BOUTLENDJ M. IEE Proc. A138, 37-43, (1991).

4. IEC 60-1 (1989) High voltage test techniques.

5. PHELPS C.T. and GRIFFITHS R.F., J. Appl. Phys. 47, 2929-2934.

6. TANG T.M. "A study of positive corona streamers up to breakdown in quasi-uniform fields". PhD Thesis, UMIST, Manchester, U.K. (1982).

7. ALLEN N.L. and GHAFFAR A (a) J. Phys.D:Appl. Phys.28 331-337 (1995).
(b) J.Phys.D: Appl.Phys. 28, 338-343, (1995).

8. JAKST A. and CROSS J. Can.Elec.Eng.J. 6, 14-20, 1981.

9. VERHAART H.F.A, TOM J, VERHAGE A.J.L and VOS C.S. "Avalanches near solid insulators" Proc. 5th Int.Symp. on High Voltage Engineering, Braunschweig Paper 31.01, 1987.

10. BOYLETT F.D.A. and WILLIAMS B.G. Brit. J. Appl.Phys.18, 593-595, (1967).

11. RAETHER H., "Electron avalanches and breakdown in gases" Butterworth, London, Chap.3. (1964).

12. GALLIMBERTI I, MARCHESI G and NIEMEYER L, "Streamer corona at an insulator surface". Proc. 7th Int.Symp. on High Voltage Engineering, Dresden, Paper 41.10, (1991).

13. GHAFFAR A "The effects of temperature on streamer propagation in air and along insulator surfaces". Ph.D. Thesis, University of Leeds, U.K. (1994).

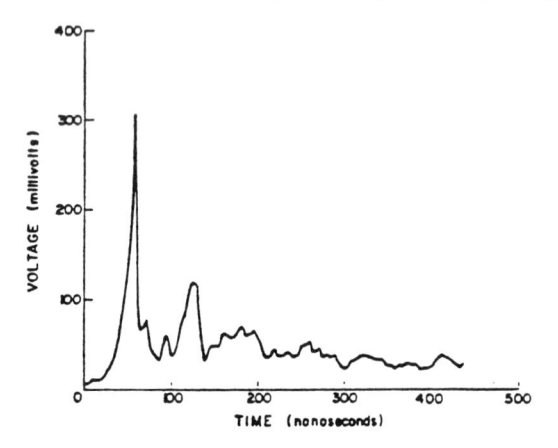

Figure 1. Avalanche electron current in a uniform field bridged by a P.T.F.E. spacer. Subsidiary avalanche at "A". [8].

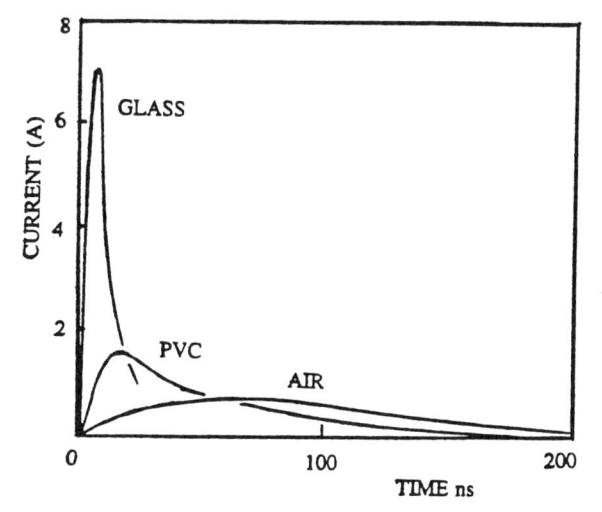

Figure 2.

Current pulses injected into rod-plane gap bridged by PVC and glass, compared with the case of the air gap [9].

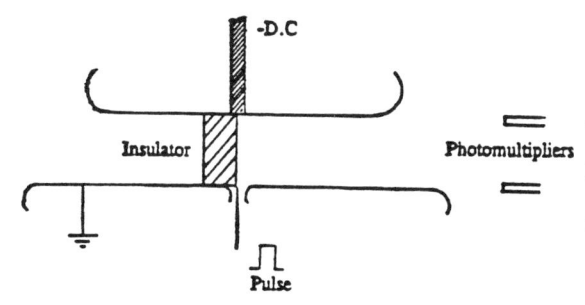

Figure 3.

Apparatus for study of streamer propagation in a quasi-uniform field.

Figure 4.

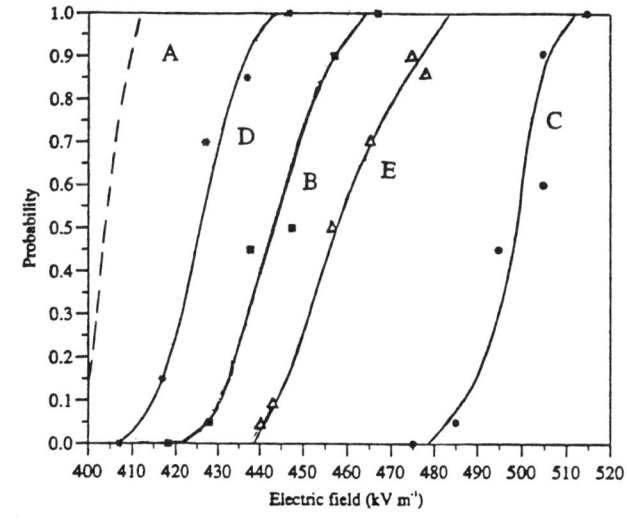

Probability of streamer propagation as a function of field near threshold.

A Air   B P.T.F.E.
C Fibreglass
D Glazed porcelain
E Silicone rubber.

Figure 5.

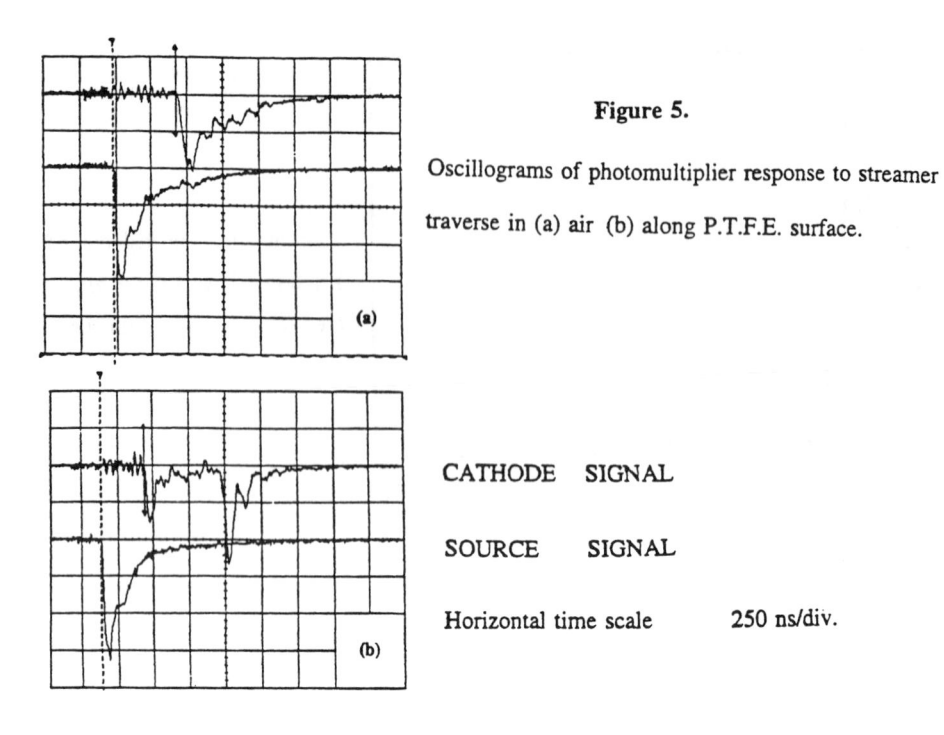

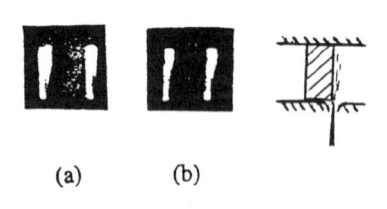

Oscillograms of photomultiplier response to streamer traverse in (a) air (b) along P.T.F.E. surface.

CATHODE SIGNAL

SOURCE SIGNAL

Horizontal time scale   250 ns/div.

Figure 6.

Photographs of streamer systems along surfaces of (a) P.T.F.E. (b) glazed porcelain.

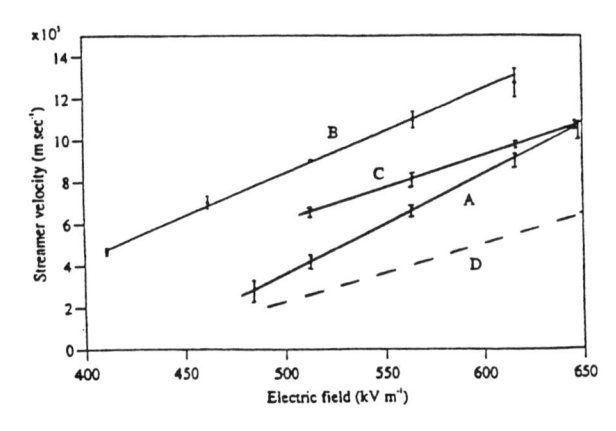

Figure 7.

Speed of propagation of surface component of different materials.

A: P.T.F.E.
B: Silicone rubber
C: Fibreglass
D: Air (reference).

Microwave Discharges used as Excitation Sources for Spectro-Chemical Analysis

Antonio Gamero

Department of Applied Physics
Faculty of Science. University of Cordoba
E-14071 Cordoba, Spain

Abstract. This paper presents the merits of two different microwave discharges as excitation sources for atomic emission spectroscopy. The first part shows the excellent analytical results obtained by using a plasma produced by surfatron in the determination of halogens, even using low power and argon as plasma gas. The last part of the paper presents the first results of a helium plasma produced by a new microwave plasma torch (TIA design). From them, this plasma seems to be a good candidate to extend the application range of microwave plasmas used as excitation sources.

INTRODUCTION

At present, atomic spectrometry is the most common method of elemental analysis used in analytical laboratories. During the last several decades atomic emission, atomic absorption, atomic fluorescence and mass spectrometries have been widely used with a large number of different atom generating devices (flames, arcs, electrothermal atomizers and plasmas). From the seventies, the development of plasma sources as atomization and excitation devices caused plasma based atomic emission and mass spectrometry to become the most extensively used tools to perform elemental analysis. The use of various plasma sources for atomic emission spectroscopy has grown immensely in recent years because of the wealth of emission lines produced. So, although not possessing the inherent selectivity of atomic absorption spectrometry, the use of a suitable high-resolution monochromator reduces the possibility of spectral interference and enables inter-element selectivity. In addition, this wealth of such emission lines also permits multi-element detection.

Real samples can be present in any physical form (solid, liquid or gas) and can be introduced into the plasma directly in their natural state or transformed previously by some physical or chemical process. Not depending on the liquid or solid form of the sample, the first step involves the transfer of the sample in a gas phase, which can be carried out by an external device or by the plasma itself. In this situation, the next step will be the atomization of the sample so that the analyte will be converted into free atoms on which the analytical spectrometries can be performed. In all cases, the plasma source acts as an energetic medium that simultaneously produces ionization of neutral species and excitation of neutral and ionized species.

Schematically, we can assume that the different reactions occurring in the plasma source pertain to three principal types: a) atomization/atomic recombination, b) excitation/de-excitation of the neutral atoms and ions, and c) ionization/electron recombination. In these reactions there is an energy transfer between the plasma and the sample and, under equilibrium conditions, the corresponding temperature characterizes the available energy in the plasma for the reactions. Thus, the gas temperature will be associated with the desolvation, volatilization and atomization processes, while the excitation and ionization temperatures will characterize the excitation and ionization processes respectively. Simultaneously with other considerations involved in a given analytical problem, the fundamental properties of the different plasma sources must be discussed in order to evaluate their relative merits in spectrochemical analysis.

MICROWAVE PLASMA SOURCES

Historically, among different plasma sources the inductively coupled plasma (ICP) has been the most extensively used in analytical spectrometry and is now the method of choice for a wide variety of elemental analyses (1). This is a radio frequency induced plasma produced mainly at atmospheric pressure in argon or argon mixed with another gas. In any case, the analyst should not neglect the merits of other plasma sources which are continuously being developed. Particularly, the microwave induced plasmas (MIPs) possess some advantages as excitation sources in atomic emission spectroscopy (AES). An important reason for this advantage of the MIPs is the efficient excitation of the halogens and other non-metals not readily accessible to ICP detection with adequate sensitivity. Argon or helium microwave induced plasmas at atmospheric pressure can be obtained by operating at powers less than 200 W (e.g. ICP operates at 0.7-5 KW). Hence, they are safe to use and the power systems are simple and less expensive than those of the other plasmas. Moreover, this low power requirement of the MIP produces both low background emission and minimum need for cooling and the support gas flow rate can be lower than 1 l/min.

The most usual microwave plasma sources used in spectrochemical analysis are produced in resonant cavities (2,3), but more recently they are also being produced by surface wave launching devices (surfatron (4) and Ro-Box(5)). From the comparative study of the analytical results existing in the bibliography, there are not any dramatic analytical differences among the various microwave plasma sources (6). For similar conditions, these possible differences depend more on the sample introduction system than on the microwave plasma type itself.

As an additional advantage, stable and easy to operate surface-wave plasmas can be produced in a wide range of operating conditions (microwave power and frequency, gas pressure, discharge tube diameter). In fact, surface-wave plasmas (SWPs) permit a good control of the discharge characteristics which facilitates the search for better conditions of the plasma sources. Besner et al. (7) have studied the effect of the frequency in low power helium surface wave plasma produced in the range from 200 to 2450 MHz. They have reported that there is little or no influence on the fundamental properties of the plasmas (electron density and excitation and gas temperatures), the major difference being the plasma geometry. In the same way, no important changes in the emission intensity of several non-metal lines have also been reported as a function of the frequency (8).

However, from another point of view, the MIPs have a relatively low gas temperature which limits the capacity of such plasmas to volatilize solid or liquid samples, or atomize the analyte species. This fact, together with most of the MIPs (typically resonant cavity) are easily perturbed and even extinguished by small changes in impedance, causes fundamental problems with sample introduction and has limited the general implementation of MIPs. Many of these problems can be avoided if the sample is presented in the form of gas. So, a wide variety of sample introduction techniques has been developed for this purpose, including ultrasonic nebulizers, electrothermal vaporization and chemical vapour generation (recent reviews in (1) and (9)). Any way, up to date the most extensive use of MIPs has been in combination with gas chromatography.

HALOGEN DETERMINATION BY SURFATRON-GENERATED ARGON PLASMA

The excitation processes in a helium microwave plasma lead to strong elemental emission lines, making the sensitive and selective detection of both metal and non-metal elements possible. However, it is generally accepted that argon plasma provides insufficient energy for the sensitive detection of non-metals (9). Although improvement with respect to signal-to-background ratio and power of detection in some instances can be obtained with helium plasma sources, this paper shows that excellent performances in non-metal detection are also possible using argon microwave plasmas.

The argon plasma is produced by surfatron at atmospheric pressure, within a capillary fused silica tube (1 mm inner diameter), which protrudes 1 cm from the surfatron gap. Emission spectroscopy techniques have been used for

TABLE 1. Argon plasma generated by Surfatron, gas flow-rate 0.5 l/min.

Power (W)	n_e (m^{-3})	T_{exc} (K)	T_{gas} (K)
50	0.6×10^{21}	3600	900
100	1.0×10^{21}	6300	1300
150	1.1×10^{21}	6700	1700

determining different plasma parameters (10): The electron density, n_e, by means of the Stark broadening of the H_β line; the excitation temperature, T_{exc}, via the Boltzmann-plot method for higher levels of the atomic energy diagram; and the gas temperature, T_{gas}, from the rotational temperature obtained with the Q_1-branch of the OH (0-0) band. The spectroscopic measurements are axially performed by using a simple silica lens to produce a 1:1 image of the flame at the entrance slit of the monochromator. A computer connected to the system permitted us to obtain the emission line intensities in digital form.

These results show that, for this argon plasma, the levels upper than 4p are in partial LTE and the Saha-Boltzmann equilibrium is established (11). The values obtained for different microwave powers and 0.5 l/min. of gas flow-rate are given in table 1. The low volatilization and atomization capabilities of these plasmas can be seen from the low gas temperatures measured. However, the values of electron density and excitation temperature are high enough to produce an efficient excitation of most of the analytes.

Particularly, this section describes some results obtained by our group relative to the detection capability of the argon plasma produced at low power by surfatron for halogen determination based on two different chemical vapour generation methods.

Continuous Vapour Generation of Br or I

The bromide (or iodide) samples are introduced into the plasma as gaseous bromine (or iodine) chemically generated in a continuous mode by using a gas-liquid separation device (figure 1). The sample solution (or blank solution), the oxidant solution (hypochlorite for bromide or hydrogen peroxide for iodide) and the diluted sulfuric acid solution are pumped via a three channel peristaltic pump. The three streams merge into a "T-piece", where oxidation and thus the bromine (or iodine) generation take place. The resulting aqueous solution goes into the gas-liquid separator, where a mild stream of argon carries the volatile bromine (or iodine) out of the aqueous phase, which is drained to waste via the "U" tube. In order to eliminate residual water load, the vapour is desiccated by passing it through concentrated sulfuric acid before entering the plasma.

Among the several very intensive emission lines in the UV-VIS region, the atomic lines at 827.24 and 206.16 nm for bromine and iodine respectively

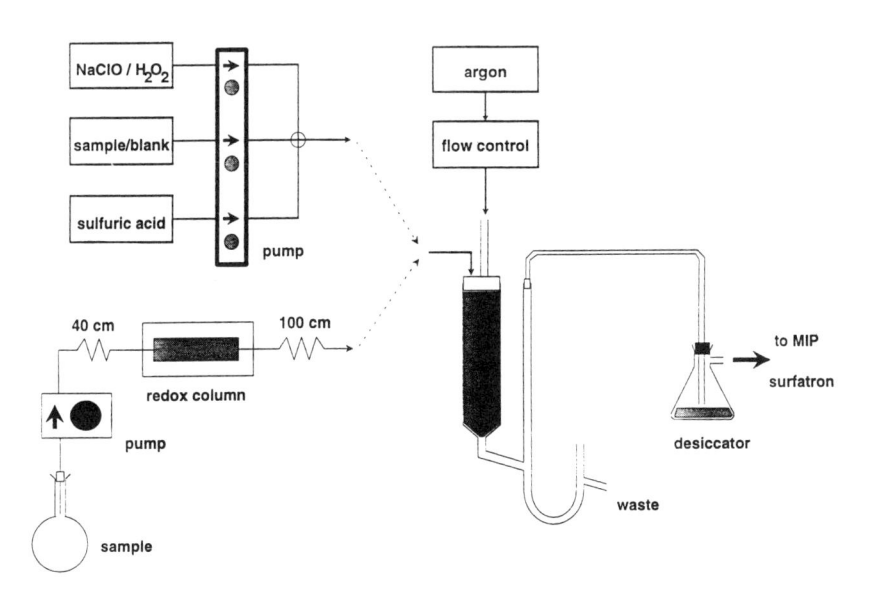

FIGURE 1. Sample introduction systems used for both vapour generations.

were finally selected because they produced the most intense emission and highest signal to the background ratio. The chemical parameters and the experimental plasma conditions affecting the atomic emission have been investigated in order to optimize the analytical performance of the entire system (12,13). When the microwave power is varied from 25 to 200 W at fixed values of the other parameters, the emission intensity turned out to be constant above 150 W. In this way, the microwave power of 100-150 W is enough for the analytical determination, thus avoiding a higher heating of the system and a more pronounced background level.

In addition to other interesting analytical features, table 2 shows the obtained detection limits (calculated as 3 times the standard deviation of the background of the blank). As can be seen from the comparation with other recent

TABLE 2. Comparison among different analytical determinations of Halogens.

	Sample introduction System	Plasma Source	Detection Limits ($\mu g/l$) Cl	Br	I
(Ref. 15)	Chemical Generation	He-TM$_{010}$ Cavity	120	1.8	-
(Ref. 16)	Ultrasonic Nebulizer	He-TM$_{010}$ Cavity	400	3000	800
(Ref. 17)	Ultrasonic Nebulizer	He-Surfatron	120	230	60
(1th method)	Chemical Generation	Ar-Surfatron	-	2	20
(2nd method)	Chemical Generation	Ar-Surfatron	50	50	200

results existing in the literature, the low power argon plasma generated by surfatron is an excellent plasma excitation source in atomic emission spectroscopy for this halogen determination.

Simultaneous Vapour Generation of Cl, Br and I

The solution of chloride, bromide and iodide is continuously introduced into the oxiding column (lead dioxide powder), and chlorine, bromine and iodine are instantaneously generated and subsequently isolated by the gas-liquid separator (figure 1). Again, a mild stream of argon carries the volatile species out and this gaseous phase is desiccated by bubbling through concentrated sulfuric acid before its insertion into the discharge (14). The atomic emission lines of chlorine and bromine selected are 837.60 and 827.24 nm respectively, of the most intense lines usually reported in the literature. For iodine, however, the line at 804.37 nm is chosen even though it is not of the most intense lines generally used, in order to accomplish the simultaneous determination of the three halogens (phototube range 300-900 nm).

We have investigated the effect of introducing samples into the plasma on the excitation temperature, by measuring it with the line pair intensity ratio method. The plasma is scarcely affected by the sample introduction (up to 100 mg/l of Cl). The excitation temperature tends slightly to decrease with the increase in the chlorine concentration but within the 10% estimated experimental error.

The signal/background ratio for Cl and Br increases with the increasing microwave power up to 100 W, this increase being smaller for iodine. Powers higher than 100 W result in significantly increased background emission, so 100 W is chosen as the optimum condition of the microwave power in routine work. The results obtained for the detection limits are shown in table 2 and support the analytical potential of the microwave plasma sources in AES for the determination of halogens, even using low microwave power and argon as plasma gas.

MICROWAVE PLASMA TORCH: TIA DESIGN

To extend the application range of the microwave induced plasmas used as excitation sources for atomic emission spectroscopy, higher power could be used, since the detection of atomic species requires both high temperatures and electron density. Higher power densities in the plasma should lead to improve its atomization, excitation and ionization capabilities. Producing a high power discharge at atmospheric pressure within a dielectric tube can create problems due to the erosion of the tube and the contamination of the carrier gas. To avoid

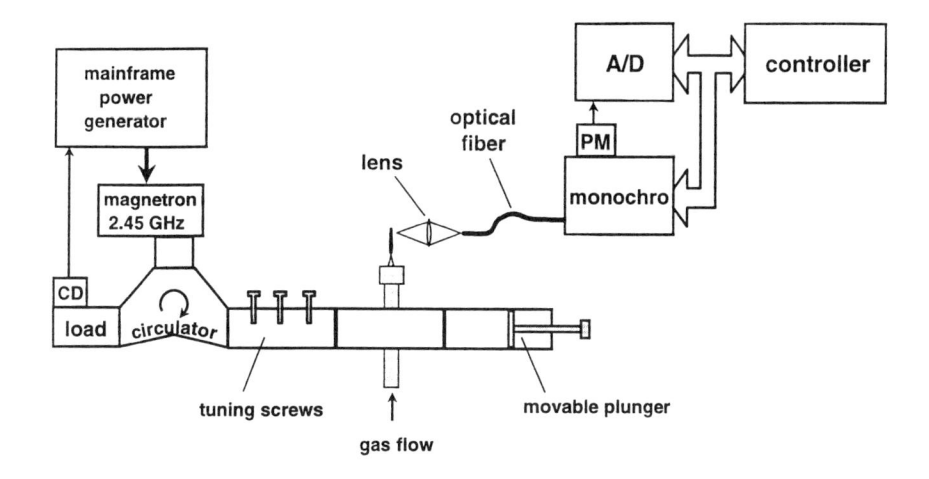

FIGURE 2. Experimental set-up used with the waveguide sustained plasma torch (TIA).

these problems, a plasma torch configuration can be used, where the discharge is excited in 'open air' at the tip of the field applicator (18,19).

This section presents several first results of an atmospheric pressure plasma produced by a new waveguide-based structure, a microwave plasma torch with axial gas injection (referred as TIA, from 'torche à injection axiale') (20). In this structure, the microwave power, at 2.45 GHz, is supplied via a rectangular waveguide and it comprises both waveguide and coaxial elements serving the purpose of wavemode conversion and impedance-matching. Using two tuning elements (waveguide and coaxial plungers) the TIA provides efficient power transfer to the plasma, typically higher than 95% of the incident power. The discharge gas flows inside the inner conductor of the coaxial line section and exits through a nozzle at its conical tip (fig 2). In this manner, the plasma is sustained at this tip, which protrudes slightly beyond the end of the outer conductor. In our case, the diameter of the nozzle is 1.5 mm and no additional cooling is necessary for incident powers up to 900 W, and a helium flow-rate as low as 3.5 l/min.

The helium plasma produced by TIA is highly stable and reproducible and reaches lengths between 10 and 20 mm, depending on the microwave power (300-900 W range) and gas flow-rate (3.5-13 l/min range) used in the discharge. Several lines arising from de-excitation of various species in the plasma are observed, both helium lines and other lines emitted by species introduced and carried by the discharge gas. The intensity of such lines significantly decreases when one moves away from the tip of the TIA nozzle (figure 3). As a result, the first few millimeters of the flame is the most interesting zone for the use as an excitation source in AES, existing a strong decrease in its excitation capability along the flame. Special mention should be made of the good results obtained in

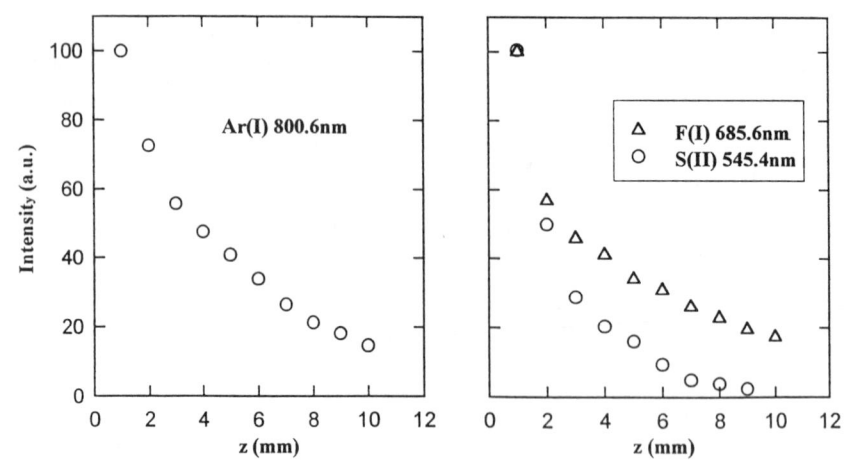

FIGURE 3. Variation of the emission intensity along the flame for several lines.

the excitation of the non-metal species, e.g. fluorine and sulphur coming from the destruction of SF_6 introduced in the discharge gas, which have a relatively high ionization potential and are therefore very difficult to excite by other conventional sources.

Various plasma parameters have been determined by using spectroscopic diagnostic techniques in terms of both the helium flow-rate and the microwave power. These preliminary results show that the plasma is not close-to-LTE, and only the excited levels very near the continuum ionization limit (higher than 4^3S) are in partial LTE (21). This deviation from the equilibrium gives rise to an overpopulation of the lower excited states with respect to their equilibrium populations, corresponding to an ionizing plasma. Because the atomic state distribution function (ASDF) cannot be described by only one excitation temperature, this magnitud is certainly not a good estimation for the electron temperature. The electron temperature is calculated by the intensity ratio of the atomic argon line at 842.46 nm and the ionic argon line at 506.20 nm. For this purpose, a small sample of argon is introduced into the plasma.

The gas temperature is obtained from the rotational temperature measured by the P-branch lines of the N_2^+ rotational band, while the electron density is calculated by using the well-known H_β stark broadening technique. The values

TABLE 3. Comparison with tipical values of ICP.

Plasma	n_e (m^{-3})	T_{exc} or T_e (K)	T_{gas} (K)
Ar-ICP (27-50 MHz, 1 KW)	~ 10^{21}	5000-6000	3000-5000
He-TIA (2.45 GHz, 300-900 W)	0.4-1.0 ×10^{21}	13000-14000	2400-2900

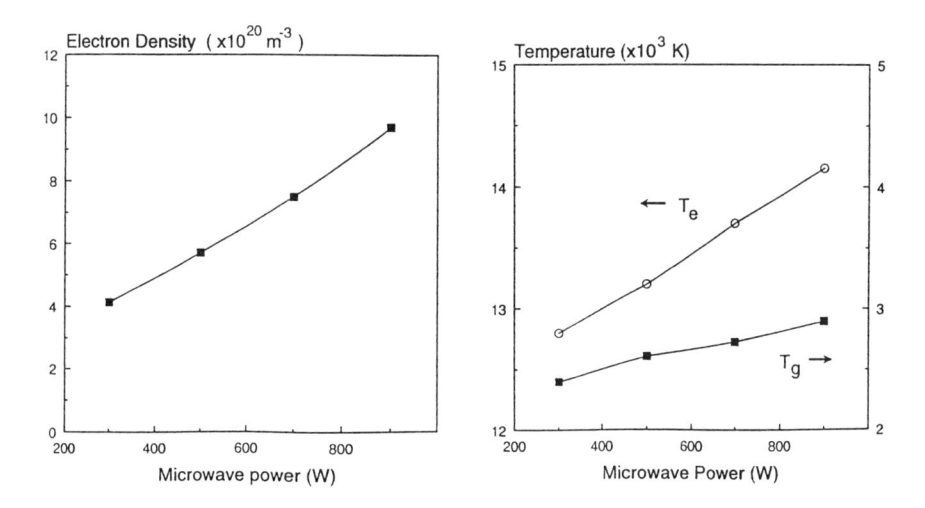

FIGURE 4. Dependence of the electron density and electron and gas temperatures on the microwave power.

of the electron density and the electron and gas temperatures, corresponding to the beginning of the plasma flame, are shown in figure 4 for a gas flow-rate of 7 l/min. In all cases, these three magnitudes increase with increasing microwave power and they similarly increase with decreasing gas flow-rate.

The properties of these plasma sources are compared with those of the ICP (22) in table 3. The gas temperatures are similar for both types of discharges. As a consequence, the microwave helium plasma produced by TIA must have high desolvation, volatilization and atomization capabilities. The electron temperature obtained in the plasma generated by TIA is significantly higher than the corresponding excitation temperature in ICP. Therefore, the helium plasma generated by TIA must produce a very efficient excitation and ionization of the analytes. For these reasons, this new microwave plasma at high pressure and high power seems like an excellent plasma source for atomic emission spectrometry, and further studies are convenient in the future.

CONCLUSIONS

The results shown here support the analytical potential of the microwave plasma produced by surfatron as an excitation source in AES for gaseous samples. Stable and easy to operate plasmas can be produced which permit a good control of the discharge characteristics and facilitates the search for better conditions for the analytical determination. Particularly, excellent analytical features have been obtained in the determination of halogens (Cl, Br and I), even using low power and argon as plasma gas.

The microwave plasma produced by TIA seems like a good candidate to extend the application range of the microwave induced plasmas used as excitation sources. As a consequence of the plasma parameters obtained, the helium plasma produced by TIA must have high volatilization and atomization capabilities, the most important limitation of MIPs up to date. So, further investigations in this direction should be convenient in the future.

ACKNOWLEDGMENTS

This work was partially supported by Project no. PB91-0847 of the DGICYT (Spanish Ministry of Education and Science).

REFERENCES

1. Sharp,B.L., Chenery,S., Jowitt,R., Sparkes,S.T. and Fisher,A., *J. Anal. At. Spectrom.* **9(6)**, R171-R200 (1994)
2. Zander,A. and Hieftje,G., *Appl. Spectrosc.* **35**, 357-371 (1981)
3. Matousek,J.P., Orr,B.J. and Selby,M. *Prog. Anal. Atom. Spectrosc.* **7**, 275-314 (1984)
4. Hubert,J., Moisan,M. and Ricard,A., *Spectrochim. Acta* **33B**, 1-10 (1979)
5. Moisan,M. and Zakrzewski,Z., *Rev. Sci. Instrum.* **58**, 1895-1900 (1987)
6. Bulska,E., Broekaert,J.A.C., Tschöpel,P. and Tölg,G., *Anal. Chim. Acta* **276**, 377-384 (1993)
7. Besner,A., Moisan,M. and Hubert,J., *J. Anal. At. Spectrom.* **3**, 863-866 (1988)
8. Hubert,J., Sing,R., Boudreau,D., Tran,K.C., Lauzon,C. and Moisan,M., in *Microwave Discharges : Fundamentals and Applications*, ed. Ferreira,C.M. and Moisan,M.; New York : Plenum Press, 1993, pp.509-530
9. Bulska,E., *J. Anal. At. Spectrom.* **7**, 201-210 (1992)
10. Mermet,J.M.,in *Inductively Coupled Plasma Emission Spectroscopy. Part II: Applications and Fundamentals*, ed. Boumans,P.W.J.M., New York : John Wiley & sons, 1987, ch.10, pp. 353-386
11. Van der Mullen,J.A.M., *Physics Reports* **191**, 109-220 (1990)
12. Quintero,M.C., Cotrino,J., Sáez,M., Menéndez,A., Sanchez Uría,J.E. and Sanz-Medel,A., *Spectrochim. Acta* **47B**, 79-87 (1992)
13. Calzada,M.D., Quintero,M.C., Gamero,A., Cotrino,J., Sanchez Uría,J.E. and Sanz-Medel,A., *Talanta* **39**, 341-347 (1992)
14. Calzada,M.D., Quintero,M.C., Gamero,A. and Gallego,M., *Anal. Chem.* **64**, 1374-1378 (1992)
15. Michlewicz, K.G. and Carnahan,J.W., *Anal. Chem.* **58**, 3122-3125 (1986)
16. Barnett,N.W., *J. Anal. At. Spectrom.* **3**, 969-972 (1988)
17. Jin,Q., Zhang,H., Ye,D. and Zhang,J., *Microchem. J.* **47**, 278-286 (1993)
18. Jin,Q., Zhu,C., Borer,M.W. and Hieftje,G.M., *Spectrochim. Acta* **46B**, 417 (1991)
19. Leprince,P., Bloyet,E. and Marec,J., *USA Patent 4611108* (1986)
20. Moisan,M.,Sauvé,G.,Zakrzewski,Z. & Hubert,J., *Plasma Sources Sci.Technol.***3**,584-592 (1994)
21. Rodero,A., García,M.C., Gamero,A., Sola,A. and Quintero,M.C., "Equilibrium separation in a high pressure helium plasma and its application to the determination of temperatures", presented in the I.C.P.I.G. XXII, Hoboken, New Jersey, USA, July 31 - August 5, 1995
22. Hasegawa,T. and Haraguchi,H., in *Inductively Coupled Plasma in Analitical Atomic Spectrometry*, ed. Montaser,A. and Golightly,D.W., New York : VCH Publishers, 1987, ch.8, pp.267-321

The Bloch-Elwert Structure as a Universal Manifestation of the Rate of "Soft" Inelastic Collisions in Plasmas: Bremsstrahlung, Electrical Conductivy, Ionization Loss

Vladimir I. Kogan

Russian Research Centre "Kurchatov Institute", Moscow

Abstract. The interrelation is traced and a mutual intrinsic identity at the "model-level" is demonstrated between varions plasma phenomena governed by distant electron-ion collisions whose rates' dependencies on the Coulomb parameter $Ze^2/\hbar v$ are described by a single universal "Bloch-Elwert structure" (BES) namely low-frequency Bremsstrahlung, electrical conductivity, and ionization loss of fast particles. In connection with the fact that the original derivation of the BES has been performed by the impact-parameter (ρ) method which in turn caused a sort of a "Bohr-Bloch controversy" the Eikonal Approximation of quantum mechanics is invoked which provides a consistent and lucid quantum-mechanical interpretation of ρ and its extremes ρ_{min} and ρ_{max} thus resolving the above-mentioned controversy.

INTRODUCTION

The Coulomb logarithm entering the cross-section of Bremsstrahlung in electron-ion collisions, the electrical conductivity of plasmas, and the rate of ionization loss of fast charged particles passing through matter has, for an arbitrary value of the Coulomb parameter $Ze^2/\hbar v$, the structure

$$\ln\frac{2mv^2}{\hbar\omega} + \psi(1) - \text{Re}\,\psi(1 + i\frac{Ze^2}{\hbar v}) \tag{1}$$

(ψ being the logarithmic derivative of the Γ-function) obtained quite independently by the impact-parameter method by Bloch (1) and from the low-frequency limit of the exact (Sommerfeld) Bremsstrahlung theory by Elwert (2); for the (static) conductivity case this "Bloch-Elwert structure" (BES), eq.(1), arises from the quantum-mechanical transport cross-section of scattering in a Debye potential (3), with the substitution $\omega \to v/D$ expressing a certain formal equivalence between static and "dynamic" screening.

© 1996 American Institute of Physics

The large difference between the "observable" physics of the above phenomena as well as between the respective derivations of BES caused a doubt as to their intrinsic (i.e. at the "model-level") mutual identity, up to an assertion on a "largely fortuitous" coincidence of the Bloch and Elwert structures (4). The latter work belonged to the pioneering once in the field of Coulomb Excitation of nuclei wherein the mathematical apparatus of the Sommerfeld Bremsstrahlung theory (including Elwert's special case) plays, on obvious grounds, a quite important part. But even after a direct demonstration of the "model-level" equivalence between the Bloch and Elwert structures in the famous Coulomb-Excitation survey (5) both of them continued to "live their own life" practically without a slightest mutual quotation.

Besides, the issue of a physical interpretation of the original Bloch result (1) (obtained in "ρ-space") in terms of a cut-off of Coulomb collisions with respect to their impact parameters (i.e. ρ_{max} and ρ_{min}) has become, for the Born domain $Ze^2/\hbar v \ll 1$, a subject of a sharp criticism by Bohr (6) directed, although somewhat implicitly, against the Bloch interpretation $\rho_{min} \sim \lambda$ (λ is the de Broglie wave-lenght of relative motion within the system: "free" atomic electron plus fast ionizing particle), see Bohr's own text under Fig.1 below.

The work at hand deals with two main issues. The first one reveals the "model-level" equivalence (more exactly, the proportionality) between the (inverse) electrical conductivity of a plasma and the intensity of its low-frequency Bremsstrahlung. This analysis is aimed at establishing a most direct proportionality relation between the (inverse) conductivity and Bremsstrahlung spectral intensity of a thermal plasma with proper (and simultaneous!) account of non-classical electron motion ($Ze^2/\hbar v \leq 1$) not only, as usual, for Bremsstrahlung but for conductivity as well which is of importance for hot plasmas. Here the main stress is made on the "transport" structures in both (mutially proportional) quantities which is, of course, a generalization of the well-known "transport theorem", see e.g. (7).

The second issue is aimed at revealing the proper physical interpretation of the impact parameter ρ. To this end we invoke the Eikonal Approximation of quantum mechanics which combines, to a large extent, the Born and quasiclasscal approximations and thus forms a natural basis not only for a unambiguous resolution of specific points like the above-mentioned "Bohr-Bloch controversy" but, much more generally, for a sound quantum-mechanical foundation of the impact-parameter method itself.

"MODIFIED TRANSPORT THEOREM" IN TERMS OF THE BLOCH-ELWERT STRUCTURE

According to (7) for electron-atom (e-a) scattering there exists an exact proportionality between the (dipole) "effective radiation" $d\kappa(\omega)/d\omega \equiv \hbar\omega d\sigma_{Br}(\omega)/d\omega$ ($d\sigma_{Br}$ being the Bremsstrahlung cross-section) in the low-frequency limit $\omega \to 0$ and the transport cross-section of elastic scattering $\sigma_{tr}^{(ea)}$; in what follows we'll denote such a proportionality as the "transport theorem" (TT). In the case of interest (Coulomb interaction, e - i scattering) the transport cross-section $\sigma_{tr}^{(ei)}$ divergess logarithmically and for this reason in (7) the Coulomb case is even removed beyond the frame of applicability of the TT.

Actually, of course, the transport theorem is **not** violated in the Coulomb case as well, rather it degenerates to an identity $\infty = \infty$. Therefore for a meaningful generalization of TT onto the Coulomb case it is sufficient, in the usual derivation of TT, to move away somewhat from the point $\omega = 0$.

To this end we start from the classical (as to electron motion) expression for $d\kappa(\omega)/d\omega$ following from (8):

$$\frac{d\kappa(\omega)}{d\omega} = \frac{2e^2}{3c^3} \int_0^\infty 2\pi\rho d\rho \left| \int_{-\infty}^{\infty} \dot{\vec{v}}(\rho,t) e^{-i\omega t} dt \right|^2 \qquad (2)$$

$\dot{\vec{v}}$ being the electron's acceleration, and ρ its impact parameter relative to the scattering centre.

For $\omega = 0$ the square of modulus in eq.(2) reduces to a "transport" structure $(\Delta\vec{v})^2 = 2v^2[1 - \cos\theta(\rho)]$, v being the electron velocity and θ its scattering angle. For ω non-vanishing but small enough the above-mentioned square reduces, as it can be shown, to a modified transport structure $2v^2[1 - \cos\theta(\rho)] \phi(\omega\rho/v)$ where $\phi(x)$ is a step-like function decreasing from the value 1 (for x <<1) to zero (for x >>1) (the explicit form of $\phi(x)$ depends on the specific form of the scattering field). Thus the difference of ϕ from unity is of importance only at large values of ρ i.e. for distant, almost rectilinear passages. In this case the acceleration $\dot{\vec{v}}$ in eq.(2) is large only within a time interval Δt_{eff} of the order of ρ/v so that the rate of oscillations in the integral over t may be described, effectively, by a factor of $\exp(-i\omega\rho/v)$-type and, hence, affects the value of the integral through a "cutting-off", monotonically decreasing function $\phi(\omega\rho/v) \leq 1$. Note that in the Coulomb case for $\omega \to 0$ in eq.(2) just the region of large ρ is of integral importance because for the point $\omega=0$ itself the divergence of eq. (2) is due to $\rho \to \infty$.

The occurrence, for small ω, of the above-mentioned modified transport structure under the integral over ρ leads, evidently, to an automatic cut-off of the integral at some $\rho_{max} \sim v/\omega$ owing to which the two cases, both non-Coulomb and Coulomb, can be encompassed by a single "modified transport theorem":

$$\omega \frac{d\sigma_{Br}(\omega)}{d\omega} = \frac{4}{3\pi}\left(\frac{v}{c}\right)^2 \frac{e^2}{\hbar c} \sigma'_{tr}(\omega), \qquad \omega \to 0 \quad (3)$$

where

$$\sigma'_{tr}(\omega) = \int_0^{\rho_{max}(\omega)} 2\pi\rho d\rho [1 - \cos\theta(\rho)], \qquad \rho_{max} = A\frac{v}{\omega} \quad (4)$$

is a modifield (by its ω-dependense) transport cross-section of elastic scattering of a classically moving electron by a force centre and A is a numerical coefficient of the order of unity, as yet indefinite.

In the non-Coulomb case the integral (4) converges at the upper limit so that in the latter one may put ω=0 and hence $\sigma'_{tr}(\omega)$ reduces to the usual quantity $\sigma_{tr}^{(ea)}$.

In the Coulomb case substituting into eq.(4) the well-known relation $\rho = a \operatorname{ctg}\frac{\theta}{2}$ ($a \equiv Ze^2/mv^2$ is the classical Coulomb lenght, m is the electron mass) we obtain:

$$\sigma'_{tr}(\omega) = \sigma'_{tr}[\theta_{min}(\omega)] = 4\pi a^2 \ln\frac{2}{\theta_{min}(\omega)} \quad (5)$$

where the minimum "taken-into-account" scattering angle $\theta_{min} = 2\operatorname{arcctg}(\rho_{max}/a) = 2A\omega/\tilde{\omega} \ll 1$ ($\tilde{\omega} \equiv v/a = mv^3/Ze^2$ is the characteristic Coulomb frequency). The substitution of (5) into eq. (3) gives, as it should be, the classical dependence for $(d\kappa/d\omega)_{Br}$ in the low-frequency limit $\omega \ll \tilde{\omega}$. For a full coincidence of the latter with the exact classical expression from (8) it remains to put $A=2/\gamma$ ($\gamma=e^C=1,78$ where $C=0,577$ is the Euler constant) so that $\theta_{min}(\omega) = \gamma\omega/\tilde{\omega}$.

The cross-section $\sigma'_{tr}(\omega)$, eq.(5), is the part of the (divergent) total Coulomb cross-section $\sigma_{tr}^{(ei)}$ corresponding to scattering angles $\theta \geq \theta_{min}(\omega)$. Thus the formulae (3) and (5) not simply express the community of the Coulomb mechanism of the processes of low-frequency Bremsstrahlung and elastic e-i scattering but "unite" two ω-dependent physical quantities accessible to direct (and independent) measurement. And what is more, though these formulae have been obtained within the framework of classical (with respect to the electron motion) theory they do not contain any quantities inherent only in it and therefore they retain their physical meaning in the general quantum-mechanical (with respect to the parameter $Ze^2/\hbar v \equiv \eta$) case as well; the transition to the quantum-mechanical case affects only the explicit form of the $\theta_{min}(\omega)$-dependence, see eq. (9) below.

In order to obtain the generalization of the Coulomb transport cross-section (5) onto the quantum (with respect to electron motion) case we use in the left-hand side of eq.(3) the low-frequency (Elwert!) limit of the exact quantum-mechanical Sommerfeld theory, see (2,7) and cf. eq. (1):

$$\omega \frac{d\sigma_{Br}(\omega)}{d\omega} = \frac{16}{3} \frac{Z^2 e^6}{m^2 v^2 c^3} \left\{ \ln \frac{2mv^2}{\hbar \omega} - C - \text{Re}\psi(i\eta) \right\}. \quad (6)$$

(let's recall that $C = -\psi(1)$, $\text{Re}\psi(1+i\eta) = \text{Re}\psi(i\eta)$).

For Re $\psi(ix)$ a quie accurate approximation is available (3):

$$\text{Re}\psi(ix) \approx -C + \frac{1}{2} \ln\left[1 + (\gamma x)^2\right] \quad (7)$$

which allows to present the expression in curly brackets in eq. (6) as a single Coulomb logarithm

$$L = \ln \frac{\omega_{quant}(v)}{\omega}, \quad \omega_{quant}(v) = \frac{2mv^2}{\hbar \sqrt{1+(\gamma\eta)^2}} = \frac{2\widetilde{\omega}}{\gamma \sqrt{1+(\gamma\eta)^{-2}}}, \quad \omega << \omega_{quant} \quad (8)$$

where $\omega_{quant}(v)$ is a characteristic "quantum" frequency which increases monotonically with increasing v and reduces to the frequencies $\frac{2}{\gamma} \frac{mv^3}{Ze^2}$ and $\frac{2mv^2}{\hbar}$ in the quasiclassical ($\eta >> 1$) and Born ($\eta << 1$) limits, respectively.

Finally, comparing the logarithms in eqs.(8) and (5) we obtain the quantum- generalized value of the electron minimum scattering angle entering the transport cross-section (5), for an arbitrary value of η:

$$\theta_{min}(\omega) = \frac{2\omega}{\omega_{quant}(v)} << 1 \quad (9)$$

It is appropriate to denote the quantity (5) with θ_{min} from eqs. (9) and (8) as the ω - modifield and quantum-generalized Coulomb transport cross-section.

THE "EFFECTIVE" IMPACT PARAMETER IN COULOMB COLLISIONS

The "pragmatic" aspect of the Bloch-Elwert structure has been exhausted by the preceding section and we shall not return to it. However the original (Bloch)

derivation of the BES has an interesting conceptual implication quite naturally incorporable into the scope of the well-known survey of Williams (9) "Application of Ordinary Space-Time Concepts in Collision Problems and Relation of Classical Theory to Born's Approximation".

In our case the role of such an "ordinary space-time concept" is played by the impact parameter ρ. The attitude towards ρ by the three members of, I would say, "the great B-Trinity" (**B**ohr, **B**ethe, **B**loch) was quite diverse: Bohr (10) treated ρ in an only possible, at that time (1913), manner i.e. purely classical. At the contrary, Bethe (11) (1930) being elated by his success in operating in "q-space" ($\vec{q} = \vec{k}_0 - \vec{k}_1$ is the collision vector of the scattered particle) was near to refusing ρ to be a "meaningful" physical quantity at all. Finally, Bloch (1) (1933) returned to the use of ρ even but operated with it in a more general manner namely for large ρ (distant passages) simply substituting,into the total ionization loss, Bohr's classical result while for close passages performing a detailed quantum-mecanical test particle-atom perturbation procedure.The resulting "Bloch contribution" to the total ionization loss plays a dominant role in the Born domain ($\eta \ll 1$) and just owing to this it assures the main property of the Bloch formula, its generality with respect to the Bohr ($\eta \gg 1$) and Bethe ($\eta \ll 1$) limits.

However despite the full validity of "Bloch's contribution (as it follows, first of all, from the above-mentioned Sommerfeld-Bremsstrahlung-Theory verification (5)) its derivation is extremely cumbersome and insufficiently clear as to physical argument. (The subsequent presentations of the Bloch theory did not add any clarification, see e.g. (12)). Apparently just this circumstance "provoked" Bohr to a sharp (although not mentioning Bloch explicitly) criticism against the non-trivial Bloch's conclusion concerning the cutt-off of the passages in the Born domain ($\eta \ll 1$) namely

$$\rho_{min} \sim \lambda, \qquad \eta \ll 1 \qquad (10)$$

Now, before quoting the corresponding paragraph from Bohr's remarkable, in all other respects, survey, we must introduce from there the figure comparing the angular distribution of scattering in a weakly screened Coulomb potential with the Rutherford formula for different values of $\eta = \dfrac{Ze^2}{\hbar v}$. The ratio ξ of the scattering cross-sections in a screened and non-screened field is given as a function of $\ln \operatorname{cosec} \dfrac{\theta}{2}$. The curves R' and R'' correspond to the quasiclassical ($\eta \gg 1$) and Born ($\eta \ll 1$) cases respectively. (The curves T and S correspond to strong screening and are therefore of no interest for us).The comparison of the curves R' and R'' and their locations ($\vartheta'_a = \dfrac{Ze^2}{mv^2 D}$, $\vartheta''_a = \dfrac{\lambda}{D}$) demonstrates, for the case of static screening, the suppression of the domain of smallest θ

accompanying the transition from η>>1 to η<<1, in full agreement with the above formulae (8), (9) for "dynamic" screening (recall the correspondence $\omega \Leftrightarrow v/\rho$).

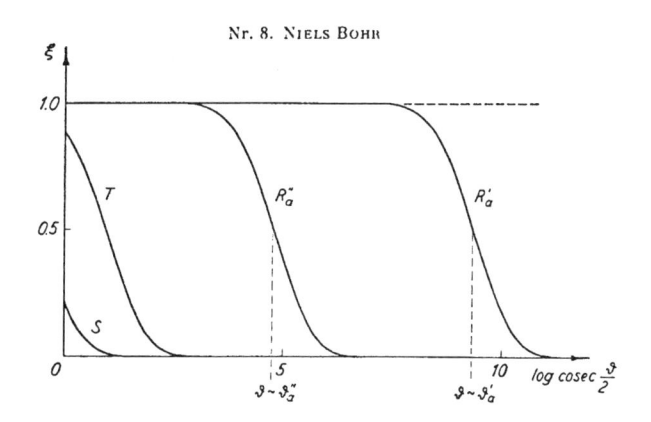

Nr. 8. NIELS BOHR

Now, the quotation from (6) reads:
<<It need hardly be stressed that the results represented in Fig. 3 cannot be interpreted by even a restricted reference to orbital pictures. Thus, **any attempt to attribute the difference between R'_α and R''_α to the obvious failure of such pictures in accounting for collisions with an impact parameter smaller than λ will be entirely irrelevant**. In fact, this argument would imply a difference between the two distributions for the large angle scattering, while the actual differences occur only in the limits of small angles. For $\eta \ll 1$, the scattering law is, in fact, determined by all parts of the field in a manner completely foreign to any ordinary mechanical analysis... >>

Here Bohr commites a qualitative mistake. Its origin lies in Bohr's "absolutization" of the meaning of ρ as an invariably well-defined (classical) quantity while in general, especially in the Born (i.e. "**anticlassical**") domain such an interpretation is centainly invalid. Postponing explicit proof of our assertion to the end of this Section, let's only mention the paradoxical circumstance that already the last sentence of the same quotation contains, essentially, the clue to the denial of the preceding content of the quotation and thus the clue to the full resolution of the "Bohr-Bloch controversion"!

In order to obtain an analytic view on the problem of the "effective" value of the impact-parameter ρ in the BES as well as its extremes ρ $_{max}$ and ρ$_{min}$ for given valus of ω (or θ for the case of static screening) and $\eta \equiv Ze^2/\hbar v$ it is appropriate to use the high-energy (or, equivalently, small-angle) Eikonal Approximation. Inteed, this approximation, first, just contains ρ as a natural integration variable and, second, encompasses, in its domain of applicability, the quasiclassical ($\eta \gg 1$) and Born ($\eta \ll 1$) approximations as two opposite extremes. It is rather obvious that since the BES describes the smooth transition between these extremes for distant (or, equivalently, "soft") Coulomb collisions, either for "dynamic" sceening (ω-depending BES) or for the static one (Debye potential, substitution $\omega \Rightarrow v/D$), see preceding Section, the application of the Eikonal Approximation to scattering, say, by the Debye potential should give, at least approximately, just the Bloch-Elwert structure. Thus for fully solving the above "ρ-problem" including the unambiquous resolution of the "Bohr-Bloch controvesion" it is sufficient to perform the "Eikonal-based" analysis of the elastic scattering amplitude for the Debye-potential up to the stage of obtaining a certain integral over ρ.

Now, let us analyse the elastic scattering amplitude $f(\theta)$ for the screened Coulomb (Debye) potential

$$U(r) = \pm \frac{Ze^2}{r} e^{-\kappa r}, \qquad \kappa \equiv 1/D \qquad (11)$$

According to the Eikonal Approximation (13)

$$f(\theta) = -ik \int_0^\infty \left[e^{2i\delta(\rho)} - 1 \right] J_0(q\rho) \rho d\rho \qquad (12)$$

where $k = \frac{mv}{\hbar}$, $q = 2k \sin\frac{\theta}{2} \approx k\theta$, J_0 is the Bessel function and the phase $\delta(\rho)$ is given by

$$\delta(\rho) = -\frac{1}{2\hbar v} \int_{-\infty}^\infty U(r) dz \qquad (13)$$

Substituting eq.(11) with $r = \sqrt{\rho^2 + z^2}$ we obtain

$$\delta(\rho) = \mp \frac{Ze^2}{\hbar v} K_0(\kappa\rho) \qquad (14)$$

where K_0 is the MacDonald function. With $Ze^2/\hbar v \equiv \eta$ we obtain, finally,

$$f(\theta) = -ik \int_0^\infty \left[e^{\mp 2i\eta K_0(\kappa\rho)} - 1 \right] J_0(2k\rho \sin\frac{\theta}{2}) \rho d\rho \qquad (15)$$

The formula (15) has the desired form of an integral over ρ and thus allows, in principle, to estimate, for a given value of η, the ρ-domain "responsible" for scattering through an angle θ, i.e. to obtain the dependence $\rho_{eff}^{(\eta)}(\theta)$ for the potential (11).

In particular, for the "Bloch-Elwert" problem as a problem of the Coulomb logarithm it is appropriate to reduce the analysis simply to the determination of the value θ_{min} restricting the domain of Coulomb scattering or, equivalently, the corresponding value of $q_{min} = 2k\sin\dfrac{\theta_{min}}{2}$, see eq. (12).

Since the purely Coulomb scattering corresponds to the Rutherford scattering amplitude $f(\theta) \sim 1/q^2$ while for $q \to 0$, according to eq.(12), $f(\theta)$ ceases to depend on q at all these exists surely a certain "de-Coulombizing" value q_{min}^{Coul}. Thus it is convenient to present the scattering amplitude in the form

$$f(\theta) = -\frac{ik}{q^2}\int_0^\infty \left[e^{\mp 2i\eta K_0(\kappa y/q)} - 1 \right] \cdot J_0(y) y\, dy \qquad (16)$$

in order to investigate the **modulus** of the integral in eq.(16) with respect to the disappearance of its q-dependence.

In the Born limit $\eta \ll 1$ expanding the expression in the square brackets eq.(16) in powers of η and performing a standard integration over y we obtain the structure $\dfrac{1}{1+(\kappa/q)^2}$ so that the "q-disappearance" condition is $q \gg \kappa$ whence

$$(q_{min}^{Coul})_{Born} = \kappa, \quad \text{or equivalently,} \quad (\theta_{min}^{Coul})_{Born} = \frac{\kappa}{k} \qquad (17)$$

In the quasiclassical limit $\eta \gg 1$ the exponential in eq. (16) becomes a strongly oscillating function so that the "effective" y-domain reduces to the neighbourhood of the "point of stationary phase" which in turn is determined by the mutual cancellation of the phase in the above-mentioned exponential with the phase in one of the two oscillating (imaginary-exponential) terms formed by the asymptotics of the Bessel function $J_0(y)$. The consistency condition of this procedure proves to be just $\eta \gg 1$ and the "point of stationary phase" corresponds as it should be to the **classical** relation between ρ and θ (14). For $(\theta_{min}^{Coul})_{class}$ we obtain

$$(\theta_{min}^{Coul})_{class} = \frac{2Ze^2\kappa}{mv^2} \ll 1 \qquad (18)$$

Combining the two "complementary" limits, eqs (17) and (18), we obtain a simple interpolation formula (up to non-important numerical coefficients)

$$\theta_{min}^{Coul} \sim \frac{\kappa}{k}\left(1 + \frac{Ze^2}{\hbar v}\right) \qquad (19)$$

Since $(d\sigma/d\Omega)^{Coul}$ is proportional to $1/\theta^4$ the small-angle transport cross-section $\sigma_{tr}^{Coul} \sim \int_{\theta_{min}}^{\theta_{max}\sim 1}(1-\cos\theta)\dfrac{d\Omega}{\theta^4} \sim \ln(1/\theta_{min})$ so that together with the "scaling factor" $\sim \left(\dfrac{Ze^2}{mv^2}\right)^2$ we obtain finally (also up to numerical coefficients)

$$\sigma_{tr}^{Coul} \sim \left(\frac{Ze^2}{mv^2}\right)^2 \ln\left(\frac{k}{\kappa}\frac{1}{1+\frac{Ze^2}{\hbar v}}\right) \qquad (20)$$

The equivalent result for for the case of dynamic screening is obtained from eq.(20) by the substitution $\kappa \equiv \frac{1}{D} \Rightarrow \frac{\omega}{v}$.

The above qualitative considerations based on the the Eikonal approximation, eq. (12), may be regarded as a rough derivation of a formula of "Bloch-Elwert" type. Besides arriving at a simple version of the quantum-generalized Coulomb logarithm they allow to give an estimate of the desired "effective" impact parameter for a given scattering angle:

$$\rho_{eff}^{Coul} \sim \frac{Ze^2}{mv^2\theta}\left(1+\frac{\hbar v}{Ze^2}\right) \qquad (21)$$

For the Born case $\eta \ll 1$ and $\theta_{max} \sim 1$ this gives $\rho_{min} \sim \lambda$ in full agreement with Bloch, eq. (10). This resolves the Bohr-Bloch controversion. Combining eq.(21) with eq. (19) we obtain

$$(\rho_{eff}^{Coul})_{max} \sim \frac{Ze^2}{mv^2\theta_{min}^{Coul}}\left(1+\frac{\hbar v}{Ze^2}\right) \sim \frac{v}{\omega} \quad \text{for arbitrary } \eta \qquad (22)$$

This verfies the validity of Bloch's procedure of direct, "phenomenologically" adoption of Bohr's classical result as the contribution of **distant** encounters to the total ionization loss.

ACKNOWLEDGMENTS

I wish to express my deep gratitude to A.B.Kukushkin and V.D.Shafranov for interesting discussions. This work has been supported by the Russian Foundation of Fundamental Research.

REFERENCES

1. Bloch, F., *Ann. der Physik*, **16**, 285-320 (1933).
2. Elwert, *Z. Naturforsch.* **30**, 477-485 (1948).
3. de Ferraris L., and Arista N.R., *Phys. Rev.*A, **99**, 2145-2159 (1984)
4. Huby R. and Newns H.C., *Proc. Phys. Soc. (Lond.)* **A 64**, 619-632 (1951).
5. Alder K., Bohr A., Huus T., Mottelson B., and Winther A., *Rev.Mod. Phys.* **28**, 432-542 (1956).

6. Bohr N. *Det. Kgl. Danske Videnskab. Selskab., Mathem.-fysic. Medd.*, **18**, N.8, 1-142 (1948).
7. Bekefi G., *Radiation Processes in Plasmas*, J. Wiley and Sons, Inc., New York, 1969.
8. Landau L.D. and Lifshitz E.M., *Classical Field Theory*, Pergamon Press, Oxford, 1975.
9. Williams E.J., *Rev.Mod. Phys.*, **17**, 217-226 (1945).
10. Bohr N., *Phil. Mag.*, **25**, 10-42 (1913).
11. Bethe H., *Ann. der Physik*, **5**, 325-400 (1930).
12. Ahlen S.P., *Rev. Mod. Phys.*, **52**, 121-173 (1980).
13. Landau L.D. and Lifshitz E.M., *Quantum Mechanics: Non-Relativistic Theory*, Pergamon Press, Oxford, 1977.
14. Galitsky V.M., Karnakov B.M., Kogan V.I., *Problems in Quantum Mecanics, Second Ed.*, Moskow, "Nauka", 1992, pp. 736-737 (in Russian).

Arc Spot Ignition on Cold Electrodes in an Ambient Gas Atmosphere

J. Mentel, R. Bayer, J. Schein, M. Schumann

Allgemeine Elektrotechnik und Elektrooptik, Ruhr-Universität Bochum, D-44780 Bochum, Germany

Abstract: Blowing an arc magnetically against a so called commutation electrode positioned perpendicularly to the discharge axis a reproducible interaction of the plasma with this electrode was achieved. Cathodic arc spots were ignited on it with a negative biased voltage and anodic spots with a positive biased voltage. Delay times of arc spot ignition on cathodes have been measured in air for 24 different materials of high purity and for several technical materials establishing an order of these materials as a function of delay time. The scattering of the delay times depends on the surface treatment. A large variety of arc traces has been found after arc spot ignition on the cathode surface by S.E.M. examination. High speed photography in pure argon revealed different forms of arc spot ignition. Partially a bright plasma ball is formed partially the cathodic surface is covered by a luminous layer initiating the formation of a bright channel between the cathode surface and the bulk plasma. Drawing conclusion in the case of cold electrodes quite different modes of current transfer occur showing different kinds of electrode erosion.

INTRODUCTION

In some applications arc discharges have to be operated on cold electrodes. Examples are switch gears of all kinds and arc heaters for chemical processes with reactive gases. In many other applications cold electrodes will be helpful for reducing the electrode consumption. However, a stationary operation of arc roots on cold electrodes is very difficult since they split up mostly in many individual spots of short lifetime. These individual spots are forming and vanishing permanently resulting in an irregular movement of the whole arc root.
Arc spots are characterized by high electric power density mostly causing the formation of craters on the electrode surface with melting and vaporization of the electrode material. This may result in a severe damage of cold electrodes and an unwanted material consumption. To control these problems a more detailed understanding of the basic processes is desirable. It is not easy by reason of the transient character of arc spots and their irregular behaviour to investigate these

processes. However when using a special experimental technique to bring a dense plasma into intimate contact with a cold electrode surface reproducible results can be achieved. This technique is quite different to the one used by Guile who reported in 1981 for the last time on this subject in an invited lecture during the ICPIG Conference in Minsk [1].

EXPERIMENTAL SET-UP

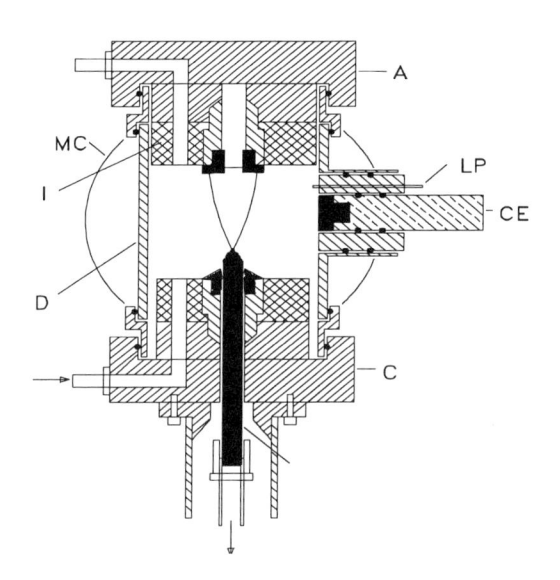

Figure 1: Discharge arrangement with A: anode; C: cathode; CE: commutation electrode; D: discharge tube; MC: magnetic field coil; LP: langmuir probe; I: insulator

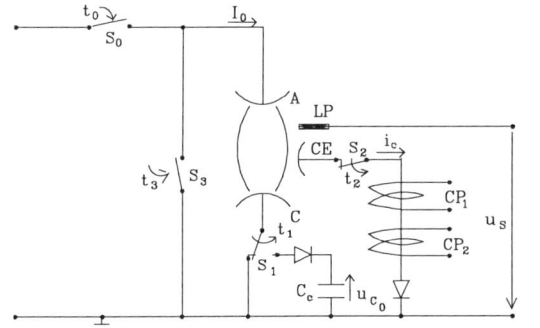

Figure 2a: Simplified electric circuit A: anode; C: cathode; CE: commutation electrode

A special experimental set-up using this technique is shown in Fig. 1. An arc of constant current is drawn by a graphite rod between a graphite anode on top and a graphite cathode on the bottom of the discharge tube within some milliseconds. The column of this arc is blown magnetically against a third electrode, so called commutation electrode CE, positioned perpendicularly to the arc axis. The subsequent electrical measurements are performed using a CE with a flat end face of 15 mm diameter. The potential related to the plasma potential near the CE is measured by a thin cone-shaped Langmuir probe (LP) inserted into the vessel close to the CE. The discharge vessel is connected to a dc power source via a thyristor network which provides current pulses of defined duration. Fig. 2 shows the simplified electric circuit and the electric signals measured at CE. S_0, S_1, and S_3 are thyristor switches, S_2 is a fast electronic switch realized by a MOSFET. The switch S_1 is activated at the time t_1 simultaneously with the magnetic blastfield. Charging the capacitor in series with the graphite cathode, at constant arc current, the voltage between the arc plasma and the CE is increased linearly. A corresponding signal is measured by the LP. At a time t_c after the application of the voltage ramp a sudden breakdown of the voltage in front of the CE is recorded and an abrupt current increase is measured in the CE. This time is called by us "arc spot formation delay time" or "commutation time". When the current in the CE has increased to several amps it is interrupted by the switch S_2. In this way the subsequent destruction of the arc traces formed during the ignition phase is prevented. At time t_3 the arc is shortened by switch S_3. A more detailed description of the experimental set-up can be found elsewhere [2, 3, 4].

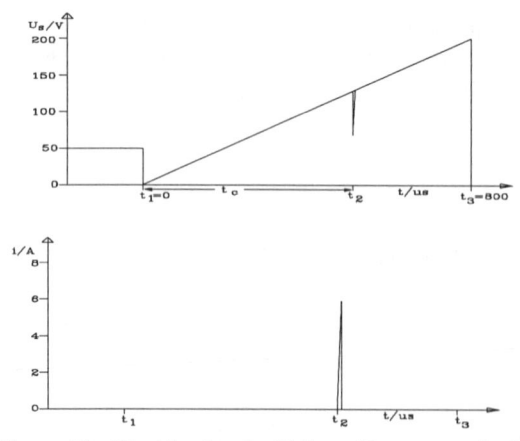

Figure 2b: Electric signals: Voltage U_s measured with LP and current i measured with current probe

ELECTRICAL MEASUREMENTS

The commutation time t_c is determined in addition to the electrode material and the structure of the electrode surface by several parameters. t_c decreases if either the rate of rise of the applied voltage (RRV) du/dt, the arc current I and the magnetic field B is increased, or if the distance d between the tube axis and the CE is lowered. These parameters are kept constant. The subsequent measurements are performed in air at atmospheric pressure with an arc operating at a current of I = 150 A at an electrode distance of 60 mm. The RRV amounts to du/dt = 340 kV/s, the magnetic blast-field to B = 8 mT and the distance of the CE from the tube axis to d = 22 mm. The magnitude of t_c scatters statistically. Therefore at least 60 individual measurements were made for each electrode type. The electrodes were polished for each ignition during the experiment. The results of t_c are represented by cumulative frequency distributions (CFDs) approximated by Weibull distributions (WDs) shown as solid curves. The results are condensed giving the average commutation times t_{ca}, the standard deviations Δt_{ca} and the

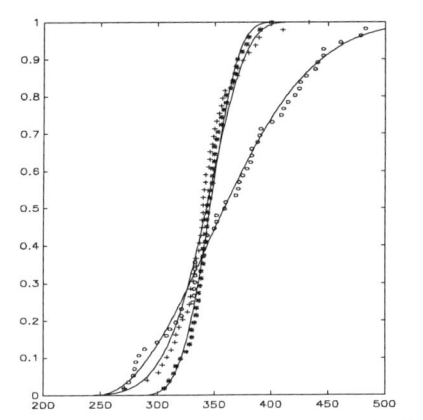

Figure 3: CFD of $t_c/\mu s$ for pure Au of 0.25µm (*,+), representing two independent measurements, and of 1.0µm(o) mechanical polished cathodes

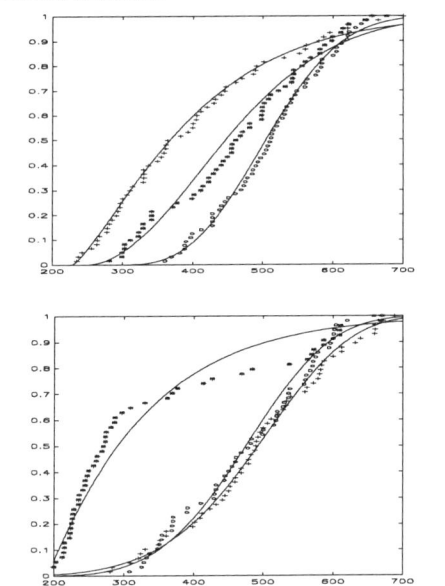

Figure 4: CFDs of 0.25µm (o), 1.0µm (+) mechanical polished and of electrolytically polished(*) cathodes
a) for pure Ni
a) for pure Pb

standard deviations of the CFD from a matched WD called Δt_{cw}. The shape of the CFDs is strongly depending on the treatment of the CE-surface. This surface was polished before every ignition with special cloths, fine grinding powder and destilled water. Diamond grinding powders with a particle size of 15µm, 7µm, 3µm, 1µm and 0.25µm were used in turn.

Having kept constant the discharge parameters and using the same surface treatment CFDs of t_c are found to be reproducible, thus characteristic for each electrode material. This is shown in Fig. 3 by two independent measurements of the CFDs with pure Au-electrodes. The degree of reproducibility can be seen quantitatively comparing the values for t_{ca}(342µs, 346µs) and its standard deviation Δt_{ca}.(28µs, 20µs) The influence of the surface treatment on the CFDs of different materials can be demonstrated if the polishing with 0.25 µm particles is omitted or if the mechanical polishing is supplemented by an additional electrolytic polishing (EP). In all figures the ordinate represents the CF and on the abscissa $t_c/\mu s$ is given. Corresponding CFDs are shown in Fig. 3 for Au, in Fig. 4a for Ni and in Fig. 4b for Pb. For pure Au the standard deviation Δt_{ca} increases with decreasing smoothness of the surface. A rougher surface produces a larger scatter of t_c whereas the average t_{ca} is not much affected. Similar results are found for Cu. For Ni a final polishing with 0.25µm particles

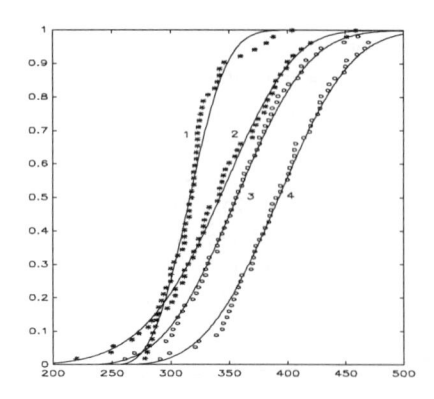

Figure 5: CFDs for cathodes made of technical Al(1), Cu(2) and of pure Al(3), Cu(4)

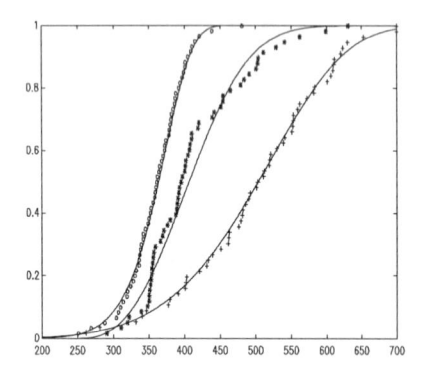

Figure 6: CFDs for cathodes made of pure Mg(o), pure Pt(*) and pure W(+). Solid curves are matched WD.

clearly increases t_{ca} and reduces Δt_{ca}. However an additional EP withdraws in parts these changes. By EP grain boundaries are laid open as shown by S.E.M. micrographs which may increase the number of favourable places for arc spot ignition. The behaviour of Fe agrees qualitatively with that of Ni. In the case of Pb additional polishing with 0.25 µm particles has nearly no effect on the CFD. The influence of the polishing procedure is less pronounced for soft materials than for hard materials. In the case of Pb an additional EP has displaced the CFD to distinctly shorter times. S.E.M. micrographs revealed that in this case EP has produced a textured surface favouring arc spot ignition. In Fig. 5 the CFDs are compared for technical materials Al(1), Cu(2), and pure materials Al(3), Cu(4). In general it can be seen that the CFDs of the technical materials are shifted to shorter times.

CFDs of several electrode materials have characteristic shapes, each one different from the other. Three typical examples are shown in Fig. 6. For Mg (as for Au, Al, Cu, Ag) a steep CFD is found which deviates only scarcely from a WD. The CFD of Pt (as of Ce, Sn, Co, Fe, Nb, Cd, Re, Pd) appears to consist of pieces of steady curves which can only insufficiently be approximated by a simple WD. The CFD of W (as those of Bi, Pb, Mo, graphite, Zr) is spread over a long time interval and deviates more statistically than in the case of Pt from a WD.

CFDs of t_c have been measured for 24 materials of high purity and for Al, Mg, Cu of technical quality. The results are summarized in Table I for cathodes mechanically polished with 0.25µm grinding powder. The sequence of elements in the table corresponds to an increasing t_{ca} given in the second column of the table. Δt_{ca} in the third column represents the width respectively steepness of the CFD. In the fourth column the standard deviation Δt_{cw} of the CFD from a matched WD is given showing the scatter of measuring points. No correlation between the properties of the bulk material and t_{ca} has been found.

TABLE I: Summary of measured materials

material	$t_{ca}/\mu s$	$\Delta t_{ca}/\mu s$	$\Delta t_{cw}/\mu s$	material	$t_{ca}/\mu s$	$\Delta t_{ca}/\mu s$	$\Delta t_{cw}/\mu s$
polished mechanically with 0.25µm grinding powder							
1 Au	342	28.3	4.7	13 Nb	476	96.1	13.0
2 Al	359	46.9	2.6	14 Pb	477	92.7	13.3
3 Mg	361	39.9	3.1	15 Mo	478	109.3	10.6
4 Zn	379	102.9	14.9	16 Cd	479	98.2	12.1
5 Cu	395	46.6	3.7	17 Re	500	108.8	11.7
6 Ce	412	67.9	12.0	18 Ti	500	108.7	9.4
7 Pt	413	71.4	11.0	19 W	502	97.4	5.4
8 Sn	429	105.1	9.3	20 C	503	115.3	7.2
9 Co	436	86.4	9.1	21 Pd	507	84.1	10.9
10 Ag	449	52.0	5.9	22 Ni	510	78.1	6.0
11 Fe	458	87.1	11.5	23 Zr	546	92.9	7.1
12 Bi	463	95.7	6.2	24 Ta	547	85.1	9.6
technical materials							
2 Al	318	27.0	3.4				
3 Mg	332	31.1	2.1				
5 Cu	346	38.7	3.8				

The CFD is steeper respectively Δt_{ca} is smaller for materials with high electrical conductivity σ than for materials with low σ. However Δt_{ca} is also reduced if the smoothness of the surface is improved. An important material parameter could be the adsorption of gases at the cathode surface. Ti, C (graphite), Pd, Ni, Zr and Ta with high values of t_{ca} are strong adsorbers of gases whereas Au, the first element in the table, scarcely adsorbs gases.

CFDs of t_c have also been measured for an anodic CE. The polarity of some elements in the circuit of Fig. 2a has to be reversed. The time t_c of anodic arc spot ignition was determined by the voltage breakdown in front of the CE. The probe current already starts to flow when the CE comes into contact with the arc plasma. In Fig. 7 CFDs for a Cu-, Al-, and Ni-anode are given. They differ scarcely showing that the anodic arc spot ignition is independent of the material properties. The $t_{ca} \approx 100$ µs represents approximately the time constant of the experiment and the $\Delta t_{ca} \approx 17$ µs the scatter of the experimental conditions from shot to shot. Moreover it turned out that different

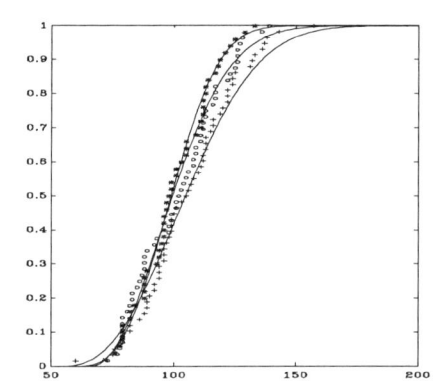

Figure 7: CFDs of anodic arc spot ignition on Al(o), Cu(+) and Ni(*)

to the cathodic arc spot ignition the anodic one is strongly influenced by the macroscopic surface structure.

S.E.M. MICROGRAPHS

Traces left behind by arc spot ignition on the electrode are investigated by scanning electron microscope (S. E. M.). The traces are produced by currents not higher than 10A flowing for several 10^{-7} s and transferring a charge of some µC. A large variety of arc traces are found formed mostly by craters. The diameters of the craters and the shape of the arc traces depend strongly on the electrode material, on the surface treatment and also on the plasma composition. Very often they are similar to the craters formed by vacuum arcs[5]. The subsequent examples are produced by arc spot ignition in air.

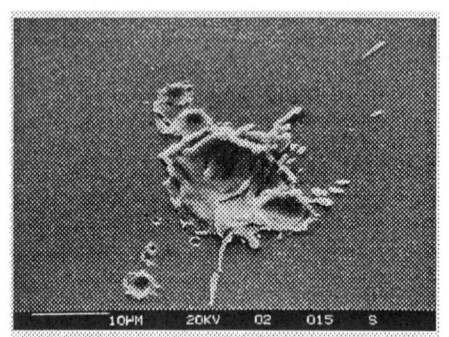

Figure 8a: Arc spot trace on pure Ni polished mechanically

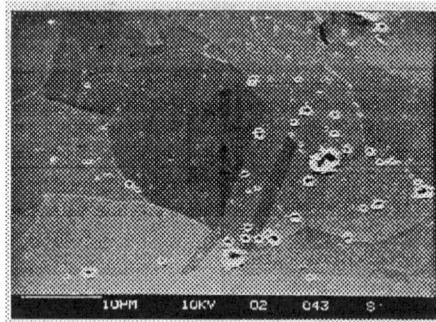

Figure 8b: Arc spot trace on pure Ni polished electrolytically

The S.E.M. micrograph of an arc trace on pure Ni polished with 0.25µm grinding powder is shown in Fig. 8a. It is made up by an agglomeration of some craters producing a gross local melting of the surface and several separate small craters in the surroundings. Similar traces have been found on Au- and Pt- cathodes. If the electrode surface is in addition polished electrolytically many small craters can be identified, often at the uncovered grain bounderies as can be seen in Fig. 8b. Even more favourable than grain bounderies are mechanical scratches along which chains of craters with diameters less than 1µm are forming. The arc traces on pure and technical materials are also very often quite different as shown for Cu and Fe [4, 6].

The S.E.M. micrographs clearly indicate for a special material a relation between surface structures produced by polishing or contaminations and arc traces. Therefore arc traces could be used to characterize the quality of surfaces.

However up to now it was not possible to relate the different values of t_{ca} of different electrode materials to special forms of arc traces.

ARC SPOT IGNITION UNDER CLEAN GAS CONDITIONS

The commutation times given in Table I may be attributed to oxide layers and adsorbed gas layers responsible for additional effects which increase the complexity of the ignition process. To remove these effects a new discharge arrangement in which the ignition process can be investigated in a clean gas atmosphere excluding contaminations by oxygen was developed. The experimental set-up shown in Fig.9 consists of a pair of horn electrodes (A, C) between which an arc is ignited by high voltage breakdown and operated for few milliseconds at 25A. After the ignition the arc is blown magnetically against a diaphragm (D) positioned horizontally above the horn electrodes.

The CE is postioned behind a small aperture (diameter 2mm) in the center of the diaphragm forcing the ignition to take place in an area of 1-2 mm^2. Up to eight CEs are fixed on the circumference of a revolver

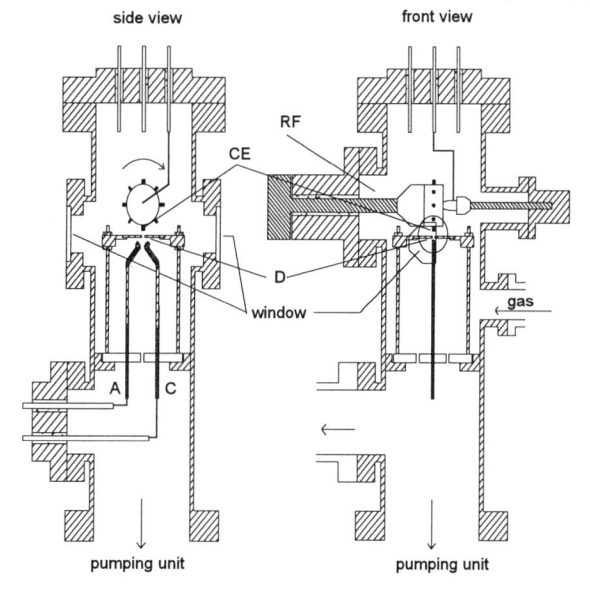

Figure 9: Experimental set-up for arc spot ignition under clean gas conditions

(RF), turned successively behind the hole. The whole discharge arrangement is enclosed in a vacuum tight chamber made of high grade steel. The chamber is equipped with windows for optical observation of the ignition, which is possible by reason of the special set-up with the high spatial resolution.

The ignition is initiated in difference to the previous experiment by applying a constant voltage between the arc plasma and the CE which is biased on mass.

The passage of the shining arc plasma through the aperture in front of the CE is recorded by a sensitive photodiode. The zero point of t_c is defined by the signal of the photodiode. The arc spot ignition in argon was recorded with high speed photography taking pictures side on by a CCD-camera with an integrated MCP shutter. The shutter making possible exposure times down to 5 ns is triggered by the current flowing through the CE. In Fig.10a,b,c different forms of arc spot ignition on W electrodes in argon are shown together with oscillograms of the current i_c transfered through the CE. In the figures the electrode is on the upper part of the picture. The picture parameters are as follows: exposure time: 100ns; picture dimensions (horizontal): 2mm; amount of i_c at time of exposure: 3A. Fig.10a corresponds to a precurrent passing over continuously into a steep current increase. The cathode surface is covered in parts with a luminous layer presumably produced by the precurrent. The layer is connected by a plasma channel forming with increasing current to the bulk plasma in front of the cathode. Similar pictures of ignitions in air are shown in [6]. Fig. 10c is related to a very steep current increase without any measureable precurrent. A bright plasma ball is formed in this case in front of the cathode spanning the gap between an unstructured bulk plasma and the cathode. Fig. 10b is characterized by a small precurrent of only 100mA. It represents a transition between the different forms of ignition in Fig. 10a and Fig. 10c showing a plasma ball and a channel. The structure in Fig. 10a is typical for the first few ignition events starting with a freshly polished electrode. The precurrent is presumeably generated by Townsend-γ-processes within the

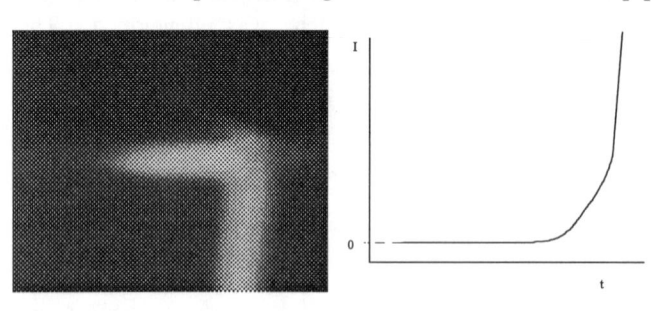

Figure 10a: High speed photograph of arc spot ignition and oscillogram of commutation current

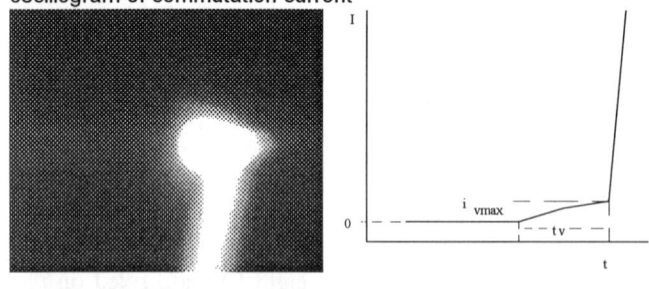

Figure 10b: High speed photograph of arc spot ignition and oscillogram of commutation current. $t_v=3\mu s$, $i_{vmax}=104mA$

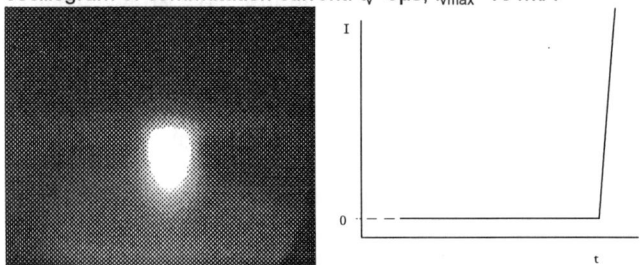

Figure 10c: High speed photograph of arc spot ignition and oscillogram of commutation current

luminous layer [7]. The plasma channel develops from a spatial instability of this layer. It is missing in some cases[6]. On the other hand the structure in Fig. 10c is dominant when the electrode surface is "arced" by numerous ignition events removing the oxide layer from the electrode surface. It seems that in this case a kind of breakdown occurs between the bulk plasma and the new cathode. The arc traces shown in Fig. 8 are footprints of the channel or of the plasma ball on the CE. Typical structures are craters. They can be avoided if the current is transfered by a luminous layer only.

The optical observations are supported by statistical investigations of t_c, the precurrent flowtime t_v and of the amount of the maximum of the precurrent i_{vmax} in dependence on the number of ignitions on one individual CE. Using 20 Au electrodes, all prepared the same way, and having 40 successive ignitions on each electrode it turned out that the average values of t_v and i_{vmax} are decreasing by a factor of 4 between the first and the 40th shot. This is shown in Fig. 11 for the

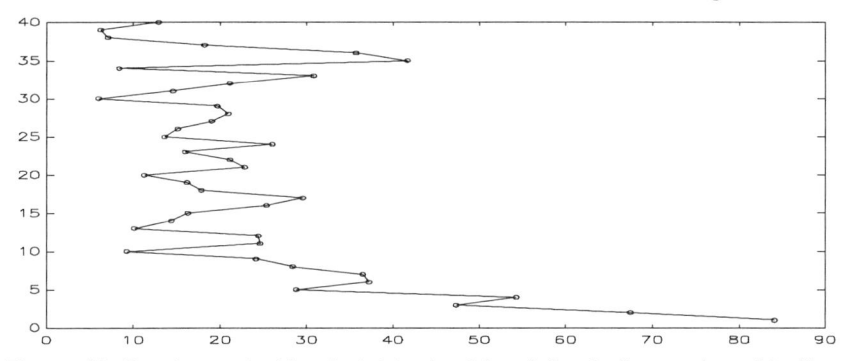

Figure 11: Development of i_{vmax}/mA (abscissa) in relation to the number of ignitions (ordinate) on a single electrode averaged over 20 electrodes for Au

maximum of the precurrent. Similar results are found for other materials. Moreover it turned out that the values of t_{ca} get closer together with increasing shot number for different materials.

Those optical and electrical measurements indicate that the ignition mechanism is changing when the electrode surface is modified by arcing.

SUMMARY

Special experimental methods based on an intimate contact between a dense plasma and a cold electrode have been developed to initiate arc spot ignition in an ambient gas atmosphere under reproducible conditions. The time needed for arc spot ignition called commutation time is an appropriate parameter characterizing arc spot ignition.

The average commutation time of the cathode arc spot ignition and its standard deviation is strongly depending on the polishing procedure.

For cathodes with well polished electrode surfaces material-specific commutation times are found. However they cannot be attributed to simple bulk properties of the electrode material (e. g. work function). The anodic arc spot ignition is independent of the electrode material, but they can be spatially fixed by macroscopic surface structures.

Cathodic arc traces respond to slight changes of the bulk and surface properties. They can be used to test surfaces. For the same electrode material smaller craters are related to shorter commutation times.

Different forms of cathodic arc spot ignition are found:

On fine polished electrodes a luminous layer in connection with a precurrent and successive formation of a plasma channel between the electrode and the bulk plasma is observed. The formation of the channel is not compelling.

On arced electrodes an ignition of a plasma ball on the surface without precurrent can be identified.

ACKNOWLEDGMENTS

This work was supported by Deutsche Forschungsgemeinschaft grant SFB 191/ A3.

REFERENCES

[1] A. E. Guile
"Electrode Processes at Non-Refractory Arc Cathodes"
Proc. of XV. ICPIG, Invited Paper, pp 165-174 (1981)

[2] K.-P. Nachtigall, J. Mentel
"Measurement of Arc Spot Formation Delay Times at Cold Cathodes in Air"
IEEE Plasma Science, vol. 19, pp 942-946 (1991)

[3] K.-P. Nachtigall, J. Mentel
"Optical Investigation of Arc Spot Formation on Cold Cathodes in Air"
IEEE Plasma Science, vol. 19, pp 947-946 (1991)

[4] R. Bayer
"Untersuchung der Zündung von Lichtbogenfußpunkten auf kalten Kathoden in Abhängigkeit von dem Material und der Oberflächenbeschaffenheit"
Ph. D. dissert. Ruhr-Universität, Bochum, Germany (1992)

[5] G. A. Mesyats, D. I. Proskurovsky
"Pulsed Electrical Discharge in Vacuum"
Springer Verlag, Berlin-Heidelberg-New York (1989)

[6] J. Mentel, R. Bayer, J. Schein, M. Schumann
"Interaction of a Dense Plasma with Cold Electrodes"
High Temp. Chem. Processes 3, pp. 627-638 (1994)

[7] W. Bötticher
"Filamentation of High Pressure Glow Discharges"
Proc. III of XXI. ICPIG, pp 128-131 (1993)

Tomography for Discharge Plasmas

N. Iwama

*Department of Electronics and Informatics, Toyama Kenritsu University
Kosugi, Toyama 939-03, Japan*

Abstract The computerized tomography for plasma measurements are reviewed from a viewpoint of numerical analysis. Interest is devoted to the methods to meet the missing of projection data, especially, the least squares fittings of series expansion models and of linear regression model. On the smoothings by parameter number reduction and with weighted objective functions, the optimizations with the statistical neg-entropies of AIC and GCV are taken as a subject of study and and discussed on experimental results in low-temperature discharge plasmas.

INTRODUCTION

The computerized tomography (CT) study for measuring the spatial parameter distributions in laboratory plasmas was, after pioneering works (1), (2) and precedent works of Abel inversion, commenced actively around the year 1980 not only for high-temperature plasmas in fusion research but also for low-temperature discharge plasmas (3)-(7). Several studies have been made on the passive and active types of sensing using radiations of the wavelengths from X-ray to microwaves and reaction particles. In comparison to the medical imaging, an important feature of plasma imaging lies in ill conditions of projection measurement: the sparseness of sampling point, the strong view-angle limitation and/or the low SN ratio, for which the standard methods of Radon inversion easily break down. The great loss of information leads us to an system of equations underdetermined, and in general, ill-conditioned on the plasma image to be reconstructed. For the discharge plasma research, a review is made in this paper on the numerical methods of determining the solution in desirable fashion. Suggestions are offered on the usage of the statistical criteria for optimizing the reconstruction.

STANDARD METHODS FOR RADON INVERSION

The projection value of a two-dimensional (2D) plasma image $E(r)$ along a line of sight specified by the perpendicular distance p from the coordinate system origin and the unit vector n along p can be written as an integral of the form

$$S(p,n) = \int E(r)\delta(p - r \cdot n)dr \quad (= R[E(r)]), \tag{1}$$

namely, the Radon transform of $E(r)$. The inversion from the samples of the function $S(p,n)$ is to solve a Fredholm equation of the first kind. The Fourier transform of $S(p,n)$ on p is associated with the 3D Fourier transform of $E(r)$ by the Fourier-slice theorem, and the theorem leads us to an FFT-aided procedure for inversion and further a real-space procedure called the filtered back projection (FBP), which are standard methods of Radon inversion. The noise enhancement by differential processings of projection data is prevented by introducing lowpass filterings of data with loss of the spatial resolution in imaging. On the sampling condition of $S(p,n)$, a suggestive study is seen in (8). Within the framework of standard methods of Radon inversion, data missing may be confronted by a preliminary interpolation of data (7). The Gerchberg-Papoulis iteration may also be useful in recovering the missed projections from the data-missing FBP/FFT image estimate corrected so as to meet physical constraints (9).

INVERSION FROM SPARSE DATA

With the view angle too much limited and/or with the data too sparse for the above FBP/FFT and related inversions, the following direct inversions from the available projection data may be advantageous.

Function Model Fitting with AIC

One approach is to fit a function model to the plasma image $E(r)$. The least squares fitting of the Radon transform of the model to projection data $S=[S_1,S_2,...,S_M]^t$ yields the estimates of model parameters and thus an estimate of $E(r)$. Particularly, when we take a series expansion model with basis functions $b_n(r)$'s, the linearity of the Radon transform $R[\ ]$ presents a series expansion model of projection with basis functions $R[b_n(r)]$'s:

$$E(r) = \sum_{n=1}^{N} c_n b_n(r), \quad R[E(r)] = \sum_{n=1}^{N} c_n R[b_n(r)]. \tag{2}$$

Then, the least squares fitting is reduced to a normal equation system to be solved on the coefficients c_n's. The solution c_n ($n=1,2,...,N$) is to be used for the expression of $E(r)$ in Equation (2) to get an estimate of $E(r)$. Some basis functions for which $R[b_n(r)]$ can be analytically calculated are known (10). If the model is suitable, the image estimate shall be a good approximation of $E(r)$ even with the parameter number N much smaller than the number of data M. Then the normal equation shall be well-conditioned.

With respect to the choice of the function model, one may use the Akaike information criterion (AIC), that is, a neg-entropy which indicates the badness of the model on the basis of Kullback-Leibler (KL) information quantity (11) and can be written as

$$AIC(N) = M\ln\hat{\varepsilon}^2 + 2(N+1), \qquad (3)$$

where $\hat{\varepsilon}^2$ is the minimized mean of squares. As N is increased, we find the decreases of $\hat{\varepsilon}^2$ and thus of the first term in Equation (3). As the projection of the image estimate approaches the particular data S in the mean squares sense, we have a tendency to get the improvement of the spatial resolution in image reconstruction and contradictorily the lowerings of both the statistical accuracy and the robustness against data missing. The lowerings appear in $AIC(N)$ as the augmentation of the second term. The resulting minimum of $AIC(N)$ is expected to indicate the best value of N, and also to decrease in magnitude for a better type of model function. The model optimization with AIC has been applied firstly to a toroidal plasma imaging from tangential projections with respect to the spline function model (12), and then to the imaging of the MHD instability of tokamak plasma with respect to the Cormack and Fourier-Bessel expansions (13).

Linear Regression Model Fitting

Another approach starts with rewriting Equation (1) to a discrete form. Dividing the plasma domain into K pixels and defining the image vector $E=[E_1,E_2,....,E_K]^t$ with sample values $E_k=E(r_k)$, we write the projection data as $S=LE+e$, where L is an M by K matrix having as its (m,k) element a weight L_{mk} which geometrically evaluates the contribution of E_k to S_m. In the usual experiment condition of $M\ll K$, L is sparse. The residual vector e represents the errors in projection measurement and also in discretizing the Radon transform as LE.

ART and Least Squares Method

With a simple model of $e=0$, we take the matrix inversion for $M=K$ and otherwise we take some useful algorithms of iteration termed the ART (algebraic reconstruction technique) (14). The algorithms can include procedures of preserving the image estimate from negative values, if necessary. The neglection of e in approximating the particular projection data, which are inaccurate and few in number, tends to give the reconstruction highly resolved, and on the other hand, statistically unreliable and easily destroyed by data missing. Taking e into account, one may take the procedure of minimizing $M^{-1}\|e\|^2$ ($=M^{-1}\|S-LE\|^2$) to get the relation $(L^tL)E =L^tS$. However, for the sparse matrix L of large scale, one usually finds the matrix L^tL ill-conditioned and singular for $M<K$. To obtain a unique and stable solution, one has to take a countermeasure of regularization.

Parameter Number Reduction with Function Model

For the purpose of regularization, one may adopt a function model of $E(r)$ on the vector E. Then the determination of model parameters by minimizing $M^{-1}\|S-LE\|^2$ may be well-conditioned when the parameter number is much smaller than M. This approach of reducing the number of unknowns in normal equation is nothing but the above-described model fitting approach, where the Radon transform of the model function is numerically calculated with the operator L. The model may be optimized with use of the AIC.

Regularization with Objective Function

Another approach of regularization is to employ an objective function $F(E)$ which measures the spatial roughness of the image $E(r)$. Then the minimization of $F(E)$ on E under the constraint that $M^{-1}\|S-LE\|^2$ should be a constant ($=\varepsilon^2$) may yield a solution $\hat{E}$ smoothed according to the value of ε^2 which assigns the distance between the projection estimate $L\hat{E}$ and the data S. The solution is obtained by minimizing the quantity $\Lambda(E)=\gamma F(E)+M^{-1}\|S-LE\|^2$, where γ is the reciproical of the Lagrangian multiplier. When a large value is given to γ, the large weight attached to $F(E)$ in minimizing $\Lambda(E)$ leads to a large attained value of ε^2, $\hat{\varepsilon}^2 = M^{-1}\|S-L\hat{E}\|^2$, with a strongly smoothed solution. The γ is a parameter to be adjusted for the optimal regularization.

Maximum Entropy Method. A useful objective function is $F(E)=-H$ with the entropy function $H=-\sum_{k=1}^{K} E_k \ln E_k$. If E_k's are properly normalized, the K-dim. convex function $F(E)$ is positive-valued and a representation of the distance between the image E and the uniform distribution; with no constraint, the distance is decreased as the image E becomes uniform. Studies have been devoted to developing the iterative algorithms of minimizing $\Lambda(E)$ (15). The use of the logarithmic function for $F(E)$ keeps the image estimate to be positive. For the purpose of regularization, the objective function can be replaced by other convex functions like $F(E)=\sum_{k=1}^{K} \ln E_k$.

Linear Algebra Methods of Tikhonov-Phillips with GCV. When the convex function $\|E\|^2=\sum_{k=1}^{K}E_k^2$ is employed, the minimization can be achieved, with no aid of SIRT-like iterative algorithms (14), by taking the linear algebra tool as far as small-scale imagings of $K \leq 900$ are concerned as usual in plasma experiments. The singular value decomposition (SVD) of the matrix L, *i.e.*, $L=U\Sigma V^t$, offers an orthogonal series expansion of the image $E(r)$ which is linear on S, that is, the Tikhonov solution

$$\hat{E} = \sum_{i=1}^{M} w_i(\gamma)[(u_i \cdot S)/\sigma_i]v_i \qquad (4)$$

with $w_i(\gamma)=[1+M\gamma/\sigma_i^2]^{-1}$ ($M \leq K$). Here σ_i's are the singular values arranged in order of magnitude as $\sigma_1 \geq \sigma_2 \geq \ldots \geq \sigma_M$, and the vectors u_i ($i=1,2,..,M$) and v_i ($i=1, 2,..,K$) are the M- and K-dim. column vectors of the orthonormal matrices U and V, respectively. With the properties of $(u_i \cdot u_j)=\delta_{ij}$ and $(v_i \cdot v_j)=\delta_{ij}$, the vector series u_i's and v_i's serve as basis vectors in representing the projection S and the image E, respectively. In Equation (4), the division of the i-th spectral component of S, $(u_i \cdot S)$, by the i-th spectral component of L, σ_i, is the Radon inversion in "Fourier" space. The inverse $(u_i \cdot S)/\sigma_i$ in Fourier space is the spectral component to be allocated to the i-th basis vector v_i with which we compose the real space solution $\hat{E}$. In composing, $w_i(\gamma)$ is a decreasing positive-valued function of i and serves as the window function (*i.e.*, the lowpass filtering function) for avoiding the enhancement of noisy components of S in the high frequency (large i) region where the values σ_i's are small and may be computing-error dominated. The idea of lowpass filtering is identical with that of the Wiener filter.

Now, in Equation (4), in the limit of $\gamma=0$ we have $w_i(\gamma)=1$ for all i, and we find that the solution is reduced to the least squares solution with norm minimum. This relation suggests a modification of the Tikhonov solution, that is, the use of other types of window function like $w_i(\gamma)=[1+M\gamma/\sigma_i^4]^{-1}$ for getting a stronger regularization. Another modification is obtained by rewriting the model as $S=(LC^{-1})(CE)+e$ with a K-dim. matrix C. With the SVD of LC^{-1}, the Tikhonov solution with respect to the objective function $\|CE\|^2$ leads to the following image estimate after the inversion C^{-1}:

$$\hat{E} = \sum_{i=1}^{M} w_i(\gamma)[(u_i \cdot S)/\sigma_i](C^{-1}v_i) \qquad (5)$$

with $w_i(\gamma)=[1+M\gamma/\sigma_i^2]^{-1}$ ($M \leq K$). When the Laplacian operator is used for C, the above image estimate reduces to the Phillips solution, which may be well regularized with basis images (vectors) smoothed by C^{-1}, $C^{-1}v_i$ ($i=1,2,...,M$). This kind of modification is possible also by using the gradient operator *etc.* for C.

The above approach of Tikhonov-Phillips is provided with a useful criterion for optimizing the value of γ. The neg-entropy of Wahba (16), namely, the generalized cross validation (GCV), $G(\gamma)$, is defined by modifying the prediction sum of squares (PRESS), $P(\gamma)$, and written as

$$G(\gamma) = \hat{\varepsilon}^2 \bigg/ \left[1 - M^{-1} \sum_{i=1}^{M} w_i(\gamma)\right]^2. \qquad (6)$$

As γ is decreased, the monotonical decrease in numerator is compensated with the approach of the denominator to 0, and as a result, $G(\gamma)$ my take a minimum for the γ value which gives the optimal image estimate. Furthermore, with respect to the same data S, $G(\gamma)$ is expected to take a lower minimum for a better type of objective funtion, in other words, for a better type of window function.

APPLICATIONS

The numerical methods of reconstruction have been applied to the measurement of low-temperature discharge plasmas: (i) 3D recovery of the ion velocity distribution $f_{3D}(v)$ from electrostatic analyzer data for the purpose of ion acoustic wave study in double plasma device; (ii) spectral-line emission CTs on the neutral particle behavior in toroidal device and on the ionization wave in positive column.

The electrostatic analyzer in plasma gives a current-voltage characteristc whose gradient is proportional to the 1D velocity distribution function $f_{1D}(u;n)$ of charged particle to be related to the 3D function $f_{3D}(v)$ as

$$f_{1D}(u;n) = \int f_{3D}(v)\delta(u - n \cdot v)dv, \qquad (7)$$

where n is the unit normal of the collector surface. The recovery of $f_{3D}(v)$ from $f_{1D}(u;n)$ measured in different directions of n is a 3D object reconstruction from planar projections (plane integrals) of the type which was studied on the measurements of electromagnetic wave scattering (2) and of laser induced fluorescence (17). The inversion by FBP needs to evaluate the 2nd differential of projection from data and therefore has to cope with a comparatively strong noise enhancement. In a double plasma device, ion current measurements on argon plasma were made in 12 directions equispaced over the range of zenithal angle $0 \leq \theta < \pi$ in the axisymmetric scheme; the total number of projection data was $M=378$ (18), (19). The results of analysis are shown in Figures 1 and 2. It is seen that the FBP was low in resolution owing to the required lowpass filtering, and that the ART was high in resolution with noisy profiles even when the iteration was stopped at an early stage, while the MEM with the algorithm of Zhuang et al. (20) gave a smoothing adequate to distinguish the ion beam. The hollow of distribution around $v=0$ is due to the electrostatic analyzer sensitivity lowered for low energy. On the other hand, the least squares fitting of the Hermite-Laguerre series expansion model gave strong smoothing, and as the term number was increased for higher resolution, the AIC failed to take a minimum for the best reconstruction as seen in Figure 2. It is under the assumption of sufficiently good estimation that the AIC is a reasonable approximation of the KL mean information. The condition number which was evaluated as 10^{15} for the normal equation of $N \geq 30$ might raise a doubt on the validity of this assumption. The failure was not prevented even when the reconstruction was facilitated by excluding the hollow part of distribution from the least squares fitting.

A similar behavior of AIC was observed on the bicubic spline function model for an Hα emisson CT in a small tokamak (CSTN-III; $T_e \leq 10$ eV), in which three

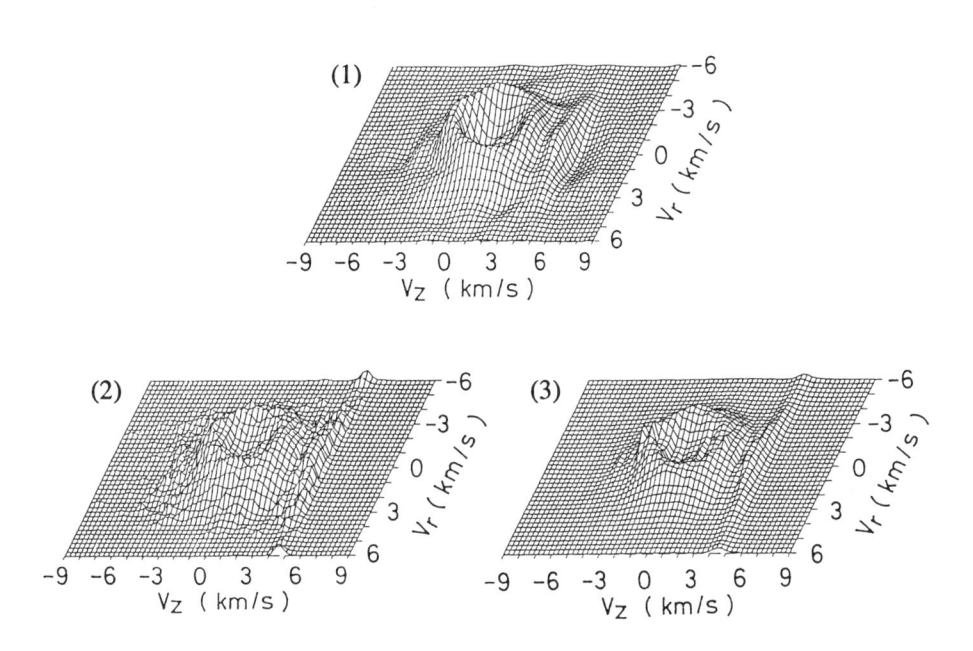

FIGURE 1. Ion velocity distribution $f_{3D}(v)$: results obtained with (1) FBP, (2) ART, and (3) MEM (displayed in v_z-v_r space with the velocity v_r perpendicular to the z axis); $K=1200$ for ART and MEM.

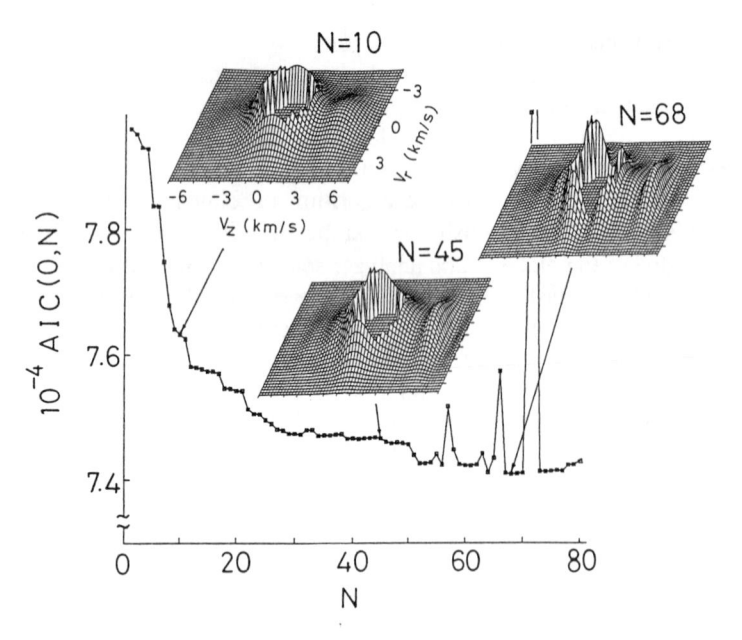

FIGURE 2. Dependences of $AIC(0,N)$ on the highest order N of Hermite polynomial when that of Laguerre polynomial was fixed to 0 for lessening the parameter number. The reconstructed $f_{3D}(v)$'s are displayed by insetting.

observation ports were available in a poloidal plane with limitations of view angle (21). These failures of AIC are in contrast with its successes on the same spline model for the tangential imaging where data missing was weak (12), and on the series expansion models for the MHD instability imaging where the models were suitable to the mode structure of plasma shape (13). When the objective function is adopted to regularize the least squares fitting of the models (22), one takes a step out of the maximum likelihood estimation framework for using the AIC.

In comparison with the behaviors of AIC, the GCV in linear algebra approach was reliable in taking the minimum for the optimal reconstruction. On the same Hα emission CT of small tokamak, the behavior of GCV in Phillips regularization was quite good as shown in Figure 3. The image estimate chosen with the minimum GCV criterion was robust on the sparseness of data and excellent in being better-smoothed in profile for the attained mean of squares that was comparable with those of the spline model fitting and of the Tikhonov regularization with modified window. The excellence over the latter was verified also by the lowered minimum of GCV (23)-(25). The GCV-aided Phillips regularization was successfully used, as shown in Figure 4, for spectral-line emission CT of the ionization wave packet of large amplitude which propagated along a cylindrical positive column of helium (26), (27).

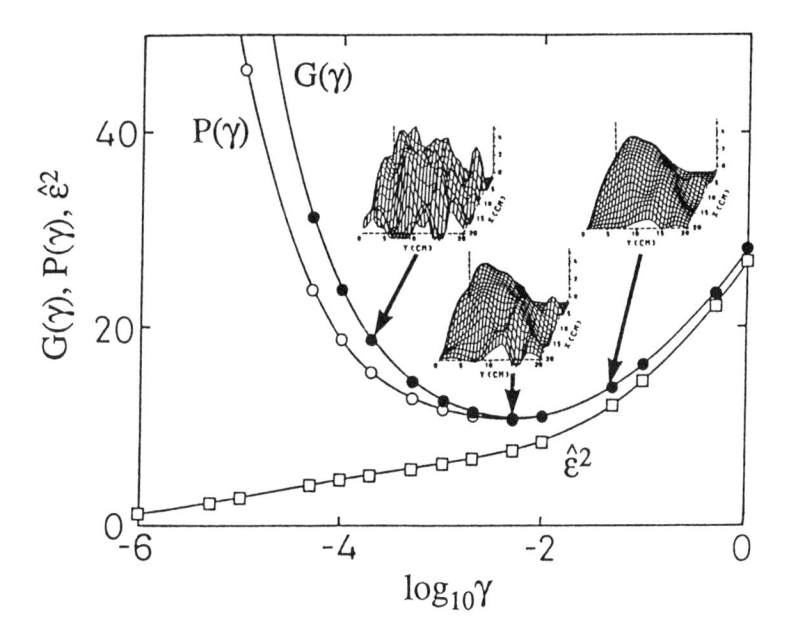

FIGURE 3. Variations of $G(\gamma)$, $P(\gamma)$, $\hat{\varepsilon}^2$ and the reconstructed emissivity distribution in poloidal plane with γ in the Hα emission CT (minor radius:10 cm; M= 600, K=900).

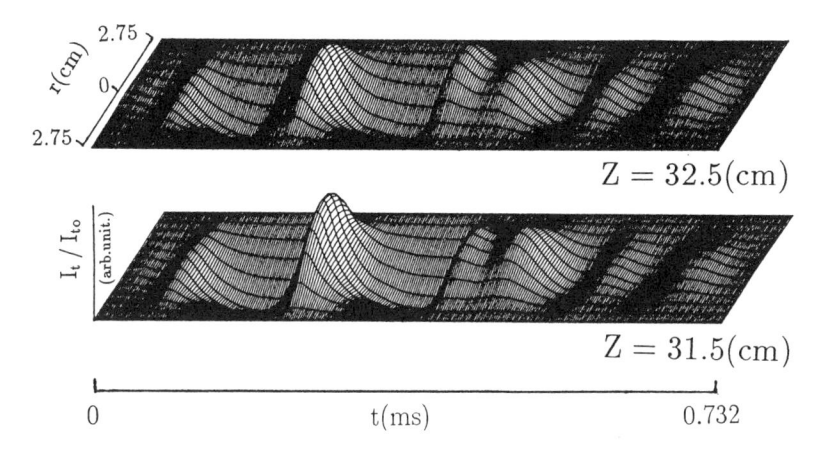

FIGURE 4. Spatio-temporal appearance of an ionization wave packet at two positions along a positive column.

ACKNOWLEDGMENTS

The author would like to thank Prof. T. Lehner of Ecole Polytechnique, Prof. S. Takamura of Nagoya University, and Prof. K. Ohe of the Nagoya Institute of Technology for their valuable supports from the standpoint of plasma experiment.

REFERENCES

1. Maldonado, C. D., and Olsen, H. N., J. Opt. Soc. Am. **56**, 1305-1313(1966)
2. Williamson J. H., and Clarke M. E., J. Plasma Physics **6**, 211-221(1971)
3. Myers, B. R., and Levine, M. A., Rev. Sci. Instrum. **49**, 610-616 (1978)
4. Sauthoff, N. R., von Goeler, S., and Stodiek, W., Nuclear Fusion **18**, 1445-1458 (1978)
5. Minerbo, G. N., Sanderson, J. G., *et al.*, Applied Optics **19**, 1723-1726 (1980)
6. Sebald, N., Appl. Phys. **21**, 221-236 (1980)
7. Melnikova, T. S., and Picalov, V. V., Beitr. Plasmaphys. **22**, 171-180 (1982)
8. Klug, A., and Crowther, R. A., Nature London **238**, 435-440 (1972).
9. Veretennikov, V. A., Koshevoi, M. O., *et al.*, Soviet J. Plasma Phys. **18**, 131-132 (1991)
10. Deans, S. R., *The Radon Transform and Its Applications*, New York: John Wiley & Sons, 1983.
11. Akaike, H., IEEE Trans. Automat. Contr., **AC-19**, 716-730 (1974)
12. Iwama, N., Takami, H., *et al.*, IEEE Trans. on Plasma Sci. **PS-15**, 609-617 (1987).
13. Nagayama, Y., J. Applied Physics **62**, 2702-2706 (1987)
14. Herman, G. T., *Image Reconstruction from Projections*, New York: Academic Press, 1980
15. Smith, C. R., and Grandy, Jr., W. T.,*Maximum Entropy and Baysian Methods in Inverse Problems*, Dordrecht: D. Reidel Pub. Co., 1985
16. Wahba, G., SIAM J. Numer. Anal. **14**, 651-667 (1977)
17. Koslover, R., and McWilliams, R., Rev. Sci. Instrum. **57**, 2441-2442 (1986)
18. Iwama, N., Lehner, T., Noziri, H., *et al.*, Jpn J. Applied Physics **27**, 1732-1742 (1988)
19. Iwasaki, H., Iwama, N., Noziri, H., *et al.*, J. Plasma and Fusion Res. **62**, 262-281 (1989)
20. Zhuang, X., *et al.*, IEEE Trans. on Acoust. Speech Signal Process. **ASSP-35**, 208-218 (1987)
21. Shen, Y., Takamura, S., Iwama, N., *et al.*, J. Plasma and Fusion Res. **59**, 30-50 (1988).
22. Zoletnik, S., and Kalvin, S., Rev. Sci. Instrum. **64**, 1208-1212 (1993)
23. Iwama, N., Yoshida, H., and Takimoto, H., *et al.*, Appl.Phys. Lett. **54**, 502-504 (1989)
24. Iwama, N., "Numerical Optimizations of Plasma Image Reconstructions with Sparce-Data Computed Tomography," in *Proceedings of the IAEA TCM on Time Resolved Two- and Three-Dimensional Plasma Diagnostics*, Nagoya, Japan, Nov. 12-22, 1991, pp. 293-298
25. Iwama, N., and Teranishi, M., *Image Processing:Theory and Applications*, Amsterdam: Elsevier (edited by Vernazza, G., *et al.*) 1993, pp. 377-380
26. Ohe, K, Naito, A., Kimura, T., and Iwama, N., Phys. Fluids **B3**, 3302-3311 (1991)
27. Ohe, K., Takeshita, S., and Kimura, T., Phys. Fluids **B5**, 2331-2333 (1993)

Thin Film Deposition on Internal Walls of Cavities and Complex Hollow Substrates

L.Soukup, M.Šícha, L.Jastrabík, M.Novák

Institute of Physics, Division of Optics, ASCR,

Na Slovance 2, 180 40 Prague 8, Czech Republic

Abstract. In the present report the radio frequency low pressure chemical system with hollow cathode is presented. This plasma chemical reactor is able to deposit the thin films on the internal walls of cavities, holes and complex shapes of hollow substrates. This system can replace the customary used electroplating technology for the surface coating technology of the machine components.

We also presented the simple theoretical model of the processes inside the low pressure RF plasma chemical reactor with the supersonic plasma jet. The results of our model are in qualitative agreement with the experimental data.

Furthemore as example of plasma jet application the deposition of germanium nitride and silicon nitride thin film is presented.

INTRODUCTION

One of the task of the surface coating technology is to deposit the thin film on the surface of the machine component with a complicated shape. One of the common problems of the two widely used technologies, i.e. electroplating and deposition where the plasma is used is depth efficiency, it means to keep thickness of the thin films even.

During the electroplating the current density i.e. the grow rate is on the edges high while in the holes is wery low. The current distribution in the electroplating technology can be influenced by two ways. The first one is the arrangement of the electrodes geometry by screening or by use of the counter electrodes. The second one is a way where we can directly influence the electrokinetics using the complexing agents. In practice the concentration of complexing agents are determined empiricaly. The electroplating technology does not include only working bath but there is a system with water rinse bath as well. Time to time it is necessary to change the content of working bath or to clean up the rinse bath. The common process for this is oxydation or reduction of metalic compounds and than precipitation like unsoluble hydroxides. But absolutely unsoluble hydroxides does not exists and as you can see from the Table 1. there exists a sufficient amount of metal compounds in the waste waters. The highest coeficient of solubility has cadmium and from this reason the use of it was forbidden. The rest of metal compounds have bad influence on the biological cleaning of waste waters and the closed technology circle must be used. For this reason the electroplating system must work like quite closed circle whith include the cleaning of the water by distillation. The costs of the closed circle technologies therefore increases and it will be convenient to replace it by other technologies. One of the possibilities is to use the plasma technologies for deposition of the thin films. The comercial available plasma-chemical reactors are not suitable for deposition of the thin films on the internal walls of cavities, holes and complex shape of hollow substrates. For this reason it is important to develop the plasma - chemical reactors which would be able to deposit the thin films in this case. The radio frequency low pressure chemical system with hollow cathode can in particular case fulfil the mentioned requirement.

TABLE 1. Solubility coefficients of hydroxides.

Metal	Solubility coefficients of hydroxides [g dm^{-3}] at 25 °C
Zn^{2+}	$4 \cdot 10^{-17} - 7 \cdot 10^{-18}$
Cu^{2+}	$2 \cdot 10^{-19} - 5 \cdot 10^{-20}$
Ni^{2+}	$5.8 \cdot 10^{-15} - 6.3 \cdot 10^{-18}$
Ca^{3+}	$3 \cdot 10^{-28} - 6.3 \cdot 10^{-31}$
Fe^{3+}	$8.7 \cdot 10^{-38} - 6.3 \cdot 10^{-40}$
Cd^{2+}	$1.3 \cdot 10^{-4} - 2.18 \cdot 10^{-4}$

RF LOW PRESSURE PLASMA CHEMICAL REACTOR WITH HOLLOW CATHODE

In the plasma-chemical reactors with hollow cathode the working gas flows customary through the hollow cathode with acts simultaneously for working gas as the nozzle. According to the working gas velocity in the hollow cathode these systems can be roughly divided into two groups : The system with supersonic working gas flow in the reactor chamber and with subsonic one. The properties of these systems also stroungly depend on the nature of the EMF by means of which the plasma inside the reactor chamber is generated.

The radio frequency plasma jet system (RPJ) with supersonic flow of the working gas inside the nozzle was developed for deposition of the thin films into the hollow substrates and substrates of complicated shapes. (see eg.[1]). The RPJ system is a new plasma chemical system in which the hollow cathode discharge is used for creation of the plasma jet channel which is added on the primary RF plasma which is created inside the reactor chamber. The advantage of this system is that the plasma jet channel exhibits the high rate plasma chemical activity.

Depending on experimental conditions the nozzle outlet can be sputtered or evaporated by the hollow cathode discharge inside the nozzle. The deposition of the thin films can be carried out by the two basic wavs:

Plasma Chemical Vapor Deposition (PCVD) when a mixture of reactive gas is activated in the plasma stream and the chemicaly activated species recombine on the substrate surface to form the thin film.

PhysicalVapor Deosition (PVD) when the components which are necessary for growth of the thin film is supplied by an errosion of the nozzle for example by reactive sputtering or by an arc evaporation.

For the process control and for the optimization of the RPJ regime a theoretical understanding and modelling of this plasma jet system is necessary.

The model of the low pressure plasma chemical reactor with hollow cathode and supersonic flow of the working gas

In its simplest form RPJ consists of the radio frequency plasma chemical reactor with two electrodes (see Fig. 1). The substrate is placed on the grounded electrode. The nozzle in the RF electrode connected with radiofrequency signal admits the working gas into the continously pumped vacuum reactor. The working gas pressure is maintained at several ten Pa and the working gas flow is adjusted in such a manner that the flow inside the reactor chamber is supersonic. At the RF power below a certain limit (for example < 20 W) an ordinary RF plasma is generated in the reactor chamber. However at conditions when the RF power exceeds this limit an intensive hollow cathode discharge is

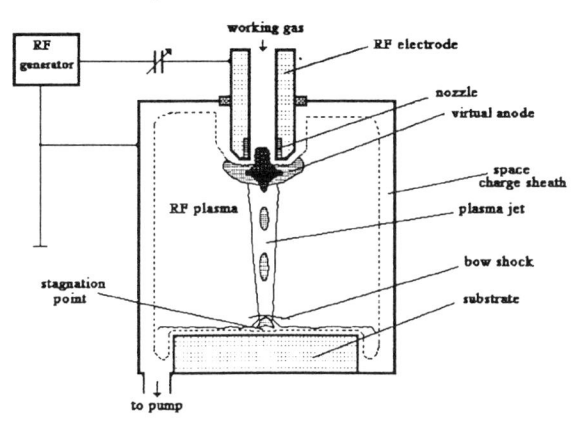

Fig.1.

generated inside the nozzle. The incoming working gas forces this discharge supersonically from the nozzle into the reactor and a well defined plasma jet is created inside the primary RF plasma.

The model consists of the following particular problems:

(i) the description of the neutral gas flow inside the nozzle,

(ii) the gas breakdown inside the cylindrical RF nozzle which acts as a hollow cathode discharge.

The aim of the present report is to present the main results of the theoretical and experimental studies dealing with the model of the described plasma system with hollow cathode [2], [3].

(i) the description of the neutral gas flow inside the nozzle

In the case in which the working gas velocity at the nozzle outlet is supersonic (RPJ system) the critical point where the velocity reaches the sonic value lies somewhere inside the nozzle near the nozzle output. At this critical point the shock wave is created on which the gas pressure drastically falls down. Downstream after shock wave the pressure in the centre of the nozzle is only about a few percent higher compared to the reactor chamber pressure and reaches the magnitude of several tens or hundreds Pa. Upstream of the shock wave the pressure of the working gas is in the order of $10^2 - 10^3$ Pa. Under described experimental conditions it is possible to assume that the working gas flow is laminar and viscous. The model of the working gas flow in the nozzle is in more detail described in [2].

(ii) The Generation of the Plasma Jet Channel

Owing to higher mobility of electrons with respect to ions in the plasma both RF electrodes are separated from the plasma by a positive space charge sheath across which the major part of the AC and the time averaged DC potential drop falls. According to our model this potential drop across the space charge sheath around the electrode with the nozzle represents the source of EMF for generation of the hollow cathode discharge which creates the plasma jet channel. At the RF power below a certain limit the potential drop across the space charge sheath is not sufficient to generate the hollow cathode discharge inside the nozzle and an ordinary RF plasma is generated in the reactor chamber. However at conditions when the RF power exceeds this limit the potential drop is sufficient to generate an intensive hollow cathode discharge inside the nozzle. The incoming working gas forces this plasma supersonically from the nozzle into the reactor and a well defined plasma jet is created inside the primary RF plasma.

Therefore we consider the gas discharge breakdown in the nozzle as a breakdown of the longitudinal hollow cathode glow discharge which is controlled by so called "hollow cathode effect". The model of processes in the hollow cathode are described in [4] where axial dependence of the cathode current is calculated.

The simplified condition for the plasma jet generation has already been described in [2]. The processes inside the nozzle are described with more precision in [5]. If we take into account the hollow cathode effect then according the paper [5] we obtain for the breakdown condition of the discharge inside the nozzle the following expression:

$$\langle K \rangle = \frac{1+\gamma}{\gamma + G} \frac{1 - \frac{G_{iS}}{\gamma}}{1 - \frac{G_{iS}}{\gamma + G}} ;$$

where

$$\langle K \rangle = \exp\left(\frac{1}{T} \int_0^T \int_0^d \alpha(x,t) dx dt\right),$$

α and γ are the first and the third Townsend coefficients, T is the period of the RF voltage, d is the thickness of the cathode fall region inside the nozzle, the coefficient G expresses the electron emission from the nozzle surface due to the hollow cathode effect and the coefficient G_{iS} expresses the particular process of electrons secondary emitted from the nozzle surface by means of ions created by pendelum high energetic electrons inside the cathode fall region.

APPLICATION

As has been mentioned above the deposition of the thin films by means of RPJ system can be carried out by the two basic ways:

Plasma Chemical vapor deposition (PCVD) when a mixture of reactive gas is activated in the plasma stream and the chemically activated species recombine on the substrate surface to form the thin film.

Physical Vapor Deposition (PVD) when the components which are necessary for growth of the thin film is supplied by an errosion of the nozzle for example by reactive sputtering or by an arc evaporation.

Typical example of PCVD thin film deposition by means of the system with the hollow cathode and the supersonic flow is the deposition of the silicon nitride thin films. The working gas mixture is composed from nitrogen and 3% silane diluted in argon [6].

Typical example of (PVD) by means of the system with the hollow cathode and the supersonic flow when one of the component is reactive sputtered from the nozzle wall is deposition of the germanium nitride thin film [7].

CONCLUSION

In the present report the radio frequency low pressure chemical system with hollow cathode is presented. Also the simple theoretical model of the processes inside the low pressure RF plasma chemical reactor with the supersonic plasma jet is presented. The results of our model are in qualitative agreement with the experimental data. This indicated that our model can serve as a good start to a more elaborated description of the processes in this type of plasma chemical reactor.

Furthermore as example of plasma jet application the deposition of germanium nitride and silicon nitride thin film is mentioned.

ACKNOWLEDGEMENTS

The research has been done in the frame of the Association for Education Research and Application in Plasma Chemical Processes and has been financially supported by GAČR 0659 grant of the Grant agency of Czech Republic.

REFERENCES

[1] L. Bárdoš, Proc. XXI ICPIG Bochum Vol. III (1993), 98.

[2] M. Šícha, L. Bárdoš, M. Tichý, L. Soukup, L. Jastrabík, H. Baránková, R. J. Soukup, J. Touš, Contributions to the Plasma Physics 34 (1994), 749.

[3] M. Tichý, M. Šícha, L. Bárdoš, L. Soukup, L. Jastrabík, K. Kapoun, J. Touš, Z. Mazanec, R. J. Soukup, Contributions to the Plasma Physics 34 (1994) 765.

[4] V. Hanzal, M. Šícha, V. Kapička, Czech. J. Phys. (1995) (in press).

[5] M. Šícha, L. Soukup, L. Jastrabík, M. Novák, M. Tichý, Surface and Coatings Technology 74-75 (1995)212

[6] L. Bárdoš, V. Dušek, M. Vaněček, J. Non Cryst. Solids 97 & 98 (1987) 181.

[7] L. Soukup, V. Peřina, L. Jastrabík, M. Šícha, P. Pokorný, R. J. Soukup, M. Novák, J. Zemek, Surface and Coatings technology (1995) (in press).

Atmospheric pressure dielectric controlled glow discharges : diagnostics and modelling

Françoise Massines*, Rami Ben Gadri*, Philippe Decomps*,
Ahmed Rabehi[†], Pierre Ségur[†], Christian Mayoux*

*Laboratoire de Génie Electrique and [†]Centre de Physique des Plasmas et de leurs Applications, Université Paul Sabatier, 118 route de Narbonne, 31062 Toulouse cedex, France

Abstract : The aim of this paper is to prove the existence of atmospheric pressure dielectric controlled glow discharge and to describe its main behavior. Electrical measurements, short time exposure photographs and numerical modelling were used to arrive at the conclusion. Experimental observations and numerical simulation agreed within a close range. Therefore, the analysis of the calculated space and time variations of the electric field together with the ion and electron densities helps to explain the discharge mechanisms involved.

INTRODUCTION

The expression "Atmospheric Pressure Glow Discharge" (APGD) is used in various fields : partial discharges (1-2), lasers (3-5) and cold plasma processing (6-12). Therefore, the features of discharges called APGD can be significantly different and depend on the domain of investigation. A partial APGD is defined as a pulseless discharge, the corresponding charge of which is lower than 1 pC. In the two other fields, APGD are pulse discharges which considerably differ from each other by the order of magnitude of the current amplitude : some tenths of mA for cold plasma processing and some A for discharge used to pump lasers. Of course, these various discharges are obtained in different configurations. Partial discharges are generated in cavities smaller than 1 mm^3. Discharge volumes for lasers are of the order of some cm^3. Gaps of some mm are used for cold plasma processing. Another major difference is the condition to get APGD. Glow partial discharge develops between aged dielectrics. In lasers, a preionisation of the gas followed by a rapid rising voltage is necessary, while in plasma processing the discharge must be controlled by a dielectric barrier and fulfil other conditions described below.

Even if most of these discharge characteristics are very different from each other, they have a common point : they do not present a radial localization i.e. there are neither filament-like discharges nor streamer-like discharges. The first stage of these discharges is controlled by Townsend mechanism (2,13-14). This phenomenon is usually observed for low values of pressure distance (Pd) product. It consists of a multiple primary avalanche build up in a space charge free-field to the point where secondary emission from the cathode provides sufficient feedback of electrons to maintain the discharge.

© 1996 American Institute of Physics

When the Pd product is higher than 10 Torr.cm (at atmospheric pressure), a streamer-like breakdown is the more probable mechanism. A single avalanche creates such a charge density that the space charge field due to ions becomes comparable to the applied field. In this case, secondary avalanches will tend to converge quickly towards the primary avalanche accelerating the development (some ns) of a conducting filamentary channel of about 100 µm radius.

If the stability of the discharges used for excimer lasers has been extensively studied (4-5), this is not the case for APGD controlled by a dielectric barrier. Previous experimental works have shown the usefulness of these discharges for cold plasma processing. It has been shown that this type of APGD is more efficient than silent or corona discharges to produce ozone (7) and to increase surface energy of polymers (9). They are also well adapted to make deposits of the same quality as those obtained with low pressure glow discharge (6). These results increase the field of application of high pressure discharges which are particularly suitable for on line treatments (8).

Obtaining APGD is not obvious. Experimental studies permit to determine two ways to stabilize the discharge: (i) working in helium using an excitation frequency higher than 1 kHz (6, 8) and (ii) replacing a plane electrode by a grid (11-12). In each case, the discharge is controlled by a dielectric layer, which means that the discharge is a transient discharge and the amount of charge as well as the energy imparted to an individual microdischarge is limited (15).

Even if there is still some ambiguity about the existence of a real glow discharge controlled by a dielectric layer, electrical (7,9,11) and acoustic (12) measurements have clearly shown that a discharge distinct from a filamentary discharge can be obtained at an atmospheric pressure. Moreover, a better comprehension of the physics of this discharge is necessary. Therefore, the aim of the present work is to determine precisely, through experiments and numerical modelling, the behavior of such a discharge. The first part of this paper presents electrical measurements and time evolution of the light distribution in the gap, the second gives results of the numerical modelling and the third, the behavior of an APGD controlled by a dielectric barrier explained from the analysis of the numerical results correlated to the experimental observations.

EXPERIMENTAL RESULTS

Experimental set-up

The experimental set-up was described elsewhere (16). The discharge is generated between two plane parallel electrodes, 4 cm in diameter, covered by an alumina plate. The power supply frequency ranges from 1 to 100 kHz. For this study, an alternative voltage of 1060V at a frequency of 10 kHz was used. Electrode gap was fixed to 5 mm and the gas used was helium at a pressure of 10^5 Pa. An electrical and an optical method of characterization were employed.

The discharge current, I_d, measured with a 50 Ω resistor in series with the grounded electrode and the external voltage, V_a, applied to the electrodes are recorded in a 100 MHz numerical oscilloscope. Values measured during one cycle are transferred to a computer and then the voltage applied to the gas, V_g, as well as the voltage applied to the alumina plates i.e. the memory voltage, Vm, are calculated from the following equations:

$$V_g(t) = V_a(t) - V_m(t) \tag{1}$$

$$V_m(t) = 1/C_{ds} \int_{t_0}^{t} I_d(t).dt + V_m(t_0) \tag{2}$$

Cds is the capacitor equivalent of the alumina plates. Assuming that the discharge has a radius of 4 cm, Cds is equal to 70 pF. Id is the discharge current and Vm(to), the voltage due to the charges accumulated on the alumina plates during the previous discharge. It is adjusted in such a way that the mean value of Vg over a cycle is equal to 0 V, which means that there is no auto-polarization.

A high speed electronic camera has been used to follow the time evolution of the spatial distribution of light during the discharge. Pictures of the gap were taken with time exposures of 10 ns or 100 ns. The intensified tube of the CCD camera was triggered by the power supply used to excite the discharge. An adjustable delay was introduced to synchronize the intensified tube excitation and the discharge current.

Electrical behavior

Figure 1 shows the different voltages and the discharge current during a single cycle. The current is characterized by one peak per half cycle. The amplitude of the peak is 90 mA. Its rise time is about 1 µs, the decrease is longer. There is at first a rapid decrease : 7 mA reached after only 2 µs and then a slower decrease for at least 20 µs. The charge density associated with a single discharge is 13 nC/cm^2.

However, the main point is that the discharge current remains exactly the same from one cycle to another i.e. it has the same periodicity as the external voltage (9).The variations of Vg and Vm during one cycle has been calculated using equations 1 and 2. At the beginning of the cycle, Va=0V and Vg=-Vm(to)=1150V. A 400 V increase of V$_a$ is sufficient to turn on the discharge. At that moment if the field applied to the gas is uniform, its value will be 3.1 kV/cm. Then, the charges,

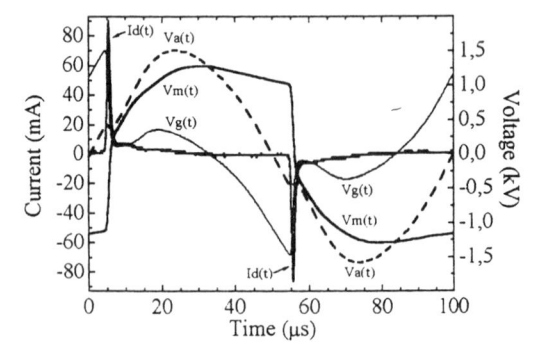

FIGURE 1. Time variation over one cycle of the measured applied voltage Va, the gas voltage Vg, the memory voltage Vm and the discharge current Id.

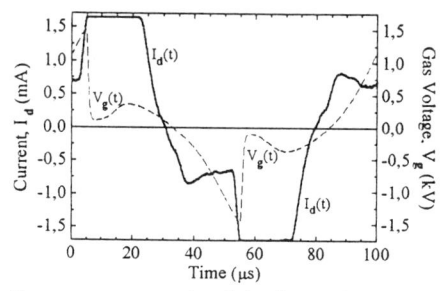

FIGURE 2. Time variation over one cycle of the low values of the measured discharge current Id, and of the gas voltage Vg.

resulting from the discharge and accumulated on both the dielectric surfaces, induce a memory voltage leading to a rapid decrease of V_g until the current becomes too small to compensate for the increase of the external voltage. A reduction of V_g results in the extinction of the discharge avoiding the formation of cathode spot (11). In the next half cycle, the applied voltage and the memory voltage increase the gas voltage which gradually becomes greater than the breakdown voltage initiating a new discharge.

The power density of this discharge is 300 mW/cm^3 increasing with frequency and voltage. Spence *et al* (12) found some tenths of mW/cm^3 and the same variations as in this work. These values are higher than those observed for filamentary discharges.

Following the discharge, the applied voltage increases by a value of 1.2 kV and since the decrease of the gas voltage during the discharge is equal to 1.4 kV, no more discharge is observed. This is very different from the normal behavior of silent discharge (15). This observation implies that at the end of the discharge the entire surface of the electrodes is uniformly charged. Okazaki *et al.* (11) present Lissajous figures of voltage-charge and define APGD as a discharge having only one peak each half cycle. This definition will be accurate if the time constant of the circuit is less than 1 ns.

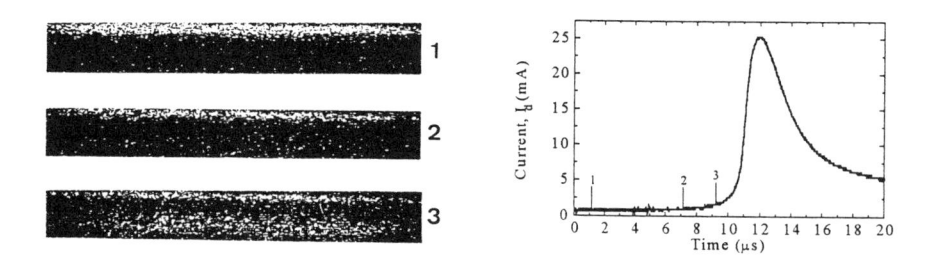

FIGURE 3. 100 ns exposure time photographs of the gap, taken during the discharge initiation. The number on the current waveform corresponds to the number on the right side of the picture and indicates the time when the picture was taken. The gap length is 5 mm and the electrode radius 4 cm. In each picture, the cathode is located at the bottom.

FIGURE 4. 10 ns exposure time photograph of the gap, taken when the discharge current is maximum.

Between two consecutive discharges the current is weak but not equal to 0. When the stationary discharge is obtained, a special behavior is always observed around the time when the polarity of the voltage applied to the gas changes (Fig. 2) : about 2 µs after the inversion of the current polarity, the variation of Id increases and what we call "residual current peak" is observed. Its order of magnitude is 0.3 mA. This current decreases when the frequency of excitation is reduced. As the electric field is low and the current variation is high, it is likely that most of the charged particles are electrons.

Time evolution of the spatial distribution of light

Photographs taken during the initiation of the discharge with a 100 ns exposure time are presented in figure 3. The development of the light intensity with time does not show a radial localization of the discharge.

The luminous area near the anode is the memory of the previous discharge. Its intensity decreases with time even when the new discharge is turned on. The light first occurs everywhere in the gap (Photograph 2) and then its intensity increases close to the cathode. This behavior agrees well with Townsend mechanism (13, 17) which presumes successive generation of avalanches. Then, since ions are less mobile than electrons, a positive space charge occurs producing a more important variation of the field near the cathode and the formation of a cathode fall leading to a glow discharge. A 4 µs delay is measured between the observation of the first luminous phenomenon and the current maximum.

At the maximum of the current, a picture was taken with an exposure time of 10 ns (Fig. 4) i.e. the order of magnitude of the life time of a streamer. No channel

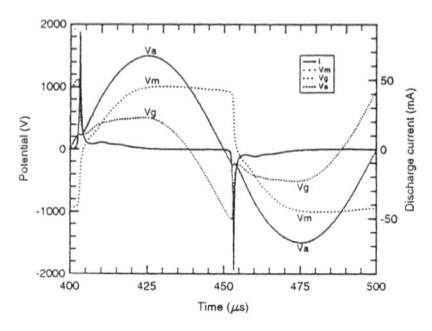

FIGURE 5. Evolution during a cycle of the calculated applied voltage V_a, gas voltage V_g, memory voltage V_m, and discharge current Id.

appears and 3 luminous domains can be distinguished : there is an area of high light intensity near the cathode followed by a dark space of about the same thickness and then a rather luminous zone of 3 mm. This is exactly the feature of a glow discharge with the cathode and the negative glow (that can not be distinguished at high pressure (18)), the Faraday dark space and the positive column.

NUMERICAL MODEL RESULTS

Numerical model

The numerical model has been described in the papers (10,19). It is a one dimensional model based on the numerical solution of the electron and ion continuity and momentum transfer equations coupled to Poisson's equation. As is usual in the case of high pressure discharges, electrons and ions are assumed to be in equilibrium with the electric field. The model is self consistent and gives the space and time variations of the electric field, the electron and ion densities. Time variations of gap voltage, memory voltage and current density are also obtained.

The charge accumulation on the dielectric as the discharge develops is carefully taken into account and the voltage boundary conditions for dielectrics are derived by considering the equivalent circuit of the gas gap in series with the equivalent capacitor of the dielectrics. The electric circuit is also considered to calculate the external voltage during the discharge pulse.

The kinetic description is very simple, only atomic ions and electrons are taken into account. Data concerning the electron transport parameters and ionization coefficient were calculated from a numerical solution of the Boltzmann equation. It was shown (19) that the current discharge calculated with pure helium data does not quantitatively agree with experimental results. Since emission spectroscopy proves (16) the presence of impurities in the gas which can play a role in an helium discharge, a model composed of a mixture of helium with 0.2 % of argon has been considered for the calculation. The mobility of ions was obtained from Ward's paper (20). The secondary cathode emission was kept constant and equal to 0.2.

Electrical behavior

The discharge is said to be in a stationary state if it is made of a succession of identical transient discharges that occur at each cycle of the applied voltage. Calculation is stopped when this condition is fulfilled. For the chosen initial conditions 5 periods are necessary.

The evolution during one cycle of the discharge current, Id, and the different voltages are plotted in figure 5. The external voltage decreases during the discharge and the transient current is similar to the measured one. Compared to the experimental observations, breakdown occurs for a lower value (1125 V) of the gas voltage. Nevertheless, the variations of voltage are comparable to those observed experimentally. The difference between experiment and calculation comes from the fact that all the collisional processes are not taken into account while calculating.

The current behavior between two discharges is exactly the same as that of figure 2. Since the calculation allows to separate the respective contribution of

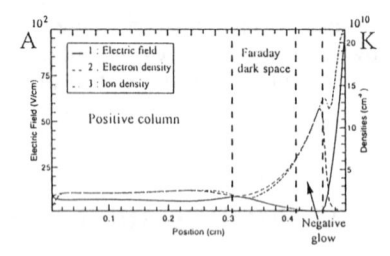

Figure 6. Spatial evolution of the electric field and the ion and electron densities at the time when the discharge current is maximum. The cathode (K) is on the right side.

electrons and ions, it appears clearly that displacement of electrons is at the origin of the residual current peak. 30 μs after the discharge switches off, there are still electrons in the gap.

A very good agreement between experimental and numerical results is achieved. Thus the analysis of the electric field, the ion and electron densities as a function of time and position will enable to explain the measurements. The first part of the following discussion shows that the discharge obtained is really a glow discharge, the others are the description of the extinction and the initiation of the discharge.

DISCUSSION

Glow discharge

We will first discuss the results found at the current maximum. The spatial distribution of the electric field as well as the ion and electron densities are presented in figure 6. These characteristics are typical and similar to those of low pressure glow discharge. Four specific regions of the glow discharge can be identified in figure 6 : (i) a strong cathode fall with a maximum field value close to

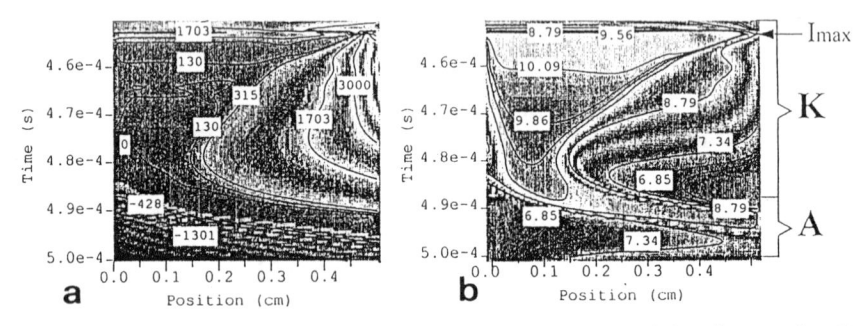

FIGURE 7. Evolution during a half cycle of the electric field, **a**, and the electron density, **b**, as a function of the position in the gap and the time. K indicates the cathode and A, the anode.

312

15 kV/cm. In this area the density of ions reaches a value of 2.10^{11} cm^{-3}. Its extension is limited to 0.4 mm, (ii) a negative glow 0.4 mm in length where the maximum density of electrons and ions is $1.2\ 10^{11}$ cm^{-3}, (iii) a Faraday dark space, 1.1 mm thick, where a small negative space charge is formed, (iv) a positive column 2.7 mm in width with a constant field equal to 700 V/cm. In this region the mobility of electrons is reduced because of their interaction with ions. The density of the two species is about 10^{10} cm^{-3}.

Considering that the light of the negative and the cathode glow can not be distinguished (18), the size of these zones is in good accordance with the light intensity measurements at the maximum of the current. This confirms that the discharge obtained is a glow one, like those obtained at low pressure. It is to be noticed that the maximum values of ion and electron densities are of the same order as those of low pressure glow discharge.

Discharge extinction

Results of figure 7 permit to understand how the discharge turns off. When the discharge current decreases, the amount of ionization becomes very small and the size of the cathode fall increases. The positive column decreases very slowly. As the electric field is almost equal to 100 V/cm, in this area, the loss of charged particles mainly occurs by recombination. 30 µs after the discharge there are still 10^9 cm^{-3} electrons near the anode and ions are more or less uniformly distributed in the gap. This is true until the polarity of the field near the anode changes. At that time, electrons quickly move to the new anode. 10 µs later, the maximum value of the electron density is less than 10^7 cm^{-3} while that of ions is 10^9 cm^{-3}.

The rapid displacement of electrons explains the particular current behavior observed around the time when the gas voltage polarity changes.

It is to be noticed that this behavior can only be obtained if 30 µs after the discharge extinction there are still electrons in the gap i. e. if a positive column of a sufficient size has been built during the discharge. The amplitude of this residual current peak depends on the number of electrons in the gas when the field polarity reversal occurs.

Discharge initiation

The literature indicates that a glow discharge occurs if the first step of the discharge is a Townsend mechanism. This will be obtained if two conditions are fulfilled : a secondary cathode emission high enough (2) and sufficient seed electrons to ensure formation of a lot of avalanches under low field (4) i. e. avoiding overvoltage that increases the probability of formation of a space charge field of a single avalanche comparable to the applied field.

Therefore, the density of electrons that are still in the gas when the breakdown happens is a major parameter. This number is related to the amplitude of the reverse current peak. Experimental as well as numerical results show that the amplitude of this peak decreases when the frequency of excitation is reduced. At 1 kHz, the stationary state can not be obtained either experimentally or numerically and there is no reverse current peak. This can be explained by the fact that the time between the end of the discharge and the polarity inversion of the gas voltage is long enough to permit the positive column to disappear.

The electron density at the breakdown, found from the numerical model results, is between $5 \cdot 10^6$ and $5 \cdot 10^7$ cm^{-3} at 10 kHz and lower than 2.10^5 cm^{-3} at 1 kHz. These values match pretty well with those necessary to get a stable discharge for lasers since a preionisation creating at least 10^6 to 10^7 electrons per cm^3 seems to be necessary (21).

Charges moving between 2 discharges should play another role. Since they arrive at the insulator surface under low field they must be trapped in shallow traps. Thus, as they are easier to be released in the next half period, they increase the value of the secondary cathode emission and so, like in the case of partial discharges (2), there is an increasing probability of obtaining a glow discharge.

According to the model, the primary avalanches appear near the anode and in the middle of the gap where the density of electrons is higher. The increase of gas ionization leads to the formation of a cathode fall that becomes more and more narrow until the current is maximum. This constriction enables to maintain the discharge even if the voltage decreases by a factor of two.

The maximum value of the current is reached 4 µs after the first avalanche. The same duration can be deduced from the photographs presented in figure 3. This is 1000 times the rise time of a streamer.

CONCLUSION

In spite of the simple kinetic description used for the model, experimental and numerical results agree within a close range. They show that a glow discharge controlled by a dielectric layer can be obtained at atmospheric pressure. This stationary transient discharge is characterized by a periodic discharge current and has the same structure as that observed for a low pressure discharge.

The APGD behavior has been described from the initiation to the extinction of the discharge : the breakdown is obtained under uniform field, no constriction is observed. As the ion density increases, a cathode fall is created. Its thickness becomes smaller and smaller until the maximum of current is reached. This decrease of the cathode fall allows to self-sustain the discharge even if the gas voltage is divided by a factor of two. The maximum values of electron and ion densities are about 10^{11} cm^{-3}. After the current maximum, the thickness of the cathode fall increases slowly and there is a decrease of the positive column length.

When the gas voltage polarity changes between two consecutive discharges, a residual electronic current is observed if a positive column has been formed and if the frequency is high enough. This current is a memory of the positive column and an indication that the number of electrons in the gap will be sufficient to produce the next breakdown under a low field which is a necessary condition to obtain an APGD. According to the numerical model, a density of electrons higher than 10^6 cm^{-3} at the breakdown voltage seems to be necessary to obtain an APGD controlled by a dielectric layer.

ACKNOWLEDGEMENTS

The authors wish to thank C. Fleurier for lending the rapid electronic camera and the L'Air Liquide company for its financial support.

REFERENCES

1. Hudon C., Bartnikas R. and Wertheimer M. R., *IEE Transaction on Electrical Insulation* **28**, 1, 1-8 (1993).
2. Morshuis P. H. F., *Partial discharge mechanisms*, Delft University Press, 1993.
3. Simon G. and Bötticher W., *J. Appl. Phys.*, **76**(9), 5036-5046, 1994.
4. Bötticher W., "Filamentation of high pressure glow discharges", in *Proceedings of the XXI ICPIG*, III, Bochum, Germany, September 19-24, 1993, pp. 128-131.
5. Stanco J., Doerk T., Ehlbeck J., Jauernik P., Schäfer J. H. and Uhlenbush J., *J. Phys. D: Appl. Phys.* **26**, 244-252, 1993.
6. Yokoyama T., Kogoma M., Kanazawa S., Moriwaki T., and Okazaki S., *J. Phys. D: Appl. Phys.*, **23**, 874-877 (1990).
7. Kogoma M. and Okazaki S., *J. Phys. D: Apll. Phys.*, **27**, 1985-1987, (1994).
8. Liu C., Tsai P. P. and Roth J. R., "Plasma-related characteristics of a steady state glow discharge at atmospheric pressure", in Proceedings of *the 20st IEEE Int. Conf. Plasma Science*, Vancouver, Canada, June 7-9, 1993.
9. Massines F., Mayoux C., Messaoudi R., Rabehi A., Ségur P., "Experimental study of an atmospheric pressure glow discharge. Application to polymers surface treatment", in Proceedings of *the X Int. Conf. on Gas Discharges and their Applications*, II, Swansea, U. K., 13-18 Septembre, 1992, pp. 730-733.
10. Rabehi A., Ben Gadri R., Ségur P, Massines F. and Ph. Decomps, "Numerical modelling of high-pressure glow discharges controlled by dielectric barrier", in Proceedings of *the Conf. on Electr. Insulation and Dielectric Phenomena*, Arlington, Texas, 23-26 October, 1994, pp.840-845.
11. Okazaki S., Kogoma M., Uehara M. and Kimura Y, J. Phys. D: Appl. Phys. **26**, 889-892 (1993)
12. Spence P. and Roth J. R., "Electrical and plasma charractetistics of a one atmosphere glow discharge plasma reactor", presented at the IEEE Conf. on Plasma Scince, Santa Fe, NM, June 6-8, 1994.
13. Koppitz J., *Z. Naturfosh.* **26 a**, 700-706 (1970).
14. Palmer J. A., *Appl. Phys. Letters* **25**,3, 138-140 (1974).
15. Eliason B., Egli W. and Kogelschatz U., *Pure & Appl. Chem.* **66**(6), 1275-1286 (1994).
16. Decomps Ph., Massines F., Mayoux C., *Acta Physica Universitatis Comenianae*, **XXXV**(1), 47-57 (1994).
17. Doran A. A., *Zeitschrift für Physik*, **208**, 427-440 (1968).
18. Raizer Y. P., *Gaz discharge Physics*, Ed. Springer-Verlag, 1990, ch 8, pp. 169-170.
19. R. Ben Gadri, A. Rabehi, F. Massines and P. Ségur, "Numerical modelling of atmospheric pressure low-frequency glow discharge between insulated electrodes", in Proceedings of XIIth ESCAMPIG, 23-26 August 1994, (Netherlands), pp. 228-229.
20. Ward A.L., *J. Appl. Phys.*, **33**, 2789-2794 (1962).
21. Taylor R.S., *J. Appl. Phys.* **B41**, 1- (1986).

Capacitively Coupled Radio Frequency Discharges

J.E. Allen

Department of Engineering Science, University of Oxford, Parks Road, Oxford, OX1 2PJ, UK.

Abstract. A review is given of research carried out on capacitively coupled radio frequency discharges in the Department of Engineering Science at Oxford University.

§ 1. Introduction

I shall not attempt, in this lecture, to give a comprehensive review of the subject of capacitively coupled radio frequency discharges. Instead I shall discuss some of the relevant physics, taking the work of my own research group to illustrate my talk. Fig. 1 sets the scene; it is a schematic diagram of a discharge showing the power supply, usually 13.56 MHz, and a blocking capacitor in series which prevents the flow of a D.C. current in the circuit. In practice almost all systems are asymmetrical and this fact is indicated in the diagram. The asymmetry may exist even if the electrodes are of the same size, due to the presence of an earthed metal chamber or pumping table. The applied voltage is alternating but D.C. potentials are automatically set up to ensure that the average electron current to an electrode is equal to the average ion current. Thus the behaviour is something like the sheath formation at the insulating wall of a discharge tube, the difference being that the sheaths in Fig. 1 have both steady and alternating voltages developed across them.

The asymmetry of the system leads to the establishment of a D.C. bias across the device. An equal and opposite voltage must, of course, be developed across the blocking capacitor. The current flowing to an electrode must be equal to the current leaving the other electrode at all times. Each current is the sum of the ion current, the electron current and the displacement current, the last of these averages out to zero over a complete cycle.

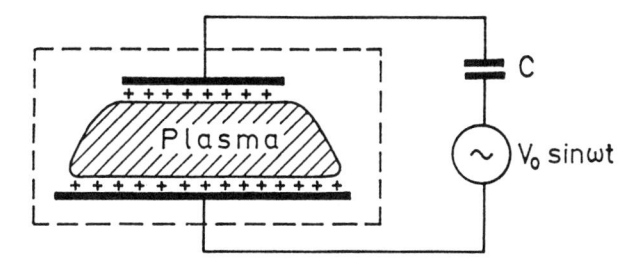

Figure 1. Schematic diagram of an asymmetric R.F. discharge.

§ 2. A Symmetrical System

Although systems are asymmetrical in general it is possible to design one which is symmetrical. Fig. 2 illustrates one employing the 'push-pull' principle (1). This was used so that Langmuir probe measurements could be made at the centre of the discharge where the potential was supposed to remain constant. Electron densities and temperatures were obtained in an argon plasma. It should be remembered, however, that the system is non-linear. By considering the symmetry of the system it is easily verified that there cannot be any odd harmonics of potential at the centre. Even harmonics may exist, however, and may lead to small errors in the probe measurements. These even harmonics were subsequently detected by Godyak and Piejak (2).

§ 3. The Self-Bias Potential Across a Sheath Adjacent to an R.F. Plasma

The calculation to be described can be applied either to the sheath surrounding a probe or to the sheaths at the electrodes. Fig. 3(a) illustrates the non-linear nature of a probe characteristic. If a probe is biassed to a certain potential, but an A.C. component of voltage also exists between the probe and the plasma, then the curvature of the plot leads to an increase in the average electron current to the probe. If the electron energy distribution is assumed to be Maxwellian it is easily shown that the increased electron current is given by the factor $I_0(eV_{RF}/kT_e)$ where $I_0(x)$ is the modified Bessel function of zero order. The quantity V_{RF} is the amplitude of the A.C. voltage which is assumed to be sinusoidal. It is a simple matter to show that the self-bias voltage for zero average current is given by (3)

$$V_{f_1} = V_{f_0} - (kT_e/e)\ln I_0(eV_{RF}/kT_e) \tag{1}$$

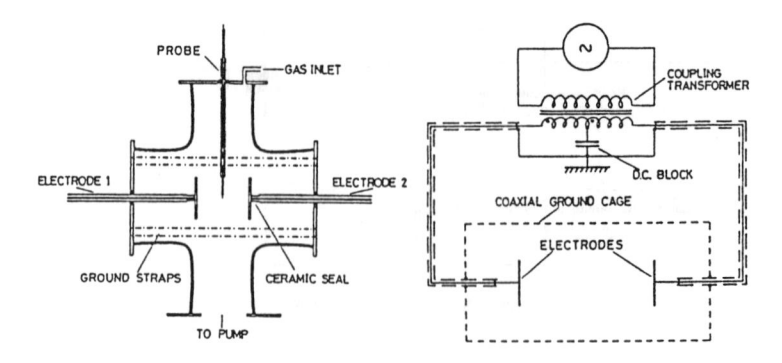

Figure 2 A symmetrical discharge employing the "push-pull" principle.

where V_{f_0} is the usual floating potential given by

$$V_{f_0} = (kT_e/e)\ln(2\pi m/M)^{1/2} \qquad (2)$$

Fig. 3 (b) shows the plot for argon, since V_{f_0} depends on the mass of the ions. Early work was carried out by Kojima et al (4), Garscadden and Emeleus (5), Boschi and Magistrelli (6) and Butler and Kino (7).

§ 4. The Driven Probe Technique

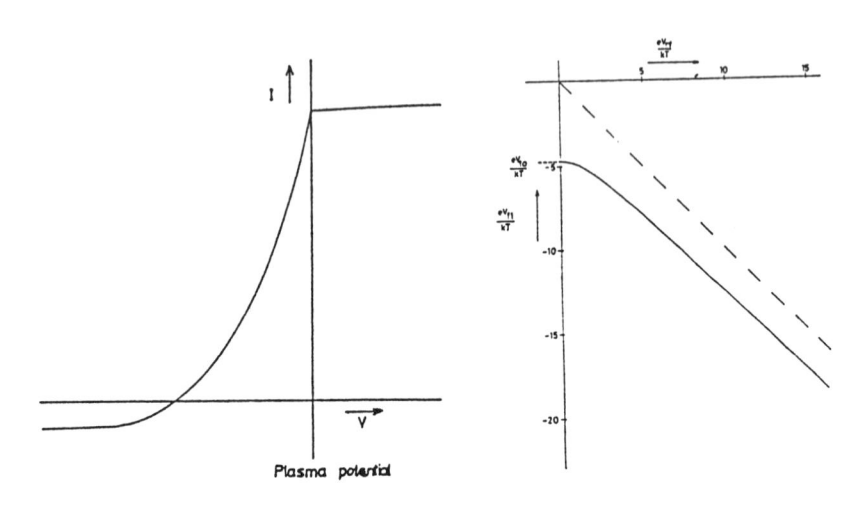

Figure 3 (a) Schematic current-voltage characteristic of a probe.
(b) The self-bias D.C. potential as a function of the R.F. amplitude (for argon)

In general the plasma potential has an alternating component, due to the asymmetry of the system. One method which enables Langmuir probes to be used is illustrated in Fig. 4 (8). An A.C. component of voltage is applied to the probe, together with the usual slow voltage ramp, in order to cancel the A.C. voltage appearing across the probe sheath. It is necessary to vary both the amplitude and phase of the applied A.C. signal. In this way errors due to the R.F. can be eliminated. An early attempt to cancel the R.F. voltage was made by Sabadil and Klagge (9), but without phase compensation. The method has more recently been extended to compensate for the second and third harmonics of the plasma potential waveform (10).

The other principal kind of probe measurement employs filters to prevent R.F. currents from flowing; relevant references can be found in (11), where other probe techniques are also described.

§ 5. The Energy Flow in R.F. Enhanced Sheaths

The early probe measurements (8) showed that, at least over a certain range of parameters, the applied A.C. voltage was developed mainly across the space charge sheaths illustrated in Fig. 1. This means that the electromagnetic energy is supplied, in the first instance, not principally to the plasma but to the space charge regions. This is shown schematically in Fig. 5. It is of interest to compare a floating electrode with one supplied with A.C. current. In the former case the electrons which arrive at the electrode lose energy on the way. The ions gain an equal amount of energy. Thus we can imagine that the electrons have given energy to the ions, via the electric field. When a current flows, however, in this case an A.C. current, another source of energy is present, that of the electromagnetic flux of energy described by the Poynting vector (12). In this

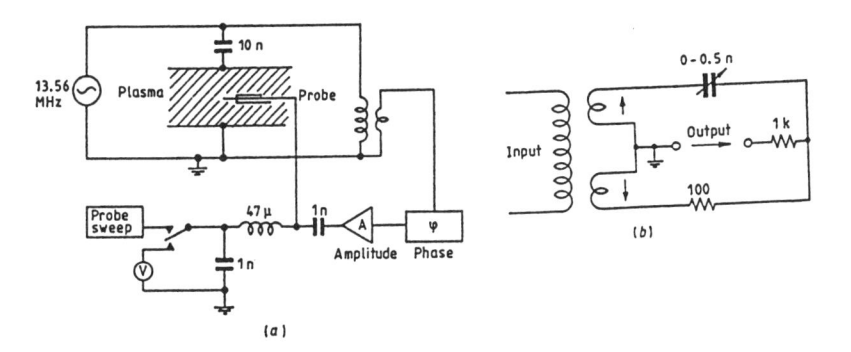

Figure 4   (a) Schematic diagram of the driven Langmuir probe circuit.
            (b) The single-frequency amplitude-independent phase shifter.

case it can be shown that most of the energy obtained by the ions is delivered to the sheath region by the Poynting Flux (13). It is clear that some energy must be supplied to the plasma and I shall return to this question later.

§ 6. Distribution of Ion Energies at an Electrode

Some measurements of the (one-dimensional) distribution of ion energies at the grounded electrode are shown in Fig. 6 (14). These were obtained using a retarding field analyser; in such an analyser the current, in this case the ion current, is given by the expression

$$I = eA \int_{v_*}^{\infty} f(v) dv \qquad (3)$$

where $\tfrac{1}{2}Mv_*^2 = eV$, V being the retarding potential. Differentiation with respect to V yields

$$f(v) = -\frac{M}{Ae^2} \frac{dI}{dV} \qquad (4)$$

The raw data are usually presented in the form $f(v)$ versus V rather than $f(v)$ against v. The latter form can, of course, be readily deduced.

At the lowest pressures collisions in the sheath are rare and all the ions have energies greater than that corresponding to the voltage drop across the sheath. A spread of energies is expected, even in the absence of collisions, because the ions are born at different positions within the plasma. This spread is expected to be of the order of kT_e/e by analogy with the (d.c.) theory of Tonks and Langmuir (15). In the one-dimensional version of that theory the distribution of ion energies is independent of the precise mechanism of ionization (i.e. single-stage or multi-stage). As the pressure increases collisions take place in the sheath and the energy distribution spreads to lower energies, as shown in Fig. 6.

§ 7. The Effect of Negative Ions on the Bohm Velocity

The flux of positive ions leaving the plasma and entering the sheath is usually calculated using the well-known Bohm condition (16). This states that the velocity of ions on entering the plasma is given by $\tfrac{1}{2}Mv^2 = \tfrac{1}{2}kT_e$. In some plasmas, however, negative ions are present in addition to positive ions,. It is clearly of interest to consider the effect of the negative ions on the Bohm condition. This can readily be carried out, assuming that both electrons and negative ion obey a Boltzmann variation with potential (17). The results are displayed in Fig. 7. which shows the variation of the potential drop in the quasi-neutral plasma versus the ratio of the negative ion density to the electron density. It is seen that the velocity acquired by the positive ions will be reduced as will the positive ion current density. A discontinuity is predicted, as the negative ion density is increased if the temperature ratio $T_-/T_e > 5 + \sqrt{24}$ (about 10). We have not yet carried out these experiments and I would be interested to know if other workers have observed a 'jump' in the plasma parameters corresponding to the predicted transition.

§ 8. Effect of the A.C. Voltages on the Ion Motion

To a first approximation the ions react to the average field in the sheath ($\omega_{pi} << \omega$) whereas the electrons react instantaneously ($\omega << \omega_{pe}$). More accurately, however, neither of these statements is exactly true (although we have calculated the sheath profiles on these assumptions, see below). In order to increase the R.F. potential across the sheath at the grounded electrode of a commercial plasma reactor the area of the 'powered' electrode was increased. The velocity acquired by an ion depends on the phase of the cycle at which it enters the sheath. The results are shown in Fig.s 8,9 and 10, where a comparison between experiment and theory is made (18). It is seen that excellent agreement is obtained at the lowest pressure (0.3 Pa) and fairly good agreement at 1 Pa. The agreement is not quite so good at 6 Pa where a multi-peaked distribution is found. This structure is due to the combination of charge-exchange collisions together with the phase of the cycle at entry into the sheath. A double-peaked structure at low pressures was first seen by Erö, but in a different context, that of ion extraction from a plasma produced in a Thonemann-type ion source (i.e. an inductively coupled plasma)(19). A well-known paper on the double-peaked structure in capacitively coupled discharges is by Coburn and Kay (20), the multi-peaked structure was first observed by Wild and Koidl (21).

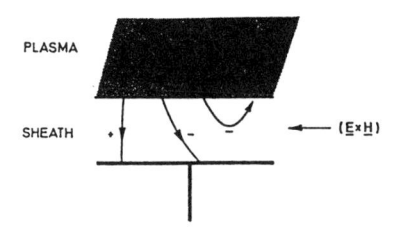

Figure 5. Schematic diagram of the sheath showing the charged particles and the Poynting flux.

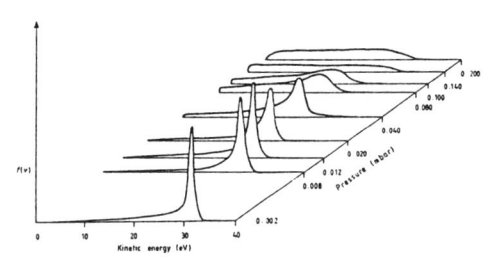

Figure 6. Ion velocities for a range of pressure. N.B. The horizontal axis is kinetic energy (½ Mv²/e).

§ 9. Stochastic Heating of Electrons

First of all the sheath theory was worked out, assuming that the ions feel only the D.C. field but the electrons can follow the potential variations in a Boltzmann-like manner. Early calculations on this basis were made by Stekolnikov et al (22). Secondly a Maxwellian distribution of electrons was injected into this sheath and their subsequent trajectories were calculated. It is clear that this theory is not entirely self-consistent. Some results are shown in Fig. 11 (23). Many electrons leave the sheath and re-enter the plasma with their energy practically unchanged. Some electrons gain energy, however, and some lose energy. There is a net gain of energy, summed over all the electrons. The diagram shows the importance of the phase at entry (a sin ωt variation is assumed, for the voltage across the sheath).

Another factor is of crucial importance; the electrons must undergo phase-mixing before they arrive at the opposite sheath. The condition for this, in the collisionless case is $p > \bar{c}_e/\omega$. In the case of collisions in the plasma region this is replaced by $p > \bar{c}_e/\omega_{coll}$. The width of the plasma is here denoted by p.

Early work on this subject is due to Godyak who considered the sheaths as oscillating rigid walls (24). A more detailed approach has been found to be necessary, however, involving the structure of the sheath.

A competing mechanism, with respect to the heating of a low pressure plasma is that of secondary emission of electrons at the electrodes, with their

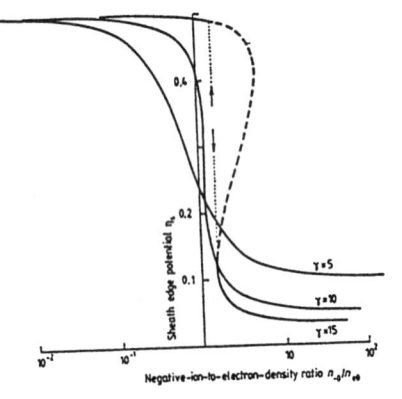

Figure 7. Plot of normalized sheath edge potential against the ratio of the negative ion density to electron density in the bulk plasma for three values of γ, the ratio of electron temperature to negative ion temperature . The arrowed dotted line indicates the necessary jump in the solution.

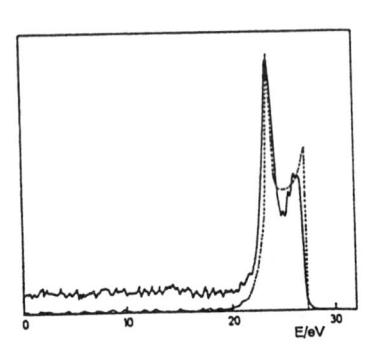

Figure 8. Ion energy distribution function (arb. units). Pressure = 0.3 Pa, n_e = 8.4 $10^{15} m^{-3} T_e$ = 2.64 eV. The solid line represents the experiment; the dashed one - simulation. The curves shown represent f(v) versus E.

subsequent acceleration across the sheath and into the plasma.

No attempt will be made, in the present lecture, to compare these two mechanisms. The secondary electrons have been detected using gridded probes (25).

§ 10. Comparison of Different Measurements of Ion Energy Distributions

Further measurements have been made of the ion energy distribution in the plasma studied above (§8). Two different instruments were employed, namely a retarding field analyser and a Cylindrical Mirror Analyser with an associated Quadrupole Mass Spectrometer (SXP300/CMX500 manufactured by V.G. Quadrupoles). Some results are shown in Fig. 12 (26); preliminary results were communicated at an earlier ICPIG Conference (27). At the lowest pressure (0.5 Pa) the agreement is excellent. At somewhat higher pressures (2.8 Pa) the agreement is fairly good at high energies, but no low energy ions were detected with the CMA/Quadrupole combination. At higher pressures the disagreement is worse. One fact to remember is that the retarding field analyser cannot discriminate between ions of different masses so that impurity ions will be present in the ion energy distribution. It would be of interest to compare results obtained with different instruments.

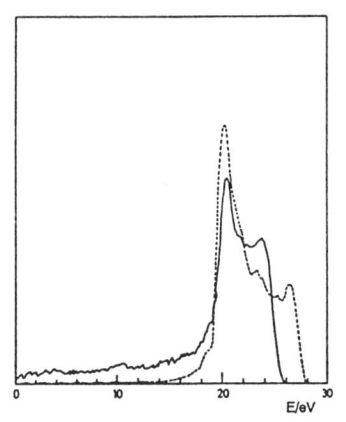

Figure 9. Ion energy distribution function (arb. units), Pressure = 1Pa, electron density $n_e = 4.1 \ 10^{16} m^{-3}$. $T_e = 2.5$ eV. The solid line represents the experimental results; dashed one - simulation.
The curves shown represent f(v) versus E.

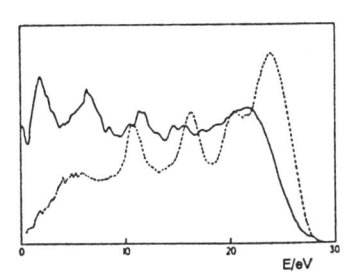

Figure 10. Ion energy distribution function (arb. units), Pressure = 6Pa, electron density $n_e = 1.8 \ 10^{15} m^{-3}$; $T_e = 2.8$ eV. The solid line represents the experimental results; the dashed one - simulation. The curves shown represent f(v) versus E.

§ 11. Theory of the Oscillating Sheaths

The basic assumptions have already been referred to above, i.e. it is assumed that $\omega_{pi} << \omega$ and $\omega << \omega_{pe}$. The initial calculations were carried out for a single sheath (28). In practice, however, the entire system must be considered since the applied voltage is developed across two sheaths, assuming that the R.F. developed across the plasma itself is negligible. The basic equation to be solved is Poisson's equation and the Bohm condition was employed to calculate the ion current density. Another boundary condition is that both sheaths carry the same (total) current. Fig. 13 shows the potential distribution within the sheaths as a function of time (29). It is seen that the potential difference developed across a single sheath is not sinusoidal, i.e. the plasma potential w.r.t. ground contains harmonics.

The voltage waveforms across the two sheaths are shown in Fig. 14(a) & (b) for different values of the electrode area ratio. Fig. 14 (c) shows the voltage across the whole device; the R.F. voltage across the plasma is assumed to be zero, as is the net D.C. voltage across the plasma. The waveform is that of a sinusoidal voltage together with a D.C. bias. Figure 14 (d) shows the current waveforms; when the electrode areas differ greatly the pulse of electron current to the smaller electrode can be seen. This occurs when the sheath voltage at that electrode drops to a low value. Fig. 15 shows the self-bias a function of the electrode area ratio. A comparison is made with two other (less detailed) models of the discharge (26). It is evident that the calculated self-bias voltage are almost independent of the model employed.

§ 12. The Bohm Criterion in the Presence of R.F. Fields

A question which arises is whether the Bohm criterion is still valid in the presence of R.F. electric fields. This question has been addressed by Riemann (30) and by Allen and Skorik (31). Riemann extended an earlier approach by Allen (32) in which the generalized sheath criterion was shown to correspond to sonic flow at the plasma boundary. Riemann considered low-frequency oscillations, i.e. frequencies much smaller than the electron plasma frequency ($\omega << \omega_{pe}$) but comparable with the ion plasma frequency ($\omega \sim \omega_{pi}$) this analysis led to the following dispersion relation:

$$\omega - kv_0 = \pm \omega_{pi} \left(\frac{k^2 \lambda_d^2}{1 + k^2 \lambda_d^2} \right)^{\frac{1}{2}} \qquad (5)$$

where v_o is the velocity of the ions, assumed to be monoenergetic. The Bohm condition corresponds to $\omega = 0$, $k\lambda_d \to 0$ and $v_0 = (kT_e/M)^{\frac{1}{2}}$. At higher frequencies the complex roots which arise from this quartic equation needed to be analysed further. The criteria may be stated as follows (33).

To decide whether a given wave with a complex k for some real ω is amplifying or evanescent, one determines whether or not k_i changes sign as the imaginary part of ω increases to a large positive value. If it does then the wave is amplifying; otherwise it is an evanescent wave.

An absolute instability is obtained whenever there is a double root of k, for some complex ω in the upper half-plane, for which the two merging roots come from different halves of the complex k plane (upper and lower) as ω_i is decreased from a large positive value.

Fig. 16 illustrates some of the plots obtained. The result is that the complex roots correspond to evanescent waves. Such (longitudinal) disturbances cannot penetrate the plasma but are reflected within a Debye distance. This means that the R.F. fields are screened out much more rapidly than the D.C. field, the pre-sheath field, which is set up automatically. The conclusion is that the Bohm criterion is unaffected by R.F. fields.

§ 13. Plasma-Sheath Resonance in a Parallel-Plate Reactor

A plasma-sheath resonance has been observed at low pressures in an argon plasma (34). This is a similar phenomenon to that associated with the R.F. resonance probe (35). Fig. 17 illustrates the principle. The plasma is represented by the parallel combination L_p and C_p which resonates at the plasma frequency. One could equally well describe the plasma as a capacitor with the appropriate permittivity. At high frequencies the sheaths behave essentially as capacitors, these are assumed to have constant capacitance in this simple linear model. Such a system has two resonant frequencies, one given by $\omega=\omega_{pe}$ and the second, the plasma sheath resonance frequency, given by

$$\omega_R = \omega_{pe} \bigg/ \left(1 + \frac{p}{2s}\right)^{\frac{1}{2}} \tag{6}$$

where p is the length of the plasma and 2s is the sum of the sheath thicknesses (not necessarily equal). The initial measurements were of the external current, but subsequently measurements were made of the distribution of R.F. potential within the plasma, for different harmonics. The resonance occured at the parameters shown in Fig. 18. A potential plot is given in Fig. 19.

This resonance can be used as a diagnostic, to measure the electron density. It could perhaps be exploited as the basis of a new plasma reactor (c.f. electron cyclotron resonance).

Further analysis (36) takes into account the non-uniformity of the plasma, assuming a profile of the form $n=n_o/(1+x^2/a^2)$ where 2a is the total width of the

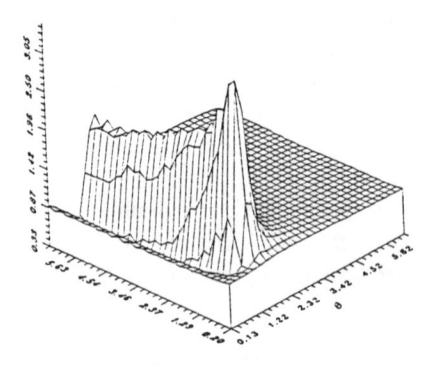

Figure 11. 3-D plot of the ratio of the electron energy on leaving the sheath to that on entering as a function of the phase angle and the initial energy; the latter is normalized in terms of kT_e.

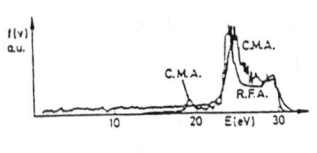

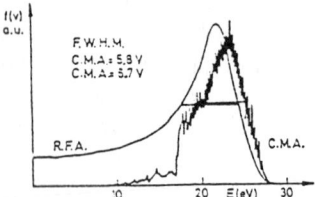

Figure 12 (a). Comparison between RFA (no boxcar) and CMA energy spectra, V_{RFpp} = 400 V, P = 0.5 Pa, V_{DCbias} = -175V. From probe measurements V_p = 26.5 V, T_e = 2.43 eV, n_e = 1.3 10^{17}m^{-3} and $V_{RFprobe}$ = 5.0V.

Figure 12 (b). Comparison between RFA and CMA energy spectra, V_{RFpp} = 400 V, P = 2.8 Pa, V_{DCbias} = -147V. From probe measurements V_p = 25.1 V, T_e = 2.97 eV, n_e = 2.8 10^{17} m^{-3} and $V_{RFprobe}$ = 4.8V.

plasma. This plasma can be represented by an infinite number of parallel circuits, each with its own resonant frequency. It was found, using experimental values for

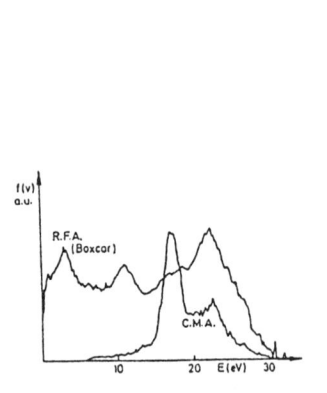

Figure 12 (c). Comparison between RFA and CMA energy spectra, V_{RFpp} = 400 V, P = 7Pa, V_{DCbias} = -136V. From probe measurements V_p = 24.4 V, T_e = 2.98 eV, n_e = 2.13 10^{17} m^{-3} and $V_{RFprobe}$ = 4.8V.

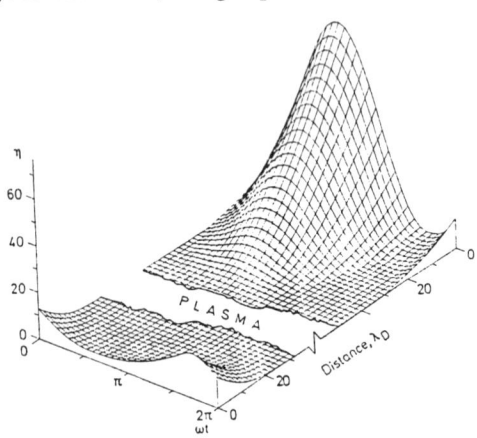

Figure 13. Normalized potential variation in the two sheath regions. The area is 0.5 and the voltage is developed mainly at the smaller electrode.

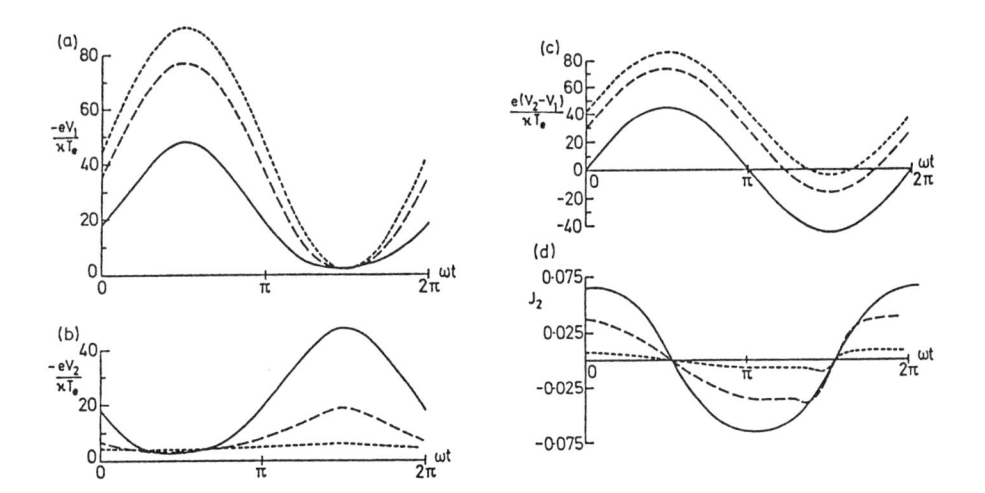

Figure 14. Normalized voltage waveforms across (a) the smaller electrode, (b) the larger electrode and (c) the whole device; the self-bias voltage is seen. The normalized current density waveforms are shown in (d). The electrode area ratios are 0.1 (...), 0.5 (---) and 1.0 (———).

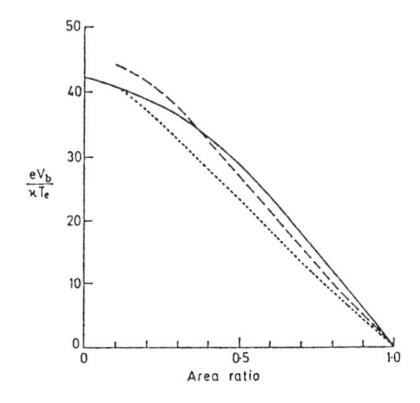

Figure 15. The self-bias voltage as a function of the ratio of electrode areas. The present calculations are illustrated by the full line. The other lines represent an electron front model due to Meijer and Goedheer (----) and another model, due to the same authors, termed the "current balance" model (...)(27).

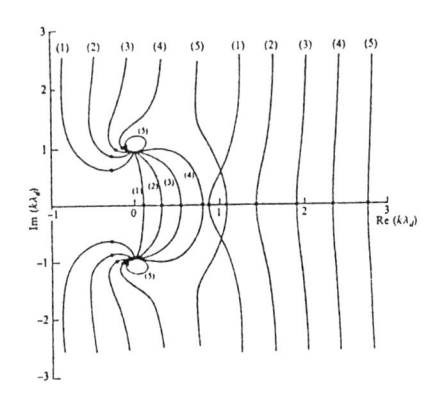

Figure 16. The locus of $k\lambda_d$ as the imaginary part of ω/ω_{pi} varies from $-\infty$ to $+\infty$. The points corresponding to real values of ω/ω_{pi} are shown. The parameters are $v_o = v_s$ and $\omega/\omega_{pi} = 0.2$ (curves 1), 0.6 (2), 1.(3), 1.4 (4), and 1.8 (5).

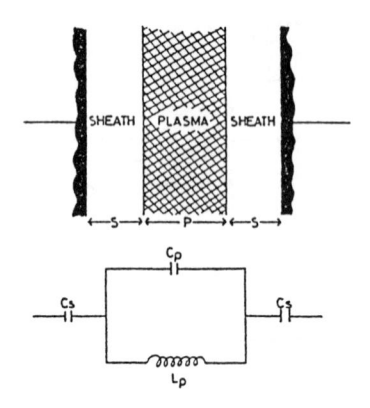

Figure 17. Schematic diagram of the plasma model.

Figure 18. Pressure and excitation voltage versus the order of the resonant harmonic, $f_o = 13.56$ MHz, for argon; results obtained with a voltage probe.

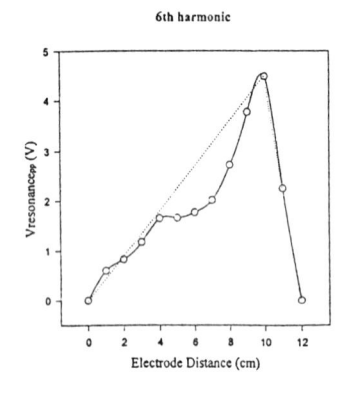

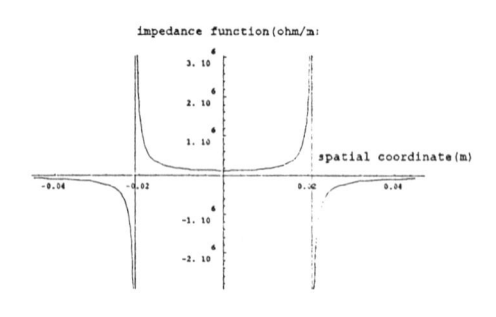

Figure 19. Spatial distribution of the resonance voltage, the driven electrode is on the right.

Figure 20. The reactance (per unit length) of the plasma versus the spatial position. The electrodes are at $\pm\ 0.06$m.

the plasma width and sheath thicknesses, that the plasma-sheath resonance coincides with the local plasma frequency at two planes within the discharge. Fig. 20 shows the distribution of reactance across the plasma region when the resonance occurs at 81.36 MHz (the 6th harmonic of the driving frequency). At two planes within the discharge the function has a pole. The theory predicts a resonant energy absorption in this region (37), the effective resistance being given by

$$R = \pi \bigg/ \epsilon_0 A \left| \frac{d\omega_{pe}}{dx} \right| \qquad (7)$$

where the gradient is evaluated at the resonant planes. The calculated damping based on this linear model is extremely high, however, so that further investigations are necessary.

It should be noted that although the resonance was observed at harmonic frequencies, in these particular experiments, this was due to the excitation of these frequencies in a non-linear system: the phenomenon itself is *not* restricted to harmonics (36).

§ 14. An Application: Etching at an Angle

Colleagues in another group, carrying out research and development in optoelectronics, wished to fabricate parallelogram-shaped gratings in SiO_2. These were for use as waveguide grating couplers and they also have potential application in grating-coupled surface-emitting lasers. Discussions led to a method of achieving the required result. A grid at the required angle (40°) was attached to one of the electrodes of a commercial reactor. The plasma boundary, parallel to the grid, was then at an angle with respect to the horizontal electrode on which the sample was placed. The grating produced in this way is shown in Fig. 21 (38). I give this as an example of the fact that basic knowledge in this field can be readily applied.

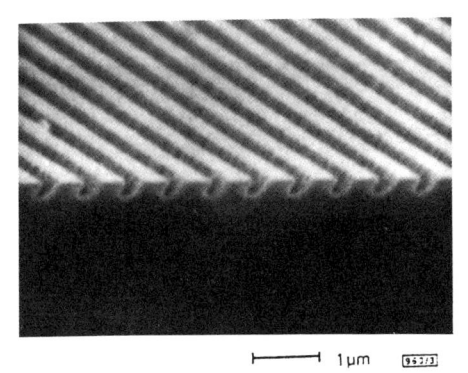

Figure 21. SEM photograph of a parallelogram-shaped grating made in SiO_2.

Acknowledgments

I have illustrated this talk with results obtained in the Department of Engineering Science at the University of Oxford. This work is ongoing in a continued fruitful collaboration with Beatrice M. Annaratone. In the early stages of the research valuable contributions were made by Nicholas St. J. Braithwaite.

References

1. Braithwaite, N. St. J., Benjamin, N.M.P. and Allen, J.E., *7th Int. Symp. Plasma Chemistry*, 564 (1985).
2. Godyak, V. and Piejak, R., *J. Appl. Phys.*, 68, 3157 (1990).
3. Braithwaite, N. St. J., Crosby, A. J. and Allen, J.E., *8th Int. Conference on Gas Discharges and their Applications*, 482 (1985).
4. Kojima, S., Takayama, K. and Shimauchi, J., *J. Phys. Soc. Japan*, 8, 55 (1953).
5. Garscadden, A. and Emeleus, K.G., *Proc. Phys. Soc. 79*, 535 (1962).
6. Boschi A. and Magistrelli, F., *Nuovo Cimento*, 29, 487 (1963).
7. Butler, H.S. and Kino, G.S., *Phys. Fluids*, 6, 1346 (1963).
8. Braithwaite, N. St. J., Benjamin, N.M.P. and Allen, J.E., *J. Phys. E: Sci. Instrum.*, 20, 1046 (1987).
9. Sabadil, H. and Klagge, S., *Proc. XVII Int. Conference on Phenomena in Ionized Gases*, 322 (1985).
10. Counsell, G. F. Braithwaite, N. St. J. and Theloham (to be published).
11. Allen, J.E., *Plasma Sources Sci. Technol.*, 4, 234 (1995).
12. Poynting, J.H., *Phil. Trans.*, 175, 343 (1884).
13. Allen, J.E., *Proc. XVIII Int. Conf. on Phenomena in Ionized Gases*, 4, 808 (1987).
14. Ingram, S.G. and Braithwaite, N. St. J., *J.Phys. D: Appl. Phys.*, 21, 1496 (1988).
15. Tonks, L. and Langmuir, I., *Phys, Rev.*, 34, 876 (1929).
16. Bohm, D., *The Characteristics of Electrical Discharges in Magnetic Fields (ed Guthrie, A. and Wakerling, R.K.)*, New York: McGraw-Hill, 1949, ch. 3.
17. Braithwaite, N. St. J. and Allen, J.E., *J. Phys. D: Appl. Phys.*, 21, 1496 (1988).
18. Annaratone, B.M., Skorik, M.A., Goruppa, A.A. and Allen, J.E., *Europhysics Conference Abstracts, 16C, III*, 1977 (1992).
19. Erö, J., *Acta Phys. Hung.*, 5, 391 (1956); *Nucl. Instrum.*, 3, 303, (1958).
20. Coburn, J.W. and Kay, E., *J. Appl. Phys.*, 43, 4965 (1972).
21. Wild, C. and Koidl, P., *J. Appl. Phys.*, 69, 2909 (1991).
22. Stekolnikov, A.F., Braithwaite, N. St. J. and Allen, J.E., *9th Int. Conf. on Gas Discharges and their Applications*, 391 (1988).
23. Skorik, M.A., Braithwaite, N. St. J. and Allen, J.E., *Europhysics Conf. Abstracts, 16C, III*, 1973. (1992)
24. Godyak, V.A., *Soviet Radio Frequency Discharge Research, Falls Church*, VA; Delphic Associates Inc., 1986.
25. Kawano, H., Annaratone, B.M. and Allen, J.E., *Proc.XX Int. Conf. on Phenomena in Ionized Gases*, 5, 1099 (1991).
26. Annaratone, B.M., Ku, V.P.T. and Allen, J.E., *Proc. XXI Int. Conf. on Phenomena in Ionized Gases*, I, 29 (1993).
27. Goruppa, A.A., Annaratone, B.M. and Allen, J.E., *Proc XX Int. Conf. on Phenomena*

in Ionized Gases, 5, 1079 (1991).
28. Skorik, M.A. and Allen, J.E., *Proc. XXI Int. Conf. on Phenomena in Ionized Gases*, I, 113 (1993).
29. Skorik, M.A., D.Phil. Thesis, University of Oxford (1995).
30. Riemann, K.-U., *Phys. Fluids*, B4, 2693 (1992).
31. Allen, J.E. and Skorik, M.A., *J. Plasma Phys.*, 50, 243 (1993).
32. Allen, J.E., J. Phys. D; *Appl. Phys.*, 9, 2331 (1976).
33. Allen, J.E. and Phelps, A.D.R., *Rep. Prog. Phys.*, 40, 1305 (1977).
34. Annaratone, B.M., Ku, V.P.T. and Allen, J.E., *J. App. Phys.*, 77, 5455 (1995).
35. Harp, R.S. and Crawford, F.W., *J. Appl. Phys.*, 35, 3436 (1964).
36. Ku, V.P.T., Annaratone, B.M. and Allen, *J.E., Proc. XXII Int. Conf. on Phenomena in Ionized Gases* 4,157 (1995).
37. Crawford, F.W. and Harker, K.J., J. Plasma Phys., 8, 261 (1972).
38. Li, M., Lin, J. C.H., Cherrill, M.J. and Sheard, S.J., *Electronics Letters*, 30, 2126 (1994).

Corona physics and diagnostics

Reidar Svein Sigmond

The Electron and Ion Physics Research Group, Physics Department, The Norwegian Institute of Technology, N-7034 Trondheim, Norway

Abstract. The criterions for oscillations in DC-fed coronas are discussed, both for the well-understood negativeTrichel pulse coronas and for the unexplained positive glow pulse coronas. Trichel-like pulses occur also in non-electron-attaching gases, due to the external circuit impedance, and this lowers the sensitivity of Trichel coronas as detectors for electronegative gas traces. Pulse excitation of positive glow coronas in argon with added trace gases show that the corona stability and resonance frequency strongly depend on the trace gas type and concentration, but the physics involved is unknown. Finally, it is shown that the low-current U(I) curve of relaxation-pulsed coronas always must have a negative slope, equal to the negative of the series resistance.

INTRODUCTION

This paper deals with the physical mechanisms of centimeter-scale, low current positive and negative corona discharges in common gases at and below atmospheric pressure, with some diagnostic methods to analyse the coronas, and with some uses of the coronas in gas diagnostics. The influence of the external electrical circuit on the corona behaviour, which often is underestimated, will be treated in some detail. Most of the paper will discuss why DC-fed unipolar negative and positive coronas usually insist in passing current as oscillations or pulses, how oscillations may be induced in stable coronas, and how the oscillations could give information about the corona mechanism, if only we understood what was going on.

The best known type of oscillating, DC fed corona is the *Trichel pulse* corona, which is usually found in air and other electron attaching gases. Figure 1B shows a scaled drawing of a Pyrex glass corona chamber with a wire loop cathode and an hemispherical anode, together with the voltage supply, series resistance, and current recording oscilloscope[1] (the equivalent circuit 1A will be discussed below). The corona current will, over a large range, consist of pulses like those shown in Figure 2, of about current independent shape and charge, and repetition frequency proportional to the current.

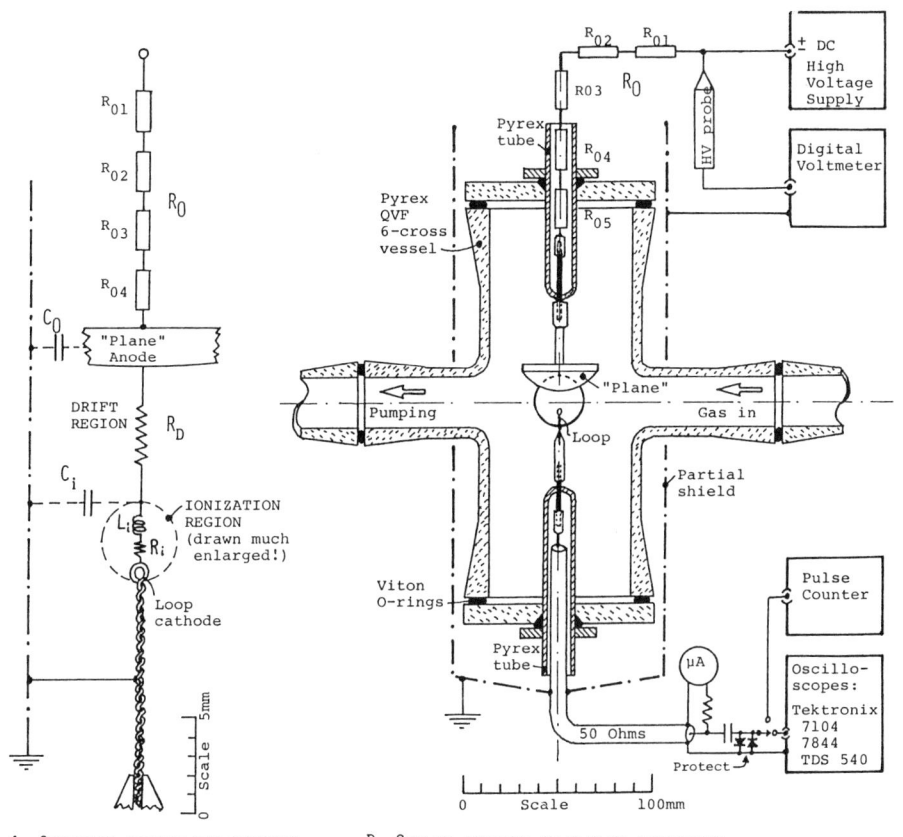

A. SIMPLEST EQUIVALENT CIRCUIT   B. SCALED DRAWING OF CORONA APPARATUS

FIGURE 1. Scaled drawings[1] of a negative corona electrode system and equivalent circuit (A), and the Pyrex-Viton corona chamber (B), with associated electrical apparatus. Mechanical supports are not drawn.

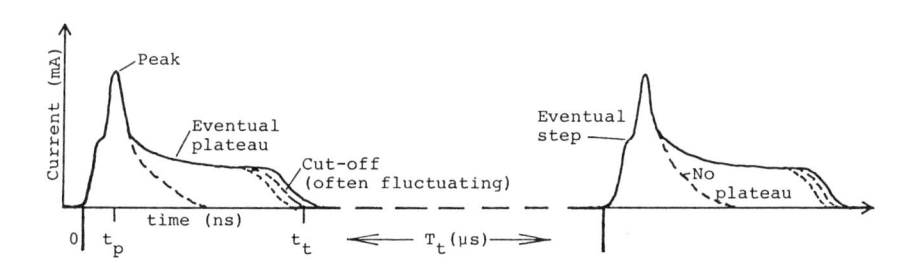

FIGURE 2. Typical Trichel pulse shapes and terminology.

However, positive point-to-plane or wire-to-plane coronas are also often oscillatory or pulsed. Figure 3 shows current and associated light pulses in a 16 mm point-to-plane gap in 5 kPa air[2]. These *positive glow pulses* are of much smaller amplitude and higher frequency than the Trichel pulses, and are often overlooked, even when the ionization is completely turned on-off by the pulses. Like Trichel pulses, these oscillations are reported to occur in electron attaching gases only. This will be discussed below.

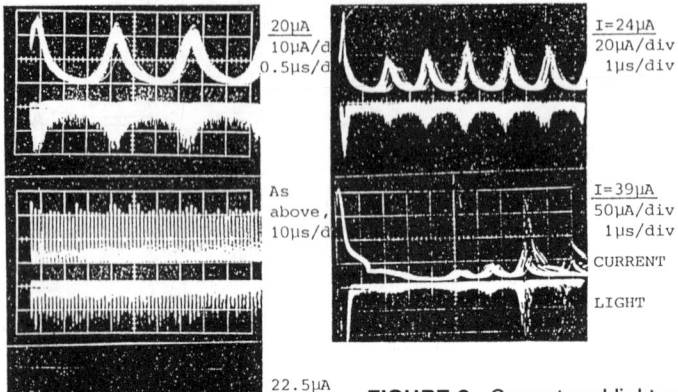

FIGURE 3. Current and light waveforms for an oscillating positive 39 μA point-to-plane corona in 5 kPa air. Point: Hyperboloid Au, R=46 μm, 16 mm from the anode plane[2].

To understand the physics behind the observed oscillations, we should look at a real or imaginary non-oscillationg corona at the same current, and discuss its stability. It is then sufficient to look at first-order current perturbations, with exponential or exponential-sinusoidal solutions. With increasing perturbation amplitudes, non-linear effects will take over, forming Trichel pulses in negative coronas and streamers in positive.

Such a constant-current corona will have an *ionization region* close to the high-field electrode, where the net ionization $\alpha' = \alpha - \eta$ is larger than zero and ion velocities are high enough to make space charge fields relatively unimportant. The ionization region can in many ways be treated like a plane parallel Townsend discharge burning near the Paschen minimum.

The *drift region* will fill the remainder of the corona gap. Here the electric field is too low to allow electron impact ionization, and electrons and ions drift, diffuse, and take part in ion-ion and ion-neutral collisions and reactions. The space charge field from the slowly drifting charges will increase the voltage fall across the drift region, much like the action of a resistance.

CORONA STABILITY

Why do coronas show oscillatory behaviour? To give oscillations, any electrical system must have two internal elements with opposite-phase current-voltage characteristics (like capacitance and inductance).

For the cases of a parallel-plate DC Townsend discharge or a corona ionization region, the first of these required circuit elements is the *capacitance* C_i of the region. For a short time this capacitance is able to sustain a discharge current different from the imposed DC current. The voltage will lag behind any current change.

The second required element for oscillations is the equivalent *inductance* L_i of the discharge or ionization region. This inductance is due to the delay between the application of a voltage step to the discharge DC voltage and the subsequent discharge current change. Any DC discharge burns with a reproduction factor $\mu=1$; i.e., each electron that leaves the cathode will on the average be replaced by exactly one new electron in the next avalanche generation. A voltage step ΔU_i added to the DC sustaining voltage will increase the reproduction factor from 1 to $1+\Delta\mu$, which will make each new avalanche generation $1+\Delta\mu$ larger than the preceding one. If the mean time between the avalanche generations is T_g (equal to the positive ion transit time in case of only positive ion feedback), then the current will change with time as:

$$I_i(t) = I_{i0} \exp\left(\frac{\Delta\mu}{T_g}t\right) \tag{1}$$

$$\left(\frac{dI_i}{dt}\right)_0 = I_{i0}\frac{\Delta\mu}{T_g} = \frac{I_{i0}}{T_g}\frac{\partial\mu}{\partial U_i}\Delta U_i \tag{2}$$

If we, for comparison, take an inductance L and impose a voltage step ΔU_i across it, then a linear current change will result:

$$\frac{dI}{dt} = \frac{\Delta U}{L} \tag{3}$$

Thus, for small voltage deviations from the DC sustainment voltage, the discharge or ionization region will act as inductances:

$$L_i = \frac{T_g}{I_{i0}\,\partial\mu/\partial U_i} \tag{4}$$

Note that L_i is inversely proportional to the DC sustainment current I_{i0}.

Will the Townsend discharge or corona ionization region also have an equivalent *resistance R_i?* Not at very low currents, because then space charges will not influence the electrical field and the reproduction factor stays equal to 1 at a given voltage, independent of the current. This means a flat $U_i(I_i)$ curve, i.e., $R_i=0$. Thus, at very low currents, the small signal characteristics of the Townsend discharge and the corona ionization region may be modelled by the gap or region capacitance C_i in parallel with the discharge inductance L_i. A voltage pulse added to the discharge sustainment voltage will then induce a train of current oscillations with the angular frequency ω_i:

$$\omega_i = \frac{1}{\sqrt{L_i C_i}} \propto \sqrt{I_{i0}/T_g} \tag{5}$$

This $\sqrt{I_{i0}}$ frequency dependence has been observed in plane parallel Townsend discharges[3] and in positive concentric cylinder coronas[4]. In the latter case the oscillations were induced by the natural stochastic fluctuations of the discharge sustainment current, and may be readily observed by triggering an oscilloscope on the fluctuation peaks.

A resistive impedance R_0 in the external circuit will damp these oscillations, as will the effective equivalent resistance R_d of a corona drift region. In no way, however, will such series impedances make the oscillations self-sustained. For this, a *negative* effective resistance must exist. In a plane parallel Townsend discharge or in the ionization region of a negative corona, such a negative resistance may be provided by the space charge field of the positive ions, which will increase the cathode field and thus the efficiency of the secondary cathode processes. This will reduce the voltage necessary to sustain the discharge, and the $U_i(I_i)$ curve will get a negative slope. In the ionization region of a positive corona, on the other hand, positive ions may lower the effective electric field, giving the $U_i(I_i)$ curve a positive slope and making the ionization region differential resistance R_i positive.

The resulting small-signal equivalent network for a corona discharge is shown in Figure 1A. For a parallel-plate Townsend discharge the value of R_i is readily measured, being the slope of the U(I) curve at small (μA) currents. For a corona ionization region, R_i will usually be masked by the drift region impedance R_d, since it is the sum R_i+R_d that is accessible as the slope of the U(I) characteristic of the corona. Only when R_i is negative and R_d+R_0 sufficiently large will R_i reveal itself, not by the slope but by making the current oscillatory.

It is easily shown that the equivalent circuit of Figure 1A has *two* possibilities of instability. Both demand negative R_i:

1. *Catastrophic:* $$R_i + R_d + R_0 < 0 \tag{6a}$$
 (when C_0 is some picofarads only), leading to spark. If C_0 is too large:

$$R_i + R_d < 0 \tag{6b}$$

leading to recurrent spark discharging of C_0.

2. *Oscillatory:*
$$R_d > \frac{L_i}{C_i |R_i|} \tag{7}$$

for C_0 greater than some ten picofarads.

At small currents in air and similar electron attaching gases, just above corona onset, R_d is usually large, and oscillations growing into Trichel pulses or positive glow oscillations occur. At increasing current, R_d diminishes, leading either to damping of the oscillations (transition into the continuous negative glow) or to catastrophic spark. Also, the ionization region boundary may become spatially unstable, leading to streamer puncture of the drift region.

TRICHEL PULSE CORONAS

Ever since the days of Trichel[5] and Loeb[6], negative ions have been considered as the necessary cause of the self-pulsing of the negative corona form known as Trichel pulse corona, see refs.[7,8] and below. This indicates that Trichel pulse coronas should be useful as detectors for electronegative trace gases in non-attaching or weakly attaching main gases. We[1] have recently confirmed this experimentally for the cases of oxygen O_2 and sulphur hexafluoride SF_6 traces in argon and air at atmospheric pressure. By measuring the Trichel pulse frequencies at given average corona currents a sensitivity better than 0.1 ppm SF_6 in Ar was obtained.

That investigation included measurements on negative coronas in "pure" argon, where the absence of Trichel pulses should give the detector a natural zero point. However, Trichel-like pulses were found even under those conditions. This phenomenon is the subject of the present chapter.

Trichel pulses are usually considered as a game between two players, the ionization and the drift region. Between pulses, negative ions in the drift region keep the ionization region voltage V_i below the onset value V_{is} necessary for a self-sustained discharge. The ions drift away to the low field electrode, and V_i rises to V_{is} (and above, if no initiatory cathode electron is available to start the discharge). When the discharge starts, the ionization region negative resistance ensures a rapid current rise, supplied by the minute ionization region capacitance C_i. At higher (mA) currents I the discharge is more effective and can exist at lower V_i. However, as discharge electrons reach the drift region and some or all form negative ions, the space charge formed lowers the V_i below the sustainment level V_{ig}, and the Trichel pulse comes to an end, often abruptly. The next pulse must wait until an equal amount of space charge has disappeared into the low field electrode. Typical Trichel pulse forms and terminology are sketched in Fig.2.

According to this idealized picture *all Trichel pulses are created equal.* Any change in the mean corona voltage U and current I should only affect the time to clear away the drift region space charge, i.e. the average pulse repetition rate f_T, and not the pulse charge Q_T or shape.

However, as one removes the electronegative constituents of a corona gas, one additional player enters the game: The external circuit, represented by the corona gap series resistance R_0 and parallel capacitance C_0. As soon as the drift region space charge becomes insignificant (only fast electrons there) R_d can no longer compensate R_i, *and C_0-R_0 take over the role of C_i-R_d* . The electrical equivalent circuit of the total system remains virtually unchanged by this, see Fig.1A. This means that *even negative coronas in free-electron gases will exhibit pulses that are (nearly) indistinguishable from "genuine" Trichel pulses,* provided C_0 is kept small enough to prevent spark formation. This has important (detrimental) consequences for the use of Trichel pulses as indicators for electronegative trace gases.

Figures 4 and 5 show[1] the Trichel pulse frequency factor (kHz per μA current) as function of contaminant concentrations in 1 bar Ar. They demonstrate that the 10 mm drift region in pure Ar has a storage ability for negative charge corresponding to 20 ppm of O_2 or 1.7 ppm of SF_6. This is simply due to the finite drift velocity of the electrons. Below 3 ppm of SF_6 in Figure 5 the change of slope indicates that the external circuit capacitance starts to determine the frequency factor, i.e. the charge Q_T per pulse. This was confirmed by varying this capacitance. 3 ppm also is where the U(I) characteristic (average corona voltage vs average current) changes initial slope from negative to positive, showing that the drift region resistance starts to dominate.

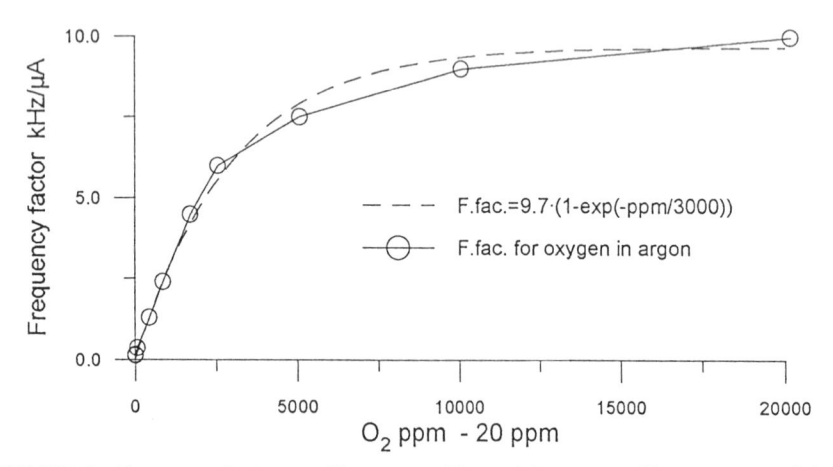

FIGURE 4. Frequency factor vs. (O_2 conc. - 20 ppm) in argon, with an exponential-type saturation curve fitted. Measured using the apparatus of Fig.1[1].

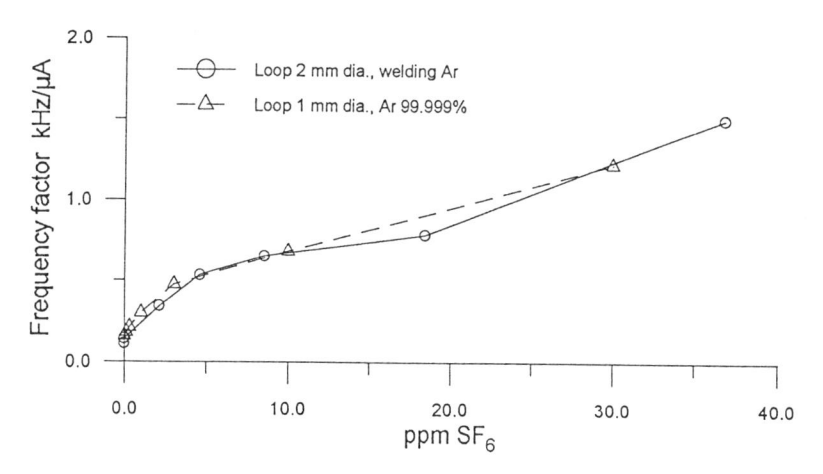

FIGURE 5. Frequency factor curves for SF_6 in two qualities of argon at 94 kPa, for cathode Pt loops 0.06/2 mm and 0.05/1 mm respectively. Note that an extrapolation to zero frequency gives -1.7 ppm on the abscissa. Measured using the apparatus of Fig.1[1].

POSITIVE GLOW CORONAS

Visually, the positive glow corona in centimeter gaps near atmospheric pressure is a uniform, stable glow closely covering the high field tip of point electrodes. In wire-cylinder coronas the whole wire surface is glowing, in contrast to Trichel wire coronas, which burns in separate, nearly equidistant points along the wire. The positive glow was studied in detail by Beattie[9], who reported it to be oscillatory in gases with electron attaching properties. In contrast to Trichel pulse coronas, however, the physical reasons for the oscillations are not at all evident. As discussed above, the positive ions will tend to decrease the ionization region field rather than to increase it. Also, positive ions are the dominating ions in positive coronas in *all* gases, but gases like Ar and N_2 do not give oscillations. On the other hand, negative ions can hardly have concentrations that affect the ionization region field, even in positive coronas in highly electronegative gases[10].

While negative Trichel current pulses always are large, in the mA range, positive glow current oscillations or pulses often are so small that they are overlooked. However, if the ionization region light is recorded instead of the current, one often finds that the ionization is turned completely off between the pulses. Nevertheless the accompanying current pulses are small, because the pulse frequency is so high that many positive ion clouds are in transit at the same time and smooth out the current.

Observations of the existence / absence of free running oscillations give only qualitative information about the ionization region stability. More detailed knowledge is obtained by pulse excitation of the burning DC corona, with observation of the induced current and light perturbations. Figures 6 and 7 show some preliminary results obtained in Trondheim, for positive wire-to-cylinder coronas in 25 kPa pure Ar and in Ar with added electron attaching and non-attaching gases. To us, the results are encouraging, but confusing.

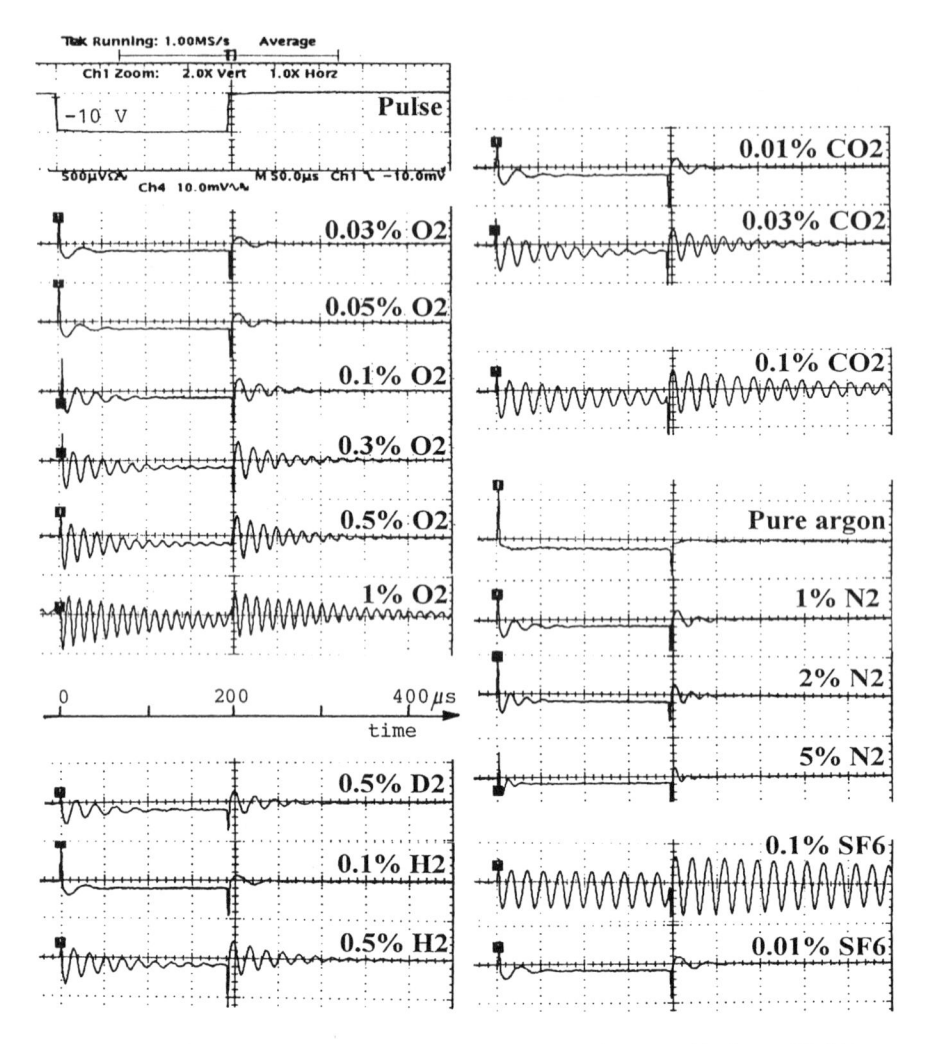

FIGURE 6. Pulse excited oscillations in various gas mixtures with Ar at 25 kPa, at 50 µA DC corona current. Positive corona between inner cylinder 1.6 mm, outer 26 mm dia., length 80 mm. Pulse -10 V to cathode. Currents increasing downwards.

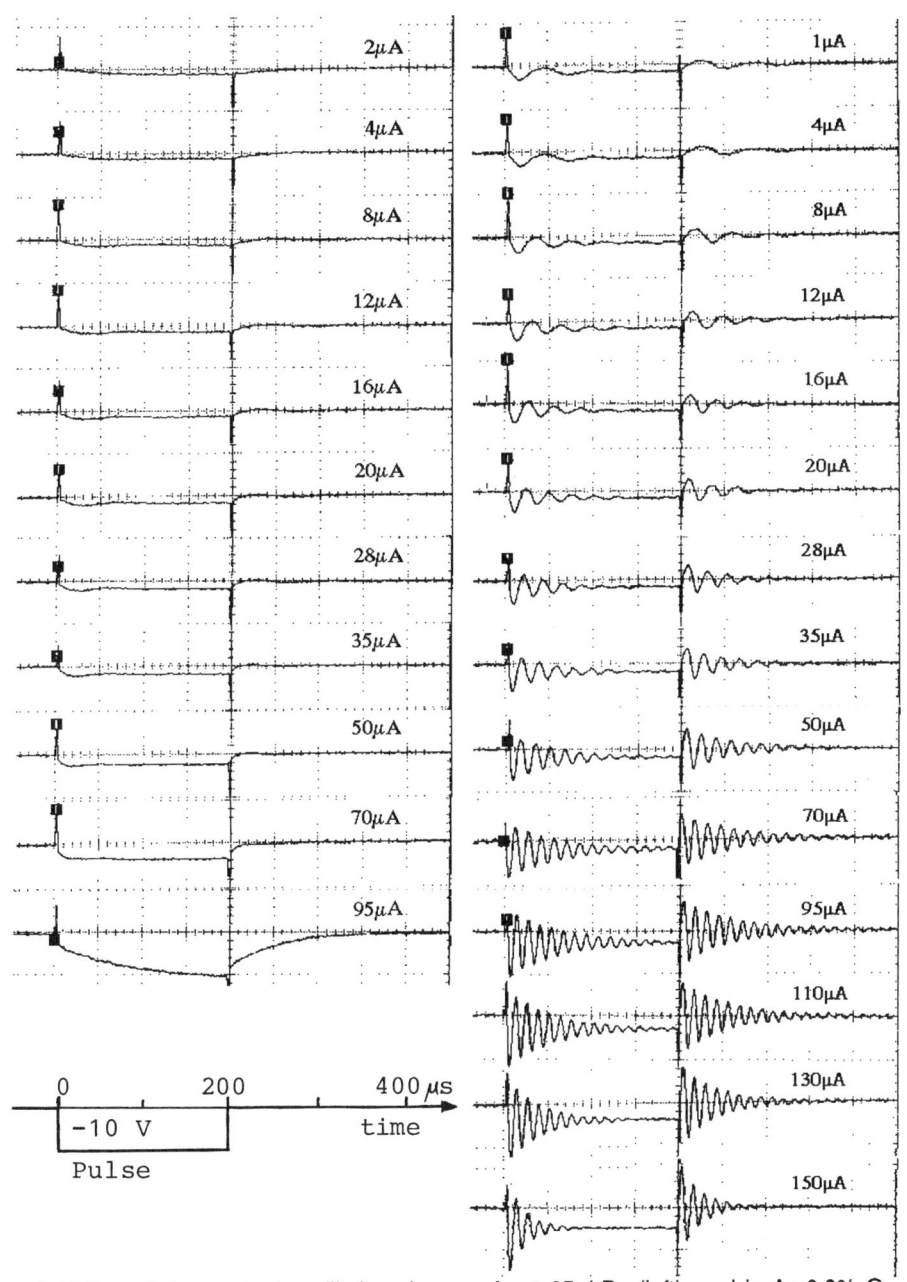

FIGURE 7. Pulse excited oscillations in pure Ar at 25 kPa (left), and in Ar+0.3% O_2 (right) at increasing corona current. Positive corona between inner cylinder 1.6 mm, outer 26 mm dia., length 80 mm. Pulse -10 V to cathode. Currents increasing downwards.

The encouragement lies in the fact that the oscillation tendency of the positive glow corona, indicated by the induced oscillation frequency and duration, certainly is very sensitive to something. The present confusion is: Sensitive to what? As evidenced by Figure 6, there is no clear correlation between the oscillation tendency and the electronagativity of the gas, compare O_2 and CO_2.

In pure Ar one hardly sees any oscillations at all, even at 95 µA DC current, where the added pulse voltage would have caused a spark if not interrupted, see Figure 7. In Ar + 0.3% O_2 the oscillation tendency first increases with the current (reflecting a decrease in the equivalent self-inductance), and then decreases (reflecting a diminishing of the drift region resistance and thus an increasing damping). The decreasing self-inductance is also shown by the increasing oscillation frequency, assuming a constant ionization region capacitance, see Eq.(5).

THE U(I) CURVE OF SELF-PULSED DISCHARGES

As pointed out above, the curve of average voltage U vs. average current I for corona discharges gives valuable information about the stability of the corona ionization region. However, at low currents near onset *all* coronas that are self-pulsed with current-independent pulse size (like Trichel pulses) must show an initial negative U(I) slope. This is illustrated by Figure 8.

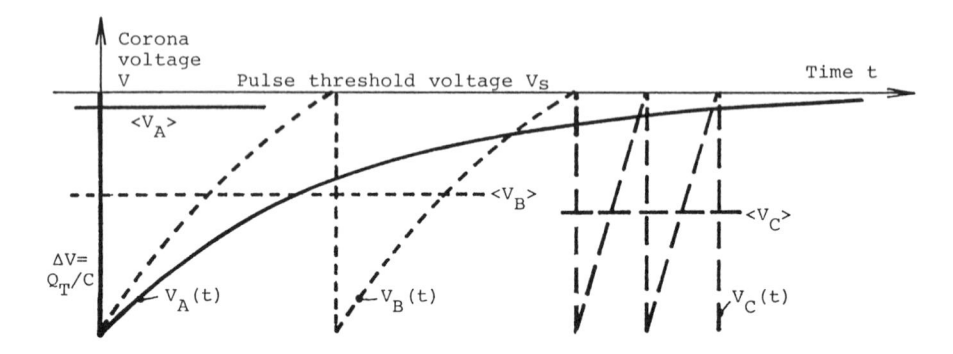

FIGURE 8. Illustrating the reason for the initial negative slope of the average U(I) curve for relaxation-type pulsed coronas.

In Figure 8 the sawtooth voltage pulses shown are all caused by discharge current pulses of equal size and shape, and so short that they cause vertical voltage drops in the figure. The rising part of the sawtooth curves are caused by recharging of the gap or ionization region capacitance, until the discharge onset voltage is reached and a new pulse occurs. The voltage observed by a slow voltage meter is the average of the sawtooth. Because of the exponential form of the recharge curve, however, this average will depend on the pulse repetition frequency, and a falling U(I) curve results. Detailed calculations[11] show that the resulting slope will be equal to the negative of the discharge series resistance. Thus, this initial negative slope has nothing to do with the size of the negative resistance of the discharge itself.

CONCLUSIONS

1. In all discharges, the ionization region is the active circuit element, which may generate oscillations for given ranges of impedances of the rest of the electrical circuit. In positive coronas, and in negative coronas in electron attaching gases, the high drift region impedance will insulate the ionization region from the external circuit, but in negative coronas in non-attaching gases the external circuit dominates and may even cause Trichel-like pulses.

2. The stability and the resonance frequency of positive glow coronas may be measures using voltage pulse excitation. The oscillations are very sensitive to the corona gas, but their physical mechanism is not known.

3. Relaxation-type self-pulsed discharges will have average U(I) curves with a negative initial slope, equal to the negative of the external series impedance.

ACKNOWLEDGEMENTS

The present paper is distilled from results obtained by visitors, staff and students in the ELION laboratory in Trondheim: A and M Goldman (Laboratoire de Physique des Décharges, Gif-sur-Yvette, France); M Laan and K Kudu (Gas Discharge Laboratory, Tartu University, Estonia); B Kurdelova, M Kurdel and K Hensel (Comenius Unicersity, Bratislava, Slovakia); B Bjones, I H (Olsen) Lågstad, R T Randeberg, and G Løfsgård (ELION).

REFERENCES

1. Sigmond R S, Goldman A, Goldman M, Laan M, Bjones B, Olsen I H, "Detection of electronegative trace gases by corona discharges", *Proc. 11th Int. Symp. on Plasma Chemistry, Loughborough UK,* 1993, pp 1315–1320.

2. Linhjell D, Sigmond R S, "Sampling image intensifier camera study of the glow pulses of positive point-to-plane coronas", *Proc. 9th Int. Conf. on Gas Discharges and Their Applications, Venezia (Benetton),* 1988, pp 403–406.

3. Sigmond R S, "On the static and quasistatic characteristics of the self-sustained low pressure Townsend discharge in hydrogen", *Proc. 5th Int. Conf. on Phenomena in Ionized Gases (North-Holland)*, 1962, pp 1359–1363.

4. Colli L, Facchini U, Gatti E, Persano A, "Dynamics of Corona Discharge between Cylindrical Electrodes", *J. Appl. Phys.* **25**, 429–435 (1954).

5. Trichel G W, "The Mechanism of the Negative Point to Plane Corona Near Onset", *Phys. Rev.* **54**, 1078–1084 (1938).

6. Loeb L B, *Electrical Coronas*, Univ. of California Press 1965.

7. Sigmond R S, Goldman M, "Corona Discharge Physics and Applications" in Kunhard E E and Luessen E H (eds) *Electrical Breakdown and Discharges in Gases*, NATO ASI Series B: Physics, Vol 89B, Plenum: New York 1983.

8. Sigmond R S, "Corona Discharges", Chapt. 4 in Meek J M and Craggs J D (eds) *Electrical Breakdown of Gases*, Wiley: London 1978.

9. Beattie J, "The Positive Glow Corona Discharge", Ph.D. thesis, Univ. of Waterloo, Canada 1975.

10. Buchet G, Goldman M, "Stability of Positive Continuous Corona Discharges in Electronegative and Non-Electronegative Gaseous Mixtures" *Proc. 9th Int. Conf. on Phenomena in Ionized Gases*, Bucharest 1969, pp 291–292.

11. Sigmond R S, "The Influence of the External Circuit on Trichel Pulse Coronas", *Proc. 4th Int. Symp. on High Pressure Low Temp. Plasma Chemistry (HAKONE IV)*, Bratislava, Slovakia 1993, pp XV–XX.

DISCHARGE PHYSICS ISSUES IN THE SCALING OF HIGH POWER CO_2 LASERS

E. Desoppere, C. Leys

Department of Applied Physics
Gent University, Rozier 44, B-9000 Gent, Belgium

Abstract. An ionization balance, a plasma chemistry model and a numerical model for the laser kinetic processes are applied in an iterative way to characterize a DC excited fast-axial-flow laser discharge. An empirical study of the discharge stability reveals a critical power density $<jE>_c$.

INTRODUCTION

The first CO_2 laser delivered 1 mW at 10 μm in 1964 (1); it is now an established commercial product, delivering from tens of watts, to multi-kilowatt of continuous output power. Careful modelling of the discharge can to a great extent replace expensive trial and error prototyping.
In the present paper the underlying physics and design principles are discussed for a fast-axial-flow (FAF) system, with special emphasis on reducing operating cost as well as footprint. A figure of merit of a laser is the maximum power output per unit length

$$\frac{P_L}{L_a} = \eta \, p_{EL} \, A_d$$

η is the electro-optical efficiency, which is a complicated function of the discharge parameters, p_{EL} is the maximum electrical power density for which the discharge remains stable, and A_d is the discharge cross section.
A CO_2 laser operates by exchanging energy between low lying vibrational-rotational levels (Fig.1); the upper (00°1) level is excited by direct excitation from the ground state and by resonant transfer from vibrationally excited nitrogen. The lower level (10°0) is depopulated by collisional transfer to the (01¹0) bending mode. Finally the (01¹0) energy is converted to heat by collisions with N_2 and He. With higher power input into the plasma, the thermal population of the lower level increases; radiation is reabsorbed by the lower laser level, halting laser action at about 150 °C. The different types of CO_2 lasers differ

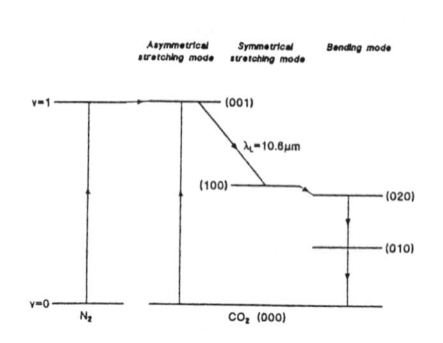

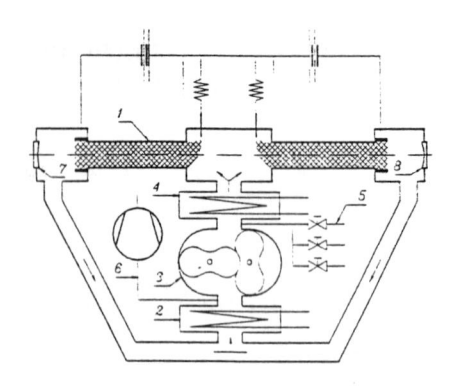

Figure 1. Low lying vibrational levels of the electronic ground state of CO_2.

Figure 2. Schematic of a FAF laser; discharge (1), heat exchanger (2,4), Roots-blower (3), gas inlet (5), replenishment (6), mirrors (7,8).

mainly in their waste heat removal scheme; in a FAF laser the hot gas itself is removed. A typical laser module is shown in figure 2. In order to investigate scaling effects, the discharge radius was increased from the commercially usual 7-10 mm to 15 mm. Typical operating parameters are V_d=10 kV, I_d=100 mA, p=60 hPa, with varying CO_2:N_2:He gas composition. The discharge is probed by a number of diagnostics listed in table I.

TABLE I. Diagnostics

Parameter		Diagnostic
E/N	reduced electric field	Langmuir probe, thermocouple
$n_p(r)$	positive ion profile	Langmuir probe
δ_{co}	degree of dissociation	mass spectrometer
α	negative ion fraction	Langmuir probe
v(r)	gas velocity	Pitot tube
T_v	vibrational temperature	IR diode laser
γ_0	small signal gain	laser power meter
I_{sat}	saturation intensity	laser power meter
$<jE>_c$	critical power density	laser power meter

CHARGED PARTICLE BALANCE

Ideally, to model a discharge, one should start from a data set of cross-sections, and solve the Boltzmann equations and the continuity equations for all charged as well as all neutral particles in a self-consistent time-dependent way. This leads to a large number of very stiff equations. The present work tries, specifically for a FAF laser discharge, to deduce a hierarchy of

processes and build in an iterative way an ionization balance, a plasma chemistry model and a vibrational kinetics model, sufficiently accurate to be used in industrial laser design.
In a discharge with gas flow the flux equations are

$$\frac{\partial n_c}{\partial t} + \nabla \cdot \Gamma_c = S_c$$

The flux density

$$\Gamma_c = - D_c \nabla n_c \pm n_c \mu_c E + n_c v$$

takes into account drift and convective flow; c=e,n,p for electrons, negative and positive ions, v is the gas velocity; other symbols take their usual significance. The source term S_c describes the loss and gain processes, which depend on the chemical composition. From the plasma chemistry model it follows that N_2^+ and CO_2^+ are the dominant positive ions, O^- and CO_3^- the dominant negative ions. Between these particles the following reactions are considered
-direct ionization

$$CO_2 + e \rightarrow CO_2^+ + 2e$$

$$N_2 + e \rightarrow N_2^+ + 2e$$

-dissociative attachment reactions 1, 3, 7, 27

$$AB + e \rightarrow A + B^-$$

-associative detachment reactions 9, 10, 11

$$AB + O^- \rightarrow ABO + e$$

-electron positive-ion recombination

$$CO_2^+ + e \rightarrow CO + O$$

$$N_2^+ + e \rightarrow N + N$$

-negative-ion positive-ion recombination

$$CO_2^+ + CO_3^- \rightarrow 2CO_2 + O$$

The numbers refer to reactions in table II. Reaction coefficients k_c are taken from literature (2)(3)(4)(5) or calculated using the ELENDIF Boltzmann solver code (6), c=i,a,d,r for ionization, attachment, detachment and recombination (ei for electron-ion, ii for negative-positive ion recombination). The steady state flux equations are solved to yield the self-sustaining E/N value in the positive column. In figure 3 calculated and experimental E/N values are compared and found to be in good agreement, except for mixtures with high $[N_2]/[CO_2]$ ratios; for these mixtures better agreement is obtained if associative ionization of N_2 is included. The model reliably predicts the self-sustaining E/N value, and from there the power input $p_{EL}=<jE>$. For the normal range of pressures (60 to 100 hPa) bulk processes dominate over diffusion and convection, which justifies the assumption of an infinite plasma. A model including wall losses shows that diffusion is the dominant loss mechanism for pressures below 20 hPa.

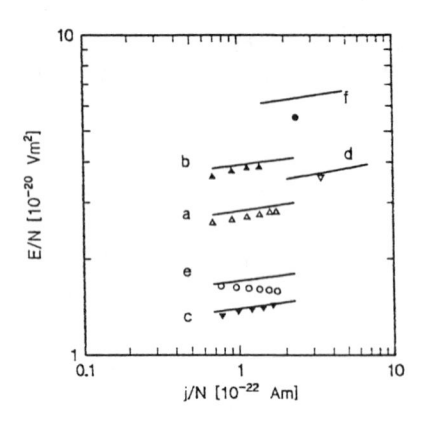

Figure 3. Calculated and experimental vol-tage-current characteristics for 4:27:69 (a), 4:48:48 (b), 15:0:85 (c), 100:0:0 (d), 0:15:85 (e) and 15:85:0 (f) $CO_2:N_2:He$.

PLASMA CHEMISTRY

The dissociation of CO_2 and the formation of negative ions affects the electro-optical efficiency η; such effects are studied in a plasma chemistry model where 30 reactions are taken into account (table II) together with a gas replenishment rate $\xi = q/Q$ (Q the volumetric gas flow, q the replenishment flow). The model calculates system averaged densities; in a closed gas recirculating system, the residence time τ_r of a particle in the discharge is only a fraction of the time τ_c a particle takes to complete the circuit loop. Therefore the rate constants for reactions in the discharge are multiplied by the factor τ_r/τ_c.

NO_x reactions are not included: in the experimental laser module mass spectrometry failed to detect nitrogen oxides (detection limit 50 ppm); Spiridonov et al. (8) measured in a similar module N_2O concentrations below 200 ppm. It is assumed that these low NO_x levels do not affect negative ion densities.

The set of 30 coupled equations is solved using a Runge-Kutta method yielding the time evolution of each species: examples are shown in figure 4 for neutral particles, and in figure 5 for charged particles in the discharge and in the gas outlet.

At steady state, all densities derive from a set of algebraic equations, solved with a Newton-Raphson method. Figure 6 compares experimental and calculated values of the dissociation degree $\delta_{co}=[CO]/([CO_2]+[CO])$.

Table II: plasma chemistry reactions; rates are taken from (2)(3)(4)(5) or calculated with ELENDIF (6)

Dissociative attachment
1. $CO_2 + e \to CO + O^-$ — $(4-7) \times 10^{-19}$ m³/s
2. $CO + e \to C + O^-$ — 3×10^{-20} m³/s
3. $O_2 + e \to O + O^-$ — 3×10^{-18} m³/s
4. $H_2O + e \to OH + H^-$ — 5×10^{-18} m³/s
5. $O_3 + e \to O_2 + O^-$ — 1×10^{-17} m³/s
6. $O_3 + e \to O + O_2^-$ — 1×10^{-18} m³/s

Three body attachment
7. $O_2 + M + e \to O_2^- + M$ — 2×10^{-42} m⁶/s (M = He)
   — 1×10^{-43} m⁶/s (M = N₂)
   — 3×10^{-42} m⁶/s (M = CO₂)
8. $O + M + e \to O^- + M$ — 1×10^{-43} m⁶/s (M = N₂)

Associative detachment
9. $CO + O^- \to CO_2 + e$ — 7×10^{-16} m³/s
10. $O + O^- \to O_2 + e$ — 2×10^{-16} m³/s
11. $CO + CO_3^- \to 2CO_2 + e$ — 5×10^{-19} m³/s
12. $O + O_3^- \to O_3 + e$ — 3×10^{-16} m³/s

Neutral dissociation
13. $CO_2 + e \to CO + O + e$ — $(4-7) \times 10^{-17}$ m³/s
14. $O_2 + e \to O + O + e$ — 1×10^{-15} m³/s
15. $H_2O + e \to H + OH + e$ — 1×10^{-16} m³/s

Neutral recombination
16. $CO + O + M \to CO_2 + M$ — 4.8×10^{-48} m⁶/s
17. $O + O + M \to O_2 + M$ — 3×10^{-45} m⁶/s
18. $O_2 + O + M \to O_3 + M$ — 5×10^{-46} m⁶/s
19. $O_3 + O \to O_2 + O_2$ — 9×10^{-21} m³/s
20. $CO + OH \to CO_2 + H$ — 1.5×10^{-19} m³/s
21. $O + OH \to O_2 + H$ — 3.3×10^{-17} m³/s

Two-body ion-neutral reactions
22. $O + O_3^- \to O_2 + O^-$ — 1×10^{-17} m³/s
23. $O_3 + O_2^- \to O_2 + O_3^-$ — 3.5×10^{-16} m³/s
24. $O + CO_3^- \to CO_2 + O_2^-$ — 8×10^{-17} m³/s
25. $O + CO_4^- \to O_2 + CO_3^-$ — 2×10^{-16} m³/s
26. $O_3 + CO_4^- \to CO_2 + O_2 + O_3^-$ — 1×10^{-16} m³/s

Three-body ion-neutral reactions
27. $CO_2 + O^- + M \to CO_3^- + M$ — 2×10^{-40} m⁶/s (M = He)
28. $CO_2 + O_2^- + M \to CO_4^- + M$ — 1×10^{-41} m⁶/s (M = CO₂)

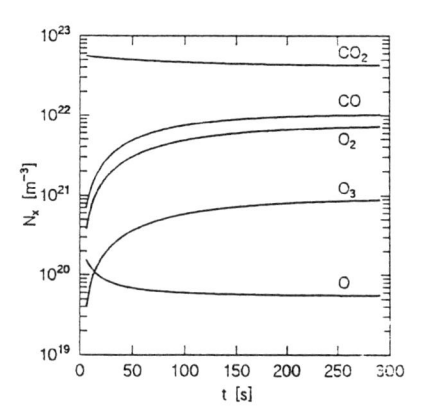

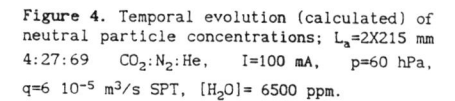

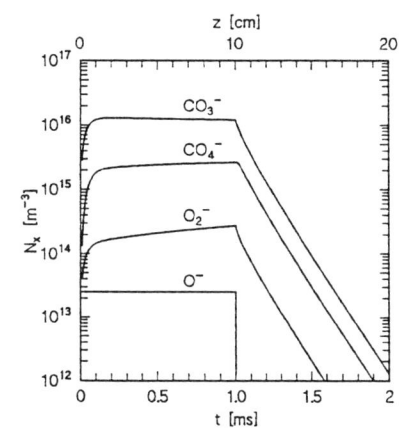

Ion-ion recombination
29. $CO_3^- + CO_2^+ \to 2CO_2 + O$ — 5×10^{-13} m³/s
30. $CO_4^- + CO_2^+ \to 2CO_2 + O_2$ — 5×10^{-13} m³/s

Figure 4. Temporal evolution (calculated) of neutral particle concentrations; L_a=2X215 mm 4:27:69 $CO_2:N_2:He$, I=100 mA, p=60 hPa, q=6 10⁻⁵ m³/s SPT, [H₂O]= 6500 ppm.

Figure 5. Temporal (spatial) evolution of charged particles in the discharge 0<t<1 ms, and in the outlet t> 1 ms; same operating conditions as in fig.4

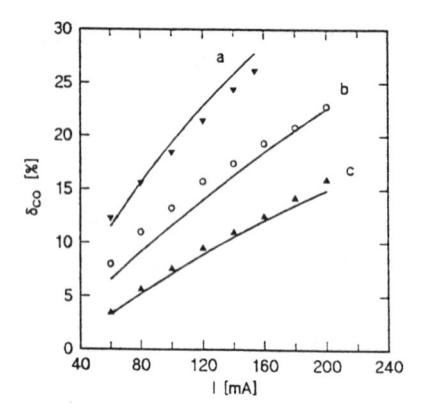

Figure 6. Calculated (line) and experimental (symbol) dissociation degree, L_a=2x215 mm (a), 1x215 mm (b), 1x120 mm (c), [H_2O]= 6500 ppm

NUMERICAL MODELLING OF A FAF CO_2 LASER

If collisions occur between vibrationally excited molecules of the same mode, little transfer occurs to other modes or to translational or rotational degrees of freedom. Due to the fast intramode relaxation time, each mode can be treated as a separate energy reservoir, in equilibrium at a specific temperature T_i; i=1,2,3 for the CO_2 symmetric stretch, bending and asymmetric stretch modes, i=4 for the N_2 vibrational mode and T is the translational and rotational temperature. A detailed discussion of this five-temperature model is found in (9). Earlier one-dimensional models (10) describe the axial laser parameter distributions; an essentially flat radial flow profile is assumed which is valid for a completely turbulent flow. However, measurements of the local flow velocity in the module demonstrate that it takes 150 mm before a flat velocity profile is established. In the transition zone a full three-dimensional description is necessary. The laser kinetics model of Sazhin et al. (11)(12) is based on the CFD (computational fluid dynamics) code FLUENT. The code uses two- and three-dimensional Navier-Stokes equations for compressible flow, and a k-ε model for turbulence (13). Mass and momentum equations are unaffected by the presence of the ionized gas; the energy equation contains additional terms for the molecular processes; these equations are complemented with transport equations for each of the vibrational modes. For the integration of the transport equations, the geometry is covered by a three-dimensional grid; differential equations are then replaced by algebraic equations. To reduce computer time, the usual five-temperature model is simplified to a three-temperature model. By comparison of relaxation times for energy transfer between vibrational modes one comes to the conclusion that $T_1 \simeq T_2 \simeq T$, as was experimentally confirmed by Spiridonov et al.(8). The three-temperature model circumvents the computational difficulties encountered with the stiff equations of the five-temperature model. Figure 7 shows the gas velocity on axis and near the wall: the parabolic profile generated by the collision of the 8 gas jets flattens out as turbulence develops. For z=10 mm the off-axis curve takes negative values, indicating that gas recirculation occurs. The combination of high residence time and poor convection favors the onset of thermal instabilities in the recirculations zones, as could be observed visually.

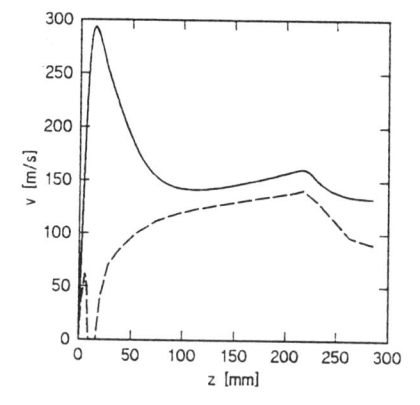

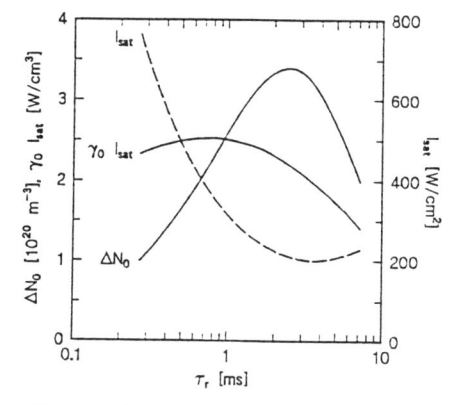

Figure 7. Calculated axial velocity profile: r=0 mm (solid line), r=12 mm (dashed line).

Figure 8. Small signal population inversion, saturation intensity and maximum optical power density $\gamma_0 I_{sat}$ as a function of residence time; I=100 mA, p=60 hPa.

As the three-dimensional simulation is computer intensive, one can compromise by solving the three-temperature kinetic equations in a one dimensinal approximation but taking the axial velocity profile from the three-dimensional simulation as input; in this way the physics of the transition zone is at least partly taken into account. Profiles of T, T_3 and T_4 are then obtained by integration of the transport equations along the axis. In the same way the variation of laser parameters can be calculated when operating parameters are changed. In figure 8 the variation of the important laser parameters on the active on gas residence time is shown. At small τ_r, the small signal population inversion ΔN_0 is low due to the convective removal of the vibrationally excited nitrogen molecules. A maximum is reached at τ_r=2.5 ms; at higher τ_r values, ΔN_0 drops due to gas heating effects. The combined effect of residence time and gas temperature on the effective relaxation time $\tau = \tau_3 \tau_r / (\tau_3 + \tau_r)$ (τ_3 the collisional relaxation time) of the upper laser level, produces a minimum for $I_{sat} (\propto 1/\tau)$. The resulting maximum available optical density $\gamma_0 I_{sat}$ (γ_0 small signal gain) is highest for $\tau_r \leq 1$ ms. Figure 9 demonstrates the flexibility of the model; a simulation was run with ($X_{-4} \neq 0$) and without ($X_{-4}=0$) vibrational deexcitation of N_2 and with varying concentrations of water. The figure clearly illustrates the effect of increased wall desorption of water as an efficient relaxator of the upper laser level.

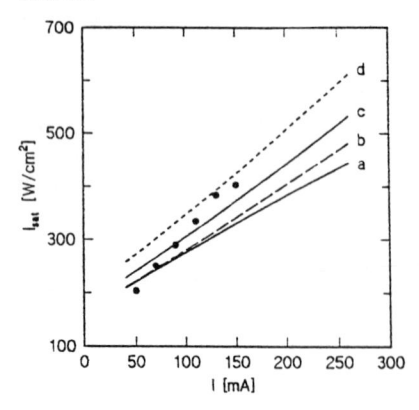

Figure 9. Saturation intensity calculated for $X_{-4}=0$ (a), $X_{-4}\neq 0$ (b), $X_{-4}\neq 0$ and $[H_2O]=6000$ ppm (c), $X_{-4}\neq 0$ and $[H_2O]=15000$ ppm (d); experimental values (full circles), p=60 hPa

DISCHARGE STABILITY

When the power density of a discharge reaches a critical value, $<jE>_c$, the column suddenly narrows. Jacobs et al.(14) suggest that in molecular laser discharges the instability arises through a positive feedback of a gas temperature perturbation. As the result

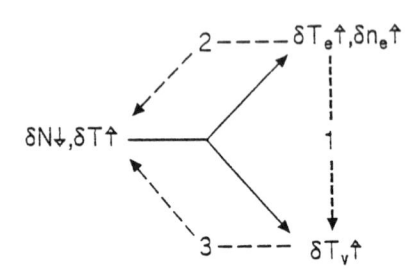

Figure 10. Feedback paths following a gas temperature or a gas density perturbation

of a local rise in temperature, T_e increases and entails a strong rise in n_e with a corresponding effect on the vibrational excitation (path 1 in fig. 10). Vibrational relaxation (path 3) and electron-molecule collisions (path2) reinforce the initial perturbation unless conduction or convection dissipates the heat. Nighan and Wiegand (15) made an analysis of the thermal stability of the collisional CO_2 laser discharge; it was revealed that such a discharge is inherently unstable and that the occurence of negative ions reinforces the instability. They calculated the time of growth of the perturbation as

$$\tau_g \simeq (10^5 \, jE)^{-1/2}$$

jE is in W/cm^2, τ_g in ms. For typical laser parameters this results in τ_g of about 1 ms, which is of the same order of magnitude as the optimum gas residence time (Fig.8). In spite of the inherent instability a discharge can be stabilised if the unstable plasma volume is removed by forced gas flow before the instability fully develops.

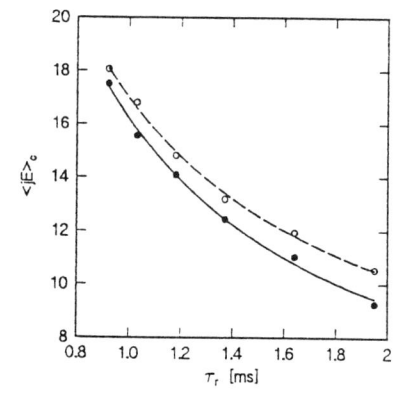

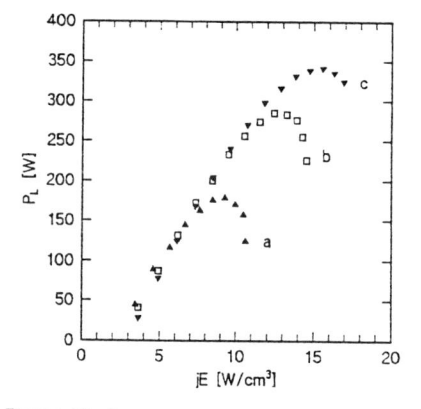

Figure 11. Critical power density as a function of gas residence time for a 4:27:69 (solid line) and a 0:30:70 (dashed line) $CO_2:N_2:He$ mixture; p=60 hPa, L_a=1x250 mm.

Figure 12. Laser power versus power density jE for different residence times: τ_r=1.95 ms (a), 1.37 ms (b), 1.03 ms (c); same conditions as in fig.11

In figure 11 the influence of the gas flow on the critical power density is shown. The adverse effect of negative ions shows through higher $<jE>_c$ values in a mixture free of negative ion source molecules. Filamentation is detrimental for the laser efficiency. If the power density exceeds the treshold value for instabilities, the laser power levels off and even drops (fig.12).

CONCLUSION

The application of an ionization balance, a plasma chemistry model and a three temperature vibrational kinetics model provides better insight in the underlying physics of fast-axial-flow CO_2 lasers. Three-dimensional modelling allows to include spatial effects relevant to R_L/L_a scaling of high power lasers.

ACKNOWLEDGEMENTS

The authors are gratefull to the European Commission for financial support under contract BREU-0516 and ERB-CIPA-CT-940183.

REFERENCES

1. C.K.N.Patel, Phys.Rev. **A136**, 1187 (1964)
2. H.Shields, A.L.S.Williams, Appl.Phys. **16**, 111 (1987)
3. W.L.Nighan, W.J.Wiegand, Phys.Rev. **A10**, 922 (1974)
4. S.R.Byron, H.Apter, J.Appl.Phys. **71**, 1976 (1992)
5. H.Hokazono, H.Fujimoto, J.Appl.Phys. **62**, 1585 (1987)
6. W.L.Morgan, B.M.Penetrante, Comp.Phys.Comm. **58**, 127 (1990)
7. C.Leys, C. van Egmond and E.Desoppere
   accepted for publication in J.Appl.Phys. **78**, 15 August (1995)
8. M.Spiridonov, C.Leys, D.Toebaert, S.Sazhin, E.Desoppere, P.Wild and S.M.P.McKenna-Lawlor
   J.Phys.D: Appl.Phys. **27**, 962 (1994)
9. W.J.Witteman, The CO_2 laser, Berlin, Springer Verlag, 1987
10. R.Rudolph, A.Harendt, P.Bisin and H.Gündel
    J.Phys.D:Appl.Phys. **26**, 552 (1993)
11. S.Sazhin, P.Wild, C.Leys, D.Toebaert and E.Sazhina
    J.Phys.D:Appl.Phys. **26**, 1872 (1993)
12. S.Sazhin, P.Wild, E.Sazhina, M.Makhlouf, C.Leys, D.Toebaert
    J.Phys.D: Appl.Phys. **27**, 464 (1994)
13. K.K.Kuo, principles of combustion, New York, Wiley, 1986
14. R.R.Jacobs, K.J.Pettipiece, S.J.Thomas
    Appl.Phys.Lett. **24**, 375 (1974)
15. W.L.Nighan, W.J.Wiegand, Appl.Phys.Lett. **25**, 633 (1974)

Nonlinear surface waves in plasmas

Slobodan M. Vuković, Najdan B. Aleksić and Dejan V. Timotijević

Institute of Physics, P.O.Box 57, 11001 Belgrade, Yugoslavia

Abstract. Evolution of electromagnetic surface waves that can propagate along the plane boundary of linear and nonlinear media, as well as in nonlinear cylindrical waveguides is reviewed and stability analysis presented. If the media are anisotropic, bisoliton solutions are found.

1. INTRODUCTION

The study of nonlinear surface waves in plasmas that has been proposed for the first time in Refs. (1) and (2) caused enormous interest over the past fifteen years in nonlinear waveguides. This was due to both theoretical and experimental advances in the early 1980's that have been applied in various optoelectronic and plasma devices (see Ref. (3)). However, some important characteristics of nonlinear modes such as polarization properties evolution and stability still remain insufficiently explained. Namely, those problems are discussed in the present paper.

The general wave equations of nonlinear electrodynamics for wave mixing that leaves the frequency unchanged shows that all field components are coupled via intrinsic nonlinearities. Consequently, decomposition of electromagnetic field into TE and TM modes even in isotropic media becomes impossible.

In fact, TE and TM modes correspond to the special limiting cases of more general nonlinear modes which represent, to some degree, a mixture of the two polarizations. This is in contrast to great majority of the papers that have appeared in the literature studying scalar wave equations, i.e. the nonlinear Schrödinger equation (see Ref. (4)).

Here, we will present results of our recent investigations that concern numerical and analytical solutions of vector wave equations assuming cubic nonlinearity approximation. That means the dielectric permittivity of the nonlinear medium can be written in the form $\varepsilon + \delta |\vec{E}|^2$, and we confine ourselves to the case $\varepsilon > 0, \delta > 0$.

Here, $\vec{E}$ is the electric field of the wave. Essentially nonlinear modes in both infinite and bounded media can be found for wave propagation constants that are of the order $\eta^2 = (kc/\omega)^2 \cong \varepsilon$. Such modes do not have linear analogs, i.e. there is a threshold power flux for their propagation. It can be shown, that longitudinal field

© 1996 American Institute of Physics

component of such modes E_z is always much smaller than the transverse ones (Refs. (5) and (6)). Moreover, within the cubic nonlinearity approximation that components must be neglected because, they are of the higher order in field intensity. Therefore, the modes studied in the present paper can be considered as quasi-transversal waves with only two field components.

Mathematically speaking, the problem reduces to the system of two coupled nonlinear Schrödinger equations. At the beginning we have found the stationary finite solutions in an infinite media. Both isotropic and anisotropic cases have been studied. Then, such solutions have been matched with the corresponding well known solutions in the linear media. Thus, by using standard boundary conditions we obtain nonlinear surface waves of hybrid polarization. In the presence of anisotropy, we use formulation of the problem given for the first time in the Ref. (6). Bi-soliton solutions found in Ref. (8) for an infinite nonlinear medium are now obtained in the form of surface waves. Our numerical analysis of evolution equations fully confirm the stability criteria proposed for the first time in Ref. (9) and generalized afterwards in Ref. (10).

Finally, a new class of finite cylindrically symmetric solutions in an infinite nonlinear media are found. With help of those solutions, nonlinear modes in cylindrical waveguides (Ref. (11)), with nonlinear core and linear cladding are now investigated. The stability criteria of Refs. (9) and (10) remain valid in this case, too.

2. PLANE GEOMETRY. ISOTROPIC MEDIA

Let us consider an electromagnetic wave that propagates along the z-axes with electric field vector of the form $\vec{E} = \vec{\mathcal{E}}_{(x)} e^{ikz - i\omega t}$. The region $x < 0$ is occupied by the linear dielectric described by a dielectric constant ϵ, while $x < 0$ contains nonlinear media (plasma) with a dielectric permittivity $\varepsilon + \delta |\vec{E}|^2$, $\delta > 0$. The solution of the Maxwell's equations in the linear medium is trivial, while in the nonlinear medium we get the system of the two equations coupled via nonlinear terms:

$$2i \frac{\partial \mathcal{U}}{\partial \tau} + \frac{\partial^2 \mathcal{U}}{\partial \zeta^2} = \left(1 + \beta^2 - |\mathcal{U}|^2 - |\mathcal{V}|^2\right) \mathcal{U},$$
$$2i \frac{\partial \mathcal{V}}{\partial \tau} + \frac{\partial^2 \mathcal{V}}{\partial \zeta^2} = \left(1 + \beta^2 - |\mathcal{U}|^2 - |\mathcal{V}|^2\right) \mathcal{V}, \qquad (1)$$

where: $\mathcal{U}, \mathcal{V} = \sqrt{\delta/(\epsilon - \varepsilon)} \mathcal{E}_{x,y}$, $\beta^2 = (\eta^2 - \epsilon)/(\epsilon - \varepsilon)$, $\zeta = (x\omega/c)(\epsilon - \varepsilon)^{1/2}$ and $\tau = \omega t (\epsilon - \varepsilon)(\varepsilon + 0.5\omega \partial \varepsilon / \partial \omega)^{-1}$.

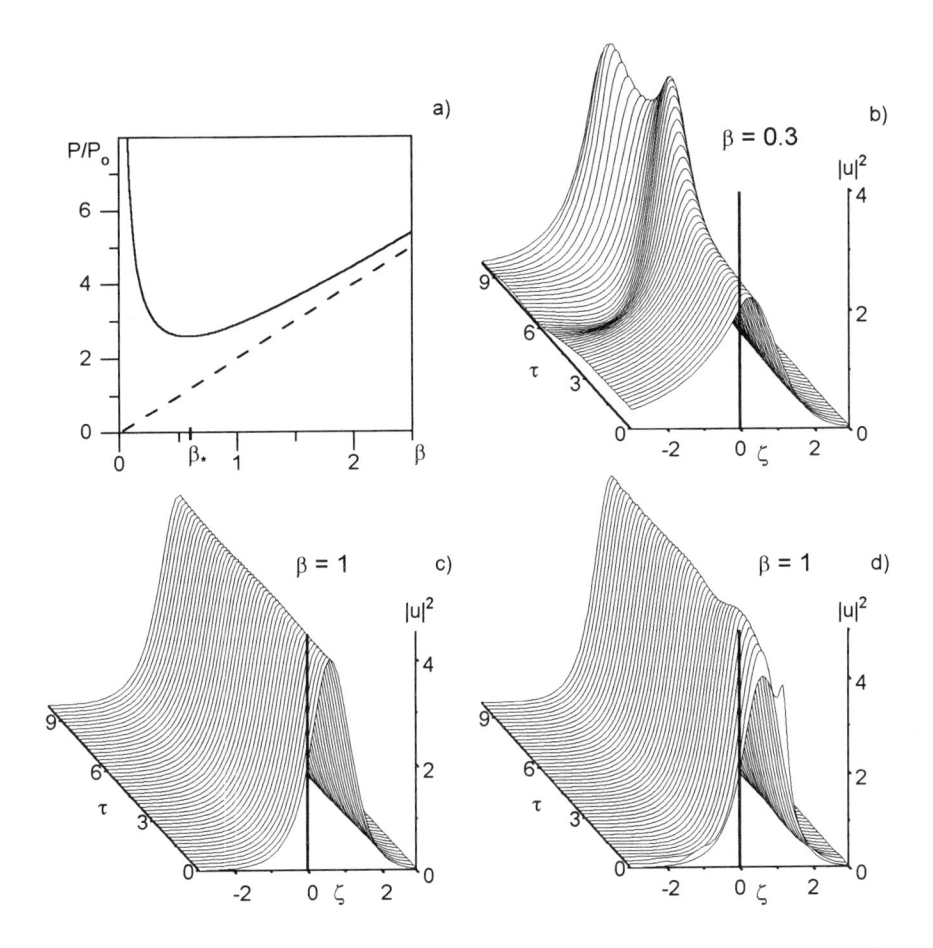

FIGURE 1. a) Dispersion of a nonlinear surface wave guided by single plane interface between linear and nonlinear media; b) evolution of unstable branch $\beta < \beta_*$; c) evolution of stable branch $\beta > \beta_*$; d) evolution of stable branch in presence of transverse perturbation. The vertical line at $\zeta = 0$ in b), c) and d) indicates the boundary interface.

The stationary solution can be written in the form:

$$\mathcal{U} = \mathcal{U}_0 e^{\beta \zeta}, \quad \zeta < 0; \quad \mathcal{U} = \frac{1}{\sqrt{1+p^2}} \frac{\sqrt{2}\alpha}{\text{ch}[\alpha(\zeta - \zeta_0)]}, \quad \zeta > 0 \qquad (2)$$

and $\mathcal{V} = \pm p\mathcal{U}$, where p is the polarization parameter $0 < p < \infty$. In the limiting cases $p = 0$ and $p \to \infty$ the mode appears to be linearly polarized: TE and TM respectively. Circular polarization is obtained for $p=1$, while in all other cases the wave is elliptically polarized. $\zeta_0 = (\beta^2+1)^{-1/2} \ln\left[(\beta^2+1)^{1/2} + \beta\right]$ is to be determined from the boundary conditions. The knowledge of the spatial distribution of the field components allows us to calculate the integrated power flux $P = \frac{1}{2\mu_0}\int_{-\infty}^{+\infty} \Re\{E_x B_y^*\} dx$ of the wave as a function of refractive index η (or β).

If the soliton is guided by the single interface between linear and nonlinear media:

$$\mathcal{P} = P/P_0 = 2\left(\sqrt{\beta^2+1} + \beta\right) + \frac{1}{\beta}, \qquad P_0 = \frac{\eta(\epsilon-\varepsilon)^{1/2}}{2\mu_0\delta\omega} \qquad (3)$$

This remarkably simple formula shows that power flux depends on a single parameter β for arbitrary polarization. Taking into account the terms that are dependent on the polarization parameter p leads to overprecision in cubic nonlinearity approximation and, consequently, can be neglected. The dispersion characteristic is shown in Fig.1 a). Our stability analysis shows that modes with $\partial P/\partial \beta > 0$, i.e. $\beta > \beta_*$ are stable, while instability occurs for $\beta < \beta_*$. This is shown in Figs. 1b), 1c). Our numeric analysis is in full agreement with the theoretical predictions of Ref. (7). In Fig. 1d) we show evolution of the stable branch in presence of the transverse perturbation. It can be seen that the mode evolves to stable state by energy leaking into the linear medium.

3. PLANE GEOMETRY. ANISOTROPIC MEDIA

We now consider an electromagnetic wave of the form $\vec{E} = \vec{\mathcal{E}}_{(x)} e^{ikz - i\omega t}$ that propagates along the z-axes in the boundary plane between anisotropic linear $(x<0)$ and anisotropic nonlinear $(x>0)$ media, i.e. magnetized plasma. We use the geometry that was studied for the first time in Ref. (7) where the external magnetic field is assumed to be oriented along y-axes. That means parallel to the boundary and normal to the wave propagation direction.

If anisotropy is sufficiently small, dielectric tensor can be written in a diagonal form with only two components in the game. In this case, we get the following system of equations in nonlinear media:

$$2i\frac{\partial \mathcal{U}}{\partial \tau} + \frac{\partial^2 \mathcal{U}}{\partial \zeta^2} = \left(1 + \beta^2 - |\mathcal{U}|^2 - |\mathcal{V}|^2\right)\mathcal{U},$$

$$2i\frac{\partial \mathcal{V}}{\partial \tau} + \frac{\partial^2 \mathcal{V}}{\partial \zeta^2} = \left(1 + \beta^2 + q - |\mathcal{U}|^2 - |\mathcal{V}|^2\right)\mathcal{V}, \qquad (4)$$

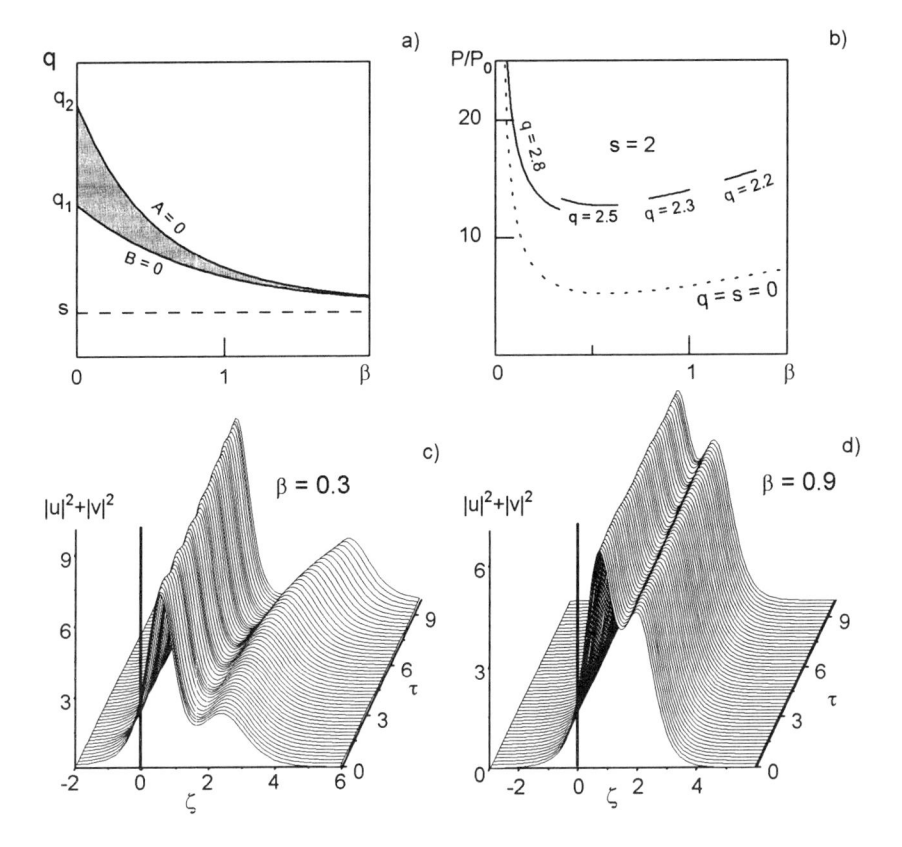

FIGURE 2. a) Illustration of existence region (shaded) for bisoliton solutions $q_1 = 0.5\sqrt{s(s+4)}$, $q_2 = (1+\sqrt{s})\sqrt{s}$; b) dispersion of bisolitons (solid lines) for various values of q; c) evolution of unstable branch; d) evolution of stable branch. The vertical line at $\zeta = 0$ in c) and d) indicates the boundary interface.

where: $\mathcal{U}, \mathcal{V} = \sqrt{\delta/(\epsilon_\| - \epsilon_\|)} \mathcal{E}_{x,y}$, $\beta^2 = (\eta^2 - \epsilon_\|)/(\epsilon_\| - \epsilon_\|)$, $\zeta = (x\omega/c)(\epsilon_\| - \epsilon_\|)^{1/2}$ and $q = (\epsilon_\| - \epsilon_\perp)/(\epsilon_\| - \epsilon_\|)$.

In the stationary case $(\partial/\partial\tau = 0)$ we get bisoliton solutions with mixed polarization:

$$\mathcal{U} = \frac{2\sqrt{2A}\,\alpha_{\|}\,e^{\alpha_{\|}\zeta}\left(1+\gamma B e^{2\alpha_{\perp}\zeta}\right)}{1+A e^{2\alpha_{\|}\zeta}+B e^{2\alpha_{\perp}\zeta}+\gamma^2 AB\,e^{2\alpha_{\|}\zeta+2\alpha_{\perp}\zeta}},$$

$$\mathcal{V} = \frac{2\sqrt{2B}\,\alpha_{\perp}\,e^{\alpha_{\perp}\zeta}\left(1-\gamma A e^{2\alpha_{\|}\zeta}\right)}{1+A e^{2\alpha_{\|}\zeta}+B e^{2\alpha_{\perp}\zeta}+\gamma^2 AB\,e^{2\alpha_{\|}\zeta+2\alpha_{\perp}\zeta}},$$
(5)

and $\mathcal{U} = \mathcal{U}_0 e^{\beta\zeta}$, $\mathcal{V} = \mathcal{V}_0 e^{\beta_\perp \zeta}$ for $\zeta < 0$. Here $\gamma = (\alpha_{\|} - \alpha_{\perp})/(\alpha_{\|} + \alpha_{\perp})$, $\beta_\perp^2 = \beta^2 + s$, $\alpha_{\|}^2 = \beta^2 + 1$, $\alpha_\perp^2 = \beta^2 + 1 + q$, and $s = (\epsilon_{\|} - \epsilon_\perp)/(\epsilon_{\|} - \varepsilon_{\|})$.

Analysis of the boundary conditions reveals the range of parameters where essentially nonlinear modes do exist. These are presented in Fig.2a). However, in the lower region we find solutions in the form of bisolitons. Similar type of solitons in an infinite media have been found in Ref.(8).

The integrated energy flux of the surface wave is given by:

$$\mathcal{P} = 2(\alpha_{\|} + \alpha_\perp)\left(\frac{\alpha_\perp - \alpha_{\|}}{\beta_\perp - \beta} + 1\right) + \frac{1}{\beta\beta_\perp}\frac{(\alpha_\perp^2 - \alpha_{\|}^2)}{(\beta_\perp - \beta)^2} \times$$

$$\times \left\{ \frac{(\alpha_\perp^2 - \alpha_{\|}^2)^2}{(\beta - \beta_\perp)} + 2(\alpha_\perp^2 - \alpha_{\|}^2)(\beta + \beta_\perp) - (\beta_\perp^2 - \alpha_{\|}^2)\beta_\perp - (\alpha_\perp^2 - \beta^2)\beta \right\}$$
(6)

which is illustrated in Fig. 2b). Evolution of this type of solitons in time is shown in Fig. 2c) and Fig. 2d). It can be seen that for $\partial\mathcal{P}/\partial\beta > 0$ we have stable bisoliton (Fig. 2d)), while in the opposite case we get unstable solitons (Fig. 2c)) due to bisoliton decay into two separate solitons. One remains coupled to the boundary, while the other one moves away deeply into the nonlinear medium. In the process of evolution two linearly polarized solitons appear.

4. CYLINDRICAL GEOMETRY. ISOTROPIC MEDIA

Let us now consider two-dimensional electromagnetic wave $\vec{E} = \vec{\mathcal{E}}_{(\tilde{\rho})} e^{ikz - i\omega t + im\varphi}$ that propagates along z-axes of cylindrical, isotropic nonlinear plasma waveguide of the radius ρ_0. The region $\rho > \rho_0$ is assumed to be occupied by linear dielectric with the permittivity ϵ. Then, the following coupled equations can be obtained:

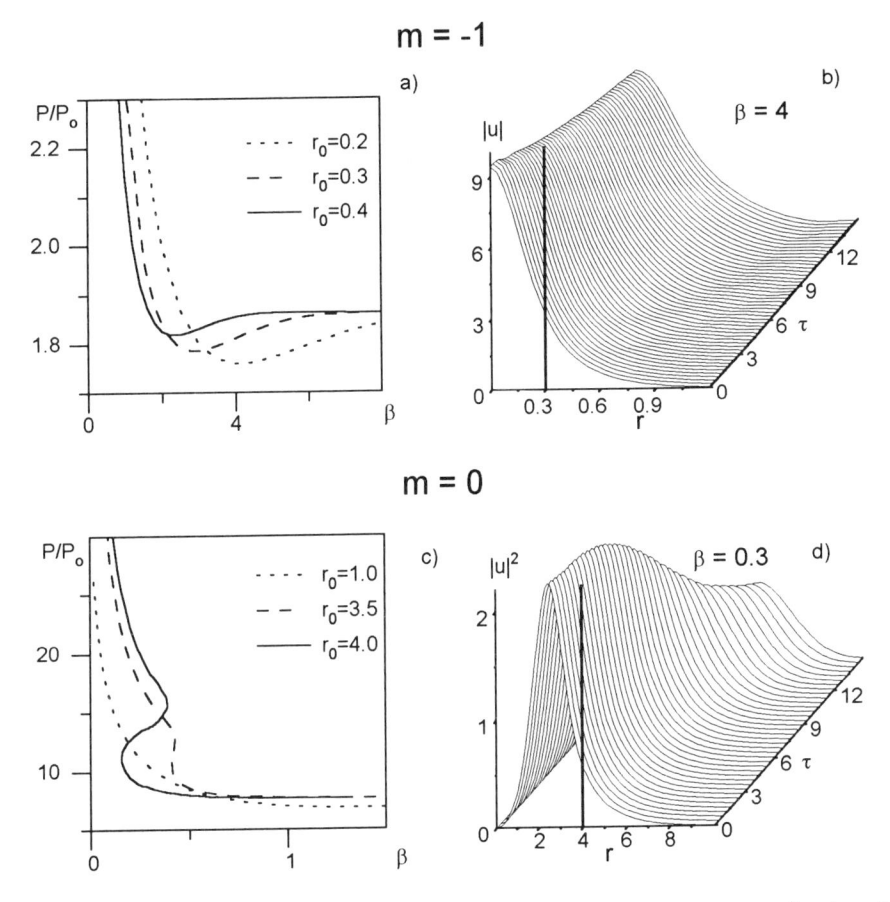

FIGURE 3. a) Dispersion of $m = -1$ mode; b) evolution of $m = -1$ mode for $\beta = 4$ and $r_0 = 0.3$; c) dispersion of $m = 0$ mode; d) evolution of mode $m = 0$ for $\beta = 0.3$ and $r_0 = 4$. The vertical line at $r_0 = 0.3$ in b) and $r_0 = 4$ in d) indicates the boundary interface.

$$2i\frac{\partial U}{\partial \tau} + \frac{1}{r}\frac{\partial}{\partial r}\left(r\frac{\partial U}{\partial r}\right) + \frac{(m+1)^2}{r^2}U = \left(1+\beta^2 - |U|^2 - |V|^2\right)U,$$
$$2i\frac{\partial V}{\partial \tau} + \frac{1}{r}\frac{\partial}{\partial r}\left(r\frac{\partial V}{\partial r}\right) + \frac{(m-1)^2}{r^2}V = \left(1+\beta^2 - |U|^2 - |V|^2\right)V,$$
(7)

where: $U, V = (\mathcal{E}_r \pm i\mathcal{E}_\varphi)\sqrt{\delta/2(\epsilon - \varepsilon)}$, $\beta^2 = (\eta^2 - \epsilon)/(\epsilon - \varepsilon)$, $r = (\omega\rho/c)\sqrt{\epsilon - \varepsilon}$.

As well known, in the linear media solutions can be expressed through McDonald's functions $K_{m\pm1}(\beta r)$, while in the nonlinear media numerical solutions are required. When $m = 0$, the system of equations (7) has a solution $\mathcal{U} = \pm p \mathcal{V}$, where p is an arbitrary polarization constant, as in the case of plane geometry (see Sec. 2). These modes, in an infinite nonlinear media, have been studied for the first time in the Ref. (12). For $m \neq 0$ only circularly polarized solutions do exist: $\mathcal{U} = 0$, or $\mathcal{V} = 0$. Integrated power flux as a function of propagation constant for $m = -1$ and $m = 0$ is presented in Fig. 3a) and Fig. 3c). The hysteresis behavior in dispersion characteristics of $m = 0$ modes are evident for $r > r_* \cong 3.5$, while such behavior is absent for $m = -1$ modes. Evolution of $m = -1$ mode is presented in Fig. 3b), and stability is evident. This is not surprising since $\beta = 4$ corresponds to the stable branch. In Fig. 3d) we present evolution of $m = 0$ mode which appears to be unstable due to energy leaking into the linear cladding.

5. CONCLUSIONS

We have analyzed polarization properties and evolution of nonlinear surface waves that can propagate along the plane boundary between linear and nonlinear media, as well as in cylindrical waveguides. Single solitons of arbitrary polarization in the isotropic case are obtained and simple analytical formula for their dispersion outlined. In the presence of weak anisotropy bisoliton solutions have been found with generally mixed polarization. For cylindrical waveguides with nonlinear core and linear cladding we have found dispersion characteristics of $m = -1$ and $m = 0$ modes, with hysteresis behavior in the latter case. In all cases, our numerical analysis of mode evolution confirms the stability criteria that involve simple derivations of the mode power-flux-density as function of refractive index.

REFERENCES

1. Alanakyan, Yu. R., *Zh. Tekh. Fiz.* **37**, 817 (1967).
2. Litvak, A. G., and Mironov, V. A., *Izv. Vyssh. Uchebn. Zav. Radiofiz.* **11**, 1911 (1968).
3. Vuković, S., ed., *Surface Waves in Plasmas and Solids*, Singapore, World Scientific, 1986.
4. Kuznetsov, E. A., Rubenchik, A. M., and Zakharov, V. E., *Phys. Rep.* **142**, 103 (1986).
5. Vuković, S., and Dragila, R., *Opt. Lett.* **15**, 168 (1990).
6. Vuković, S., and Aleksić, N., in *Proceedings of the EQEC '91*, Edinburgh, 1991, p. NDFr 23.
7. Litvak, A. G., *Izv. Vyssh. Uchebn. Zaved. Radiofiz.* **9**, 900 (1966).
8. Tratnik, M. V., and Sipe, J. E., *Phys. Rev. A* **38**, 2011 (1988).
9. Kolokolov, A. A., *J. Appl. Mech. Tech. Phys.* **11**, 426 (1975).
10. Mitchell, D. J., and Snyder, A. W., *J. Opt. Soc. Am. B* **10**, 1572 (1993).
11. Boardman, A. D., Cooper, D. S., and Robbins, D. J., *Opt. Lett.* **11**, 112 (1986).
12. Litvak, A. G., and Friman, G. M., *Zh. Eksp. Teor. Fiz.* **68**, 1288 (1975).

Optical and Probe Diagnostics of RF Discharges

Vratislav Kapička[1], Miloš Šícha[2] and A.Brablec[1]

[1] *Department of Physical Electronics, Faculty of Sciences, Masaryk University, 611 37 Brno, Kotlářská 2, Czech republic*
[2] *Department of Electronics and Vacuum Physics, Faculty of Mathematics and Physics, Charles University, V Holešovičkách 2, 18000 Prague 8, Czech Republic*

Abstract. At present, many types of RF discharges have been studied. A probe diagnostics, optical and spectroscopical methods have been used frequently both at a fundamental research and at technological applications of the discharges, especially for a simplicity of these methods. In the paper, basic principles of the methods and some interesting results will be presented.

INTRODUCTION

The RF plasma has been used in many technological applications [1-2]. To determine physical conditions in the plasma one can apply many different diagnostics methods [3]. However, the most of them influence the plasma itself. This is true especially in case of the RF plasma. This is usually not valid for optical methods, which make possible to determine from radiation processes occurring in plasma a presence of individual species and their temperatures (excitation temperature, temperatures of electrons, molecules and neutral particles, resp.), the density of charged particles and toms upon the upper and lower, as well metastable, energetic levels, either the mean values, or the radial distribution of magnitudes, without any influence of the measurement methods on the processes in the plasma.

Contrary to the optical methods a crucial problem of the probe diagnostics is how to eliminate an influence of the RF voltage component on the plasma and the probe [4]. On the other side, the probe diagnostics represents a cheap and simple diagnostics method.

Both mentioned methods will be discussed in case of the RF discharges.

OPTICAL AND SPECTROSCOPICAL METHODS

Equidensitometrical Analysis

The distribution of radiation emitted by plasma can be also studied equidensitometrically [5-8]. Namely, the equidensitometry makes possible to obtain curves of the same blackening measuring photographic plates which is done in two steps

- photographic process
- measurement of the blackening.

At present, the CCD cameras have been frequently used instead of the mentioned procedure which makes possible to obtain directly intensities of the radiation emitted by plasma.

An interpretation of these curves make possible to get a needed spatial and time information on properties of the investigated plasma discharge, as shown in Fig.1. Namely, pictures of the discharges reconstructed by a computer contain an information about geometry of the discharge and the animation make possible to follow a time development of the discharges.

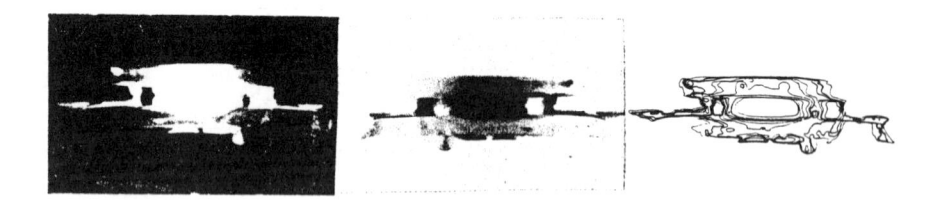

FIGURE 1. The capacitively RF discharge burning in nitrogen, its photo and obtained equidensities.

Spectral Investigation

From spectral lines emitted by plasma one can deduced onto the properties of the plasma and its state such as follows:

- presence of a given element in plasma (according to emitted lines)
- concentration of atoms upon energetic levels (from the intensity of spectral lines)
- temperature of radiating atoms (from spectral line broadenings)
- density of charged particles (from spectral line broadening)

- an average value of excitation temperature or its radial distribution excitation in the discharge (from relative or absolute spectral line intensities).

The physical parameters getting by means of the plasma diagnostics can used in order to set optimal conditions in the discharge, which is important for example in plasmachemistry, first of all, to keep the conditions constant in plasma, where the spectral lines are generated. While the temperatures of individual species in the plasma being in the thermodynamical equilibrium are the same, than in other cases it holds a relation: electron temperature > excitation temperature > vibrational temperature > rotational temperature > temperature of neutral particles.

To estimate temperature from spectral line intensities one must start with a relation for radiated - irradiated energy from unit space and time, which is defined as follows:

$$I_{ik} = h\nu_{ik} n_i A_{ik} \qquad (1)$$

where i, k denotes an upper and a lower level, respectively. ν_{ik} is the frequency of emitted radiation, h is the Planck constant, n_i is the density od atoms upon the upper level, A_{ik} is the Einstein coefficient of the spontaneous emission. Assuming the Maxwell-Boltzman distribution of electrons upon the energetic level E_i in the atom then the intensity of spectral line is given in absolute units by

$$I_{ik} = h\nu_{ik} g_i n_0 B \exp(-\frac{E_i}{kT}) A_{ik} \qquad (2)$$

where quantities g_i (the statistical weight of the given atom level), n_0 (density of atoms in the ground state), B (constant which is the same for all states in the atom) must be known [9-10]. In case of two lines, under conditions $E_i - E_1 \gg kT$ and $E_m - E_1 \gg kT$, the ratio of the intensities is a function of the temperature, i.e. $I_{ik}/I_{mn} = f(T)$. The formulae are strictly valid in case of optically thin plasma, only.

To evaluate the vibrational temperature T_v and the rotational one T_r the well known pyrometric line method has been used [11]. The intensity of vibrational band head is given by the general relation

$$I_{v'v''} = K\nu^4 A_{v'v''} i \Delta J \exp(-\frac{i}{kT_v}(E_{v'} + E_{rv'})) \qquad (3)$$

where ν is the frequency of the band edge, $A_{v'v''}$ is the Einstein coefficient for the transition from the lower vibrational state to the upper one, $E_{rv'}$ is the energy needed for the rotational transition at the band edge, i is the intensity factor and ΔJ is the number of captured rotational lines in vibrational band head recorded by a photometer with any definite width of the slit, respectively.

The intensity of any rotational line can be described by

$$I_{J'J''} = K'.\nu^4 C \exp(-\frac{hcB'}{kT_r} J'(J'+1)) \qquad (4)$$

where C is the intensity factor taking into account the multiplicity of the energy rotational state, h, c, k are the well known physical constants, J', J'' is the rotational quantum number of the upper and lower rotational state, ν is the frequency of the rotational line J' and B' is the rotational constant of the upper state [12-13], respectively.

To demonstrate a practical use of optical diagnostics in case of the RF discharge. we estimated the rotational temperature determined from the relative intensities of rotational lines OH in the 0-0 band – 306,4 nm (usually Q_1 branch) and from the R branch of the 0-0 band of the N_2 molecules.

On the contrary, the concentration of charged particles has been estimated from the broadening of the spectral lines.

Here, an influence of the spectral apparatus on the spectral line profile is omitted. Then, taking into account no other effects the spectral line width is equal to the natural spectral line width which is equal to $\gamma_i = \sum_k A_{ik}$ in case of a single emission. Of course, additional effects such as autoionization by an electric field, inelastic collisions and others increase the mentioned spectral line width.

If the radiating atom moves towards the optical device with any velocity then the frequency registered by a detector will be shifted (Doppler effect) and due to the chaotic motion of many atom the spectral line will be broadened as follows :

$$\Delta\lambda = 2\lambda/c\sqrt{2\ln 2kT/m_i} = 7,16.10^{-7}\lambda\sqrt{T/m_i}. \qquad (5)$$

In addition, the Doppler effect can excited collisions of another type, which consequently may change initial velocities of particles.

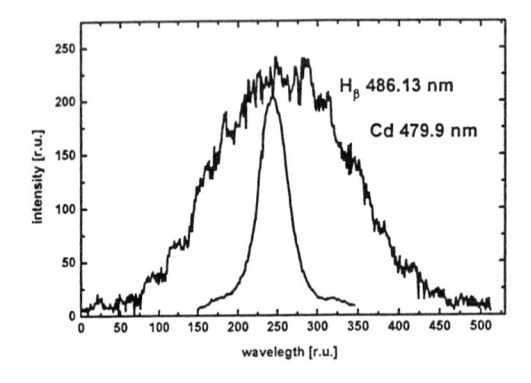

FIGURE 2. A spectral line profile emitted by RF plasma burning in humid air at reduced pressure and recorded by means of the Fabry-Perot interferometer including its apparatus function.

If the radiating atoms is not affected by constant electric field, but by a variable one, the spectral line becomes broadened. The physical basis of the spectral line broadening has been clarified by the Holtsmark statistical theory and the Lorentz collision one, respectively. Both theories have been developed intensively and various approaches have been discussed studying their interrelations limitations as to the validity of individual theories. Of course, a complex solution of the problem can be done in the framework of the quantum mechanical theory of broadenings [14-16]. to be most revealing method for establishing the spectral line broadening.

The total half-width of the line and its shift due to the quadratic Stark effect are given by

$$w_{total} = [1 - 1.75\alpha(1 - 0.75r)]w$$
$$d_{total} = [d/w \pm 2\alpha(1 - 0.75r)]w \quad (6)$$

where $w, w/d$ and d are tabulated, r is given in [14]. These expression may be applied with sufficient accuracy for $\alpha \leq 0.5$ and $r \leq 0.8$. For once ionized atoms the term $0.75r$ must be replaced by $1.2r$.

The determination of of charge particle density and temperature of neutral particles from spectral line broadenings becomes more difficult in case od low temperature plasma as the density of charged particles is usually smaller than 10^{21} m^{-3} and the temperature is less than 10^4 K. For this reason, the line widths are relatively small and it is necessary to use an apparatus with higher resolution power to take into account an influence of the apparatus function, too.

To estimate the temperature of neutral particles and density of charged particles from the Doppler and Lorentz broadening mentioned above may be also applied, even in case of plasma which isn't in thermodynamical equilibrium.

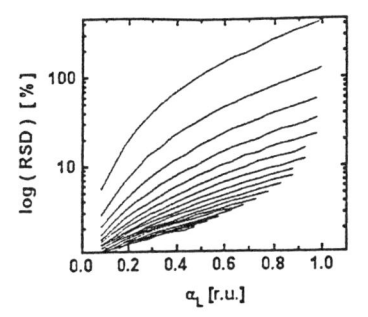

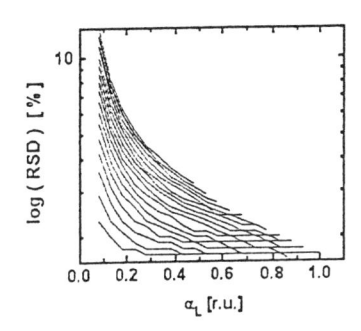

FIGURE 3. Relative standard deviation (RSD) of the Lorentz (α_L) and Doppler broadening (α_D) as a function of α_L for the fixed halfwidth of the apparatus function (0.01) and different values of α_D (starting with 0.05 and step 0.05 from left down corner) . In all cases the fixed number of points (N=101) and the constant value od $\sigma_i = 1$ for all points of the profiles were used. All profiles were normalized to 1. For more details – see [17].

The situation becomes more complicated, when the line profile is influenced by more than two independent phenomena. In the low temperature plasma the Doppler and Stark (Lorentz) broadening have been assumed only. Then the profile is described by means of the Voigt function written in relative units

$$V(x,y) = \frac{y}{\pi} \int_{-\infty}^{\infty} \frac{e^{-t^2}}{(x-t)^2 + y^2} dt, \quad x = \frac{\gamma - \gamma_0}{\alpha_D}, \quad y = \frac{\alpha_L}{\alpha_D}. \tag{7}$$

Finally, in case that it is impossible to neglect an influence of the apparatus function h_a on the initial line profile I (in our case the Voigt function) emitted by plasma one must solve the Fredholm integral of the first kind expressed as follows

$$I_p = \int_{-\infty}^{\infty} h_a(\lambda') \cdot I(\lambda - \lambda') d\lambda, \tag{8}$$

where I_a is the recorded profile, using for example the least squares method in order to estimate 4 unknown parameters, i.e. a scaling factor, the center γ_0 of the profile, α_D and α_L, resp.

Knowing statistical quantities such as confidence interval, correlation coefficients (from dispersion matrix), resp. in advance one can estimate a chance to obtain the plasma parameters with reasonable precision. An example of such approach is demonstrated in Fig.3.

The radial courses of physics parameters in discharge

The radial course of the physics parameters has been determined by the spectroscopic, interferometric and schlieren methods too. The Abel transformation expresses an relation between a radial intensity $I(r)$ and measured intensity $G(u)$ emitted by the plasma with cylindrical symmetry in the distance of u from the axes of the plasma column of the radius R:

$$G(\xi) = 2 \int_{\xi}^{1} \frac{I(\varrho)\varrho d\varrho}{\sqrt{\varrho^2 - \xi^2}}, \quad \varrho = r/R, \quad \xi = u/R. \tag{9}$$

Estimating the radial distribution of e.g. temperature in the discharge the cylindrical symmetry of the measured part has been usually considered. In this case the radial distribution of any physical parameter can be determined using methods as follows:

- high speed photography - makes possible to get an immediate visual information about transient processes which is a great advantage. There are various types of high-speed cameras. They can operate at a different speed or framing rate and they use different recording methods.

- schlieren methods - a various refraction index of the medium has been for a visualization of the plasma. The changes depend on the composition of the gas, its density, pressure and temperature, resp. [18].

- interferometric methods (including holographic methods) - the shift f of fringes is proportional to the change of the index refraction Δn given by $f.\lambda = d.\Delta n$, where λ is the wavelength and d is a the thickness of the plasma. Considering the plasma is close to the ideal medium one can derive $\Delta\rho/\rho = \Delta n/(n-1)$. However, the ratio $\Delta\rho/\rho$ depends only on the gas temperature which is known for different gases. So, the temperature can be estimated and using the Abel transformation also the corresponding radial distribution.

PROBE DIAGNOSTICS

The Langmuir probes are useful diagnostic tools in the low-pressure weekly ionized RF discharges. On the contrary to the other plasma diagnostic methods the probe technique enables the determination of the plasma parameters locally in the neighbourhood of the probe. In the Maxwellian plasma it gives the information about the electron number density, electron temperature and plasma potential. For non-Maxwellian plasma it enables to determine the electron energy distribution.

Generally in the plasma of the RF discharge the probe electron current is affected by the RF oscillations of electron density, energy and plasma potential. If the relation $2\pi f/p \geq 10^8$ [s^{-1} torr^{-1}] (f is the frequency of the driving RF signal and p is the pressure of the working gas in the RF ' discharge) is satisfied then the influence of the electron density and energy oscillations can be neglected and only the oscillations of the plasma potential should be considered [19]. For widely used frequency of the RF discharge 13,56 MHz and if the pressure of the working gas is order of several ten Pa the relation is sufficiently satisfied. In the present report we confine ourselves on the experimental conditions where the mentioned relation is valid and only the oscillations of the probe-plasma potential shouLd be taken into account.

Due to the oscillations of the probe-plasma potential the time-averaged I-V probe characteristic is modified. This modification is associated with the formation of the positive space charge sheath around the probe and with the inability of the positive ions ta response as quickly as electrons to the RF probe-plasma oscillations. For Maxwellian electron energy distribution function the average electron current $< I_e f(V_p) >$ increases due to probe-plasma oscillations according to:

$$< I_e f(V_p) > = I_{e0} \exp(-\frac{q_e V_p}{kT_e}) J_0(\frac{q_e \nu_0}{kT_e}), \qquad (10)$$

where V_p is the probe-plasma bias voltage, ν_0 is the amplitude of RF osculations across the probe-plasma junction, q_e and T_e is electron charge and temperature respectively, k is the Boltzman's constant and $J_0(x)$ is modified Bessell function of the first kind and of zero order.

Due to the modification of the time-averaged probe characteristic the determined electron temperature is higher than in reality. In the paper [20] has been on the base of the theoretical model calculated the dependence of the error in determination of the electron temperature on the amplitude of the RF oscillations.

In order to obtain reliable values of electron temperature it is essential to suppress the RF voltage oscillations across the probe-plasma junction. The methods which eliminate the influence of the RF voltage oscillations across the probe plasma junction on the probe measurement can be roughly divided into three groups:

1. In the first group the ratio Z_p/Z_c is minimized (Z_p is the impedance of the probe-plasma junction and Z_c is the impedance of the external circuit (see e.g. [21-23]). The minimization can be achieved both by increasing of Z_c and by lowering the impedance Z_p by coupling capacitively the probe to the auxiliary electrode with large surface.

   The increasing of the impedance Z_c of the external circuit can be achieved with the help of filters tuned to the fundamental and second harmonic of the driving RF frequency. It is important to notice that the stray capacitance of the probe with respect to the ground is connected parallel to these tunned filters and caused the lowering of the external circuit impedance Z_c. The effect of the stray capacitance can be minimized by placing the miniature tuned RF filter as close as possible to the probes [23].

2. In the second group the RF voltage with the proper amplitude and phase is superimposed on the probe bias voltage in order to compensate the RF oscillations over the plasma probe junction [23-27].

   In order to optimized the suppression of the RF voltage across the probe-plasma junction it is convenient to combine the both methods (e.g. [23]).

3. The RF voltage across the probe-plasma junction can also be avoided by measurement of the probe characteristic in the sharp time interval during of which the variation of the plasma due to RF oscillations is small (e.g. [28,29]). The disadvantage of this method is that the used probe technique must have sufficient time resolution. For this reason this method has been used only for RF discharge the driving frequency of which was lower than 100 kHz.

In order to estimate the error in the determination of the electron temperature from the probe characteristic it is necessary in the particular case to estimate the amplitude of the RF oscillations over the probe-plasma junction ν_0.

The normalized value $q_e\nu_0/kT_e$ has been in [30] estimated under assumption that for frequencies in MHz range the impedance of the probe-plasma junction could be considered in the first approximation to have pure capacitive character. Then the magnitude has been estimated from the probe characteristic measurement for two values of the external capacitor connected parallel with the stray capacitance of the probe i.e. between the probe and ground.

The other method of estimation of the amplitude ν_0 consists of the measurement of the probe characteristic second derivation [31]. According to paper [31] the voltage difference between the minimum and maximum of the second derivation is proportional to the quantity $2\nu_0$.

The symmetrical double probe method has been in the paper [32] used as a diagnostic method in the electrodeless RF discharge. The RF oscillation over the probe-plasma junction has been in this case suppressed by using the RF chokes the parallel resonance of which are tuned on the RF driving frequency. The auxiliary electrode was made by means of long insulating wire which was wrapped around the Pyrex glass probe envelope.

Because the wire was insulated it created the additional capacity between the probe and plasma. The additional capacity is connected parallel to the probe-plasma and causes be lowering the impedance Z_p.

ACKNOWLEDGMENTS

This work has been done in the frame of the Association for Education, Research and Application in Plasma chemical Processes and has been partially financially supported by GA CR.

REFERENCES

1. Bardos,L.,"The Supersonic Rf Plasma JET System", presented at 21th ICPIG, Bochum, Germany, September 19–24, 1993, p. 98.
2. Robin L,,Vervisch P.,and Cheron B.G., *Phys. Plasmas* **1**, 444–458 (1994).
3. Engel A., *Electric Plasmas, their Nature & Uses*, New York: Taylor Francis Ltd, 1983, ch. 1, pp. 14.
4. Jirůtka P.,Šícha M., Tichý M., Jastrabík L., Soukup L., "Langmuir Probe Measurement in the Rf Low Pressure Jet Plasma-Chemical Reactors", presented at 16th Symposium on Plasma Physics and Technology, Prague, Czech Republic, April 27–29, 1993, p. 183.
5. Dittrich K., Nieberga K., *Prog. Analyt. Atom. Spectrosc.* **7**, 315–350 (1984).
6. Petrakiev A., Koleva I., Ditrich K., Kapička V., *Folia UJEP Brno* **20**, 73–81 (1979).
7. Bušov B., Aubrecht V., Bartl S., Gross B., Jadrný P., Peška L., "Diagnostics and Modelling of Arc Behaviour", presented at XIth Symposium on Physics of Switching Arc", Brno, Czech Republic, September 26–30, 1994, p. 163–166.
8. Bušov B., Jadrný P., Aubrecht V., Gross B., Maloch J., "Switching Arc Diagnostics and CAD, presented at XXIth ICPIG, Bochum, Germany, September 19–24, 1993, pp. 42–43.
9. Corlis Ch.H., Bozman W.R., *Experimental Transition Probabilities for Spectral Lines of Seventy Elements*, Russian transl., Moscow: Mir, 1968.
10. Wiese W.L., Smith M.W., Glennon B.M., *Atomic Transition Probabilities* **I**, NSRDS–VBS 4, Washington, USA, 1966.
11. Tesař C., Janča J., "Vibrational and Rotational Temperatures of Capacitively Coupled RF Discharges", presented at 17th Symposium on Plasma Physics and Technology, Prague, Czech Republic, June 13–16, 1995, pp. 258–260.

12. Herzberg G., *Molecular Spectra and Molecular Structure I, Spectra of Diatomic Molecules*, Russian transl., Moscow: Nauka, 1949
13. Kosoruchcina A.D., Trekkov E.S., *ZhTF* **45**, 1082–84.
14. Griem H.R., *Spectral Line Broadening by Plasmas*, Russian transl., Moscow: Mir, 1978.
15. Griem H.R., *Plasma Spectroscopy*, Russian transl., Moscow: Atomizdat, 1969.
16. Fris S.E., *Spektroskopija Gazorazrjadnoj Plazmy*, Leningrad: Nauka, 1970, ch. 2, pp. 9–16.
17. Brablec A., Šťastný F., Kapička V., "Quantitative Estimates of Doppler and Lorentz Broadening from Spectral Line Profile", presented at 17th Symposium on Plasma Physics and Technology, Prague, Czech Republic, June 13–16, 1995, pp. 296–298.
18. Kapoun K., Aubrecht V., Gross B., Pešta L., "Optical Diagnostics of Discharges", presented at XIth Symposium on Physics of Switching Arc", Brno, Czech Republic, September 26–30, 1994, pp. 171–177.
19. Winkler R., Deutch H., Wilhelm J., Wilke Ch., *Beitr. Plasmaphys.* **24**, 285 (1984).
20. Klagge S., Maess M., *Beitr. Plasmaphys.* **23**, 335 (1983).
21. Cantin A., Gagne R.R.J., *IEEE Trans. AP* **16**, 279 (1986).
22. Godyak V.A., Popov O.A., *ZhTF* **47**, 766 (1977).
23. Godyak V.A., Piejak P.B., Alexandrowich B.M., *Plasma Sources Sci. Technol.* **1**, 36 (1992).
24. Gagne R.R.J., Cantin A., *Appl. Phys. Lett.* **30**, 316 (1977).
25. Sabadil H., Klagge S., *Proc. of 17th ICPIG*, Budapest, 1985, p. 322.
26. Braithwaite N.S.J., Benjamin N.M.P., Allen J.B., *J. Phys. E.: Sci. Instr.* **20**, 1046 (1987).
27. Godyak V.A., Piejak R.B., Alexandrowich B.M., *J. Appl. Phys.* **73**, 3657 (1993).
28. Anderson C.A., Graham V.G., Hopkins M.B., *Appl. Phys. Lett.* **52**, 783 (1988).
29. Hopkins M.B., Anderson C.A., Graham V.G., *Europhys. Lett*, **8**, 141 (1989).
30. Špatenka P., *Czech. J. Phys. B* **38**, 996 (1988).
31. Sabadil H., Klagge S., Kammeyex M., *Plasma Chem. Plasma Process.* **8**, 425 (1988).
32. Gagne R.R.J., Cantin A., *J. Appl. Phys.* **43**, 2639(1972).

Cluster Lamp -
A New Kind of Light Generation Mechanism

R. Scholl and G. Natour

Philips Research Labs, P.O.Box 1980, D-52021 Aachen, Germany

Abstract. Investigations on a new type of light generation mechanism have led to the development of the so called cluster lamp. Their operation is based on the regenerative formation and heating of small metal particles (clusters) in a microwave excited high intensity discharge. These lamps combine the excellent colour quality of incandescent lamps with the high efficiency of discharge lamps. Depending on the used filling, lamps with different colour temperatures for different application areas can be made.

A New Way of Generating Light

After Edison's invention of the carbon-filament lamp (1879) major steps were: first the search for new incandescent materials, i.e. tantalum, osmium and at last tungsten. Already Edison knew, that materials that could withstand higher temperatures could be used to produce more light. Further important steps were the introduction of an inert gas into the bulb, the coiling of the tungsten wire, and the introduction of the halogen cycle as applied in halogen incandescent lamps.

The use of a halogen cycle allows evaporated tungsten to return to the filament by chemical transport, which prevents blackening of the bulb. The efficiency of halogen incandescent lamps is still rather low because the tungsten filament emits mainly in the infrared and not in the visible spectral range (Fig.1). Increasing the filament temperature would led to a dramatically shortened lifetime and is at last limited by the melting point of tungsten of 3680 K.

It was the efficiency problem, that led the lighting industry to alternatives for the incandescent lamp. The innovation of discharge lamps offered light sources with much higher efficiencies by applying mechanisms to generate radiation more or less selectively in the visible spectral range. These mechanisms include selective radiation from atoms, ions, and molecules in high pressure discharge lamps or from phosphors excited by ultraviolet radiation in low pressure discharge lamps.

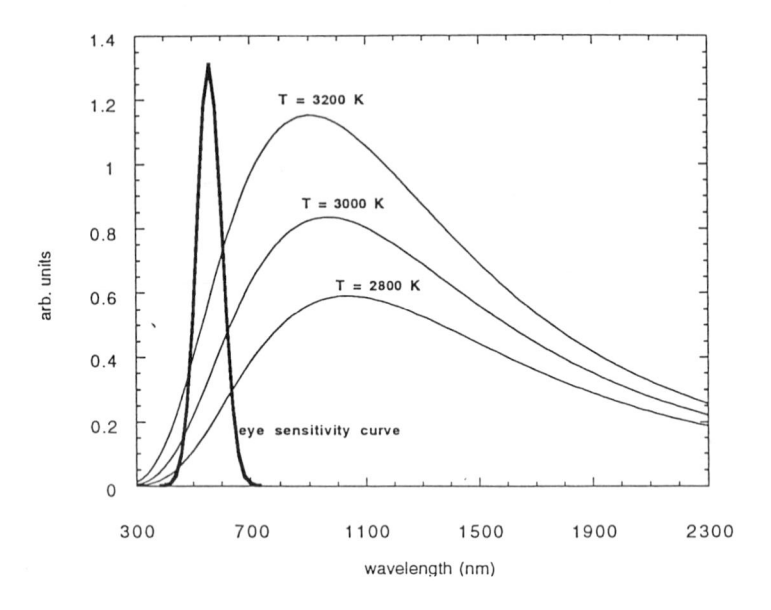

Fig. 1 Spectral distributions of black body radiators at different temperatures and spectral distribution of the eye sensitivity

In Philips Research, a new way of generating light was invented: formation and heating of clusters of metal atoms in a high-intensity discharge[1]. This cluster radiation, combines the high efficiencies of discharge lamps with the good light quality, i.e. continuous spectrum, of incandescent radiators, resulting in an efficient light source with excellent colour rendering. The use of different fillings leads to different metal clusters and hence to different colour temperatures. This make the lamp a good candidate for many application areas, where high colour rendering is needed like in TV-studio lighting and home lighting applications, requiring colour temperatures of about 5600 K and 3200 K, respectively.

Lamp Operation

The cluster lamp is a microwave excited high pressure discharge lamp (see fig.2). It consists of a cylindrical TM_{010} resonant microwave cavity in which microwave power (100 Watt) is coupled in via an antenna (frequency 2.45 GHz). The microwave power is used to support a high pressure discharge in an electrodeless quartz tube with a typical volume of about 0.3 cm³. This tube is filled with a metal compound and centred in the cavity to excite a vapour discharge. The experimental set-up has been described in more detail in references [2] and [3].

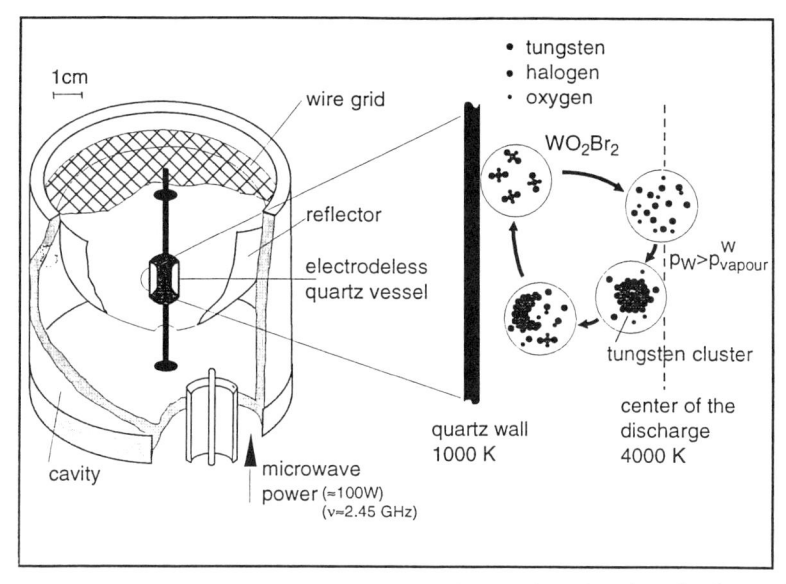

Fig. 2 Cluster lamp together with a schematic drawing showing the formation of tungsten clusters

In the discharge small metal particles (clusters), typically consisting of 1000 atoms, are generated and heated to use their thermal emission for light generation. The thermal emission guarantees a continuous spectrum with perfect colour rendering. On the other hand it is possible to heat the clusters above their melting points and to shift the maximum of the emission into the visible range.

Regenerative Formation of Clusters

There have been several attempts to replace the tungsten filament of conventional incandescent lamps by incandescent radiators. One involves injection of small particles into a gas discharge that heats them to incandescence. However, these particles must be continuously supplied by an external device, which makes the system not suitable for application as a light source. With a regenerative chemical cycle the formation of incandescent particles and the continuous delivery of them is achieved as long as the discharge is sustained.

The example of fig. 2 illustrates the regenerative formation of tungsten clusters, one of the first discovered cluster forming systems. The compound tungsten oxybromide (WO$_2$Br$_2$) evaporates inside the discharge tube at a tube wall temperature of about 1000K, leading to a pressure of 0.2-2.0 bar. Tungsten oxybromide molecules dominate the gas composition near the tube wall. Convection and diffusion transport the molecules to hotter regions of the discharge where they dissociate. Tungsten atoms are released in such high amounts that condensation of the metal vapour takes place and tungsten clusters are formed. The clusters are heated by the surrounding discharge and emit incandescent radiation until they move out of the condensation region. They finally evaporate in hotter zones of the discharge or chemically dissolve in cooler parts.

This cycle of cluster formation and decomposition proceeds stationary without any consumption of the filling. A similar regenerative chemical cycle is known from halogen incandescent lamps, invented by Zubler and Mosby [4], where the evaporated tungsten from the hot filament and which normally leads to wall blackening, is transported back to the filament by the oxygen / halogen cycle.

Experimental Results

Fig. 3 show a typical emission spectrum of a WO$_2$Br$_2$ cluster lamp. The spectrum is dominated by a continuum, which results from the incandescent radiation of the tungsten clusters. The spectrum includes contributions of molecular bands (WO) and of atomic lines of the used metal and alkali atoms (cesium is filled in the lamp and other alkali elements are presumably included by contamination).

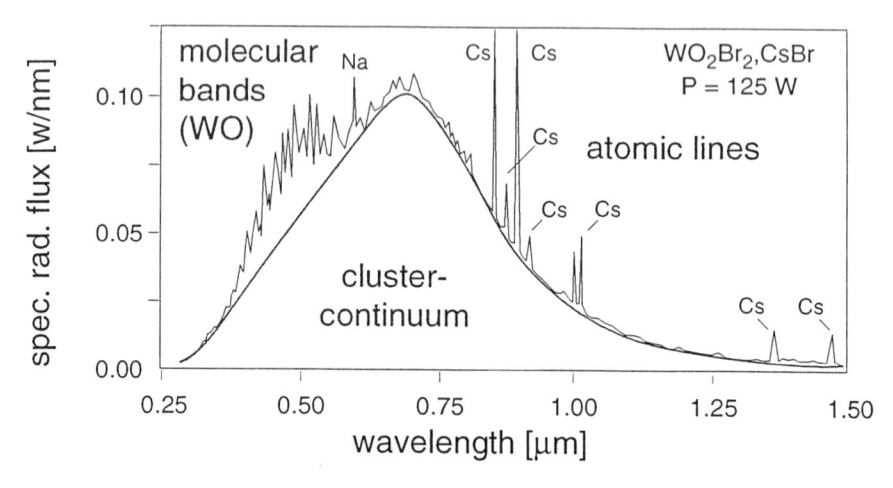

Fig.3 Measured emission spectrum of a WO$_2$Br$_2$ / CsBr discharge

The different contributions of the emission spectrum originate from different areas of the discharge. Fig.4 shows the spectral radiance observed on lines of sight through the discharge axis and 1.0 mm beside it. The emission from the axis is dominated by the molecular band spectrum of the WO molecules. The off-axis regions emit a broad structureless continuum, the incandescent radiation of tungsten clusters.

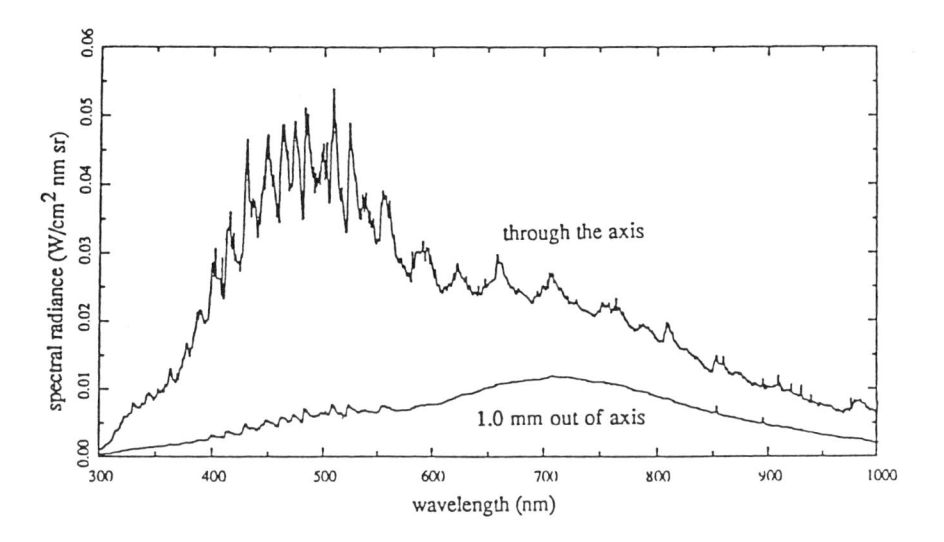

Fig.4 Spectral radiance observed on lines of sight through the discharge axis and 1.0 mm beside it (microwave power 85 W, spectral resolution 1nm)

Cluster Forming Metals

Selection of cluster forming compounds have been assisted by thermodynamic equilibrium calculations supplying the temperature and pressure criteria for the stability of the condensed phases. Fig. 5 shows an example of such a calculation. The filled compound WO_2Br_2 is stable up to about 2500 K. At higher temperature one of the Br atoms is separated and the partial pressure of WO_2Br increases. At about 3000 K the second bromine atom is separated, and so on. At very high temperatures atomic tungsten dominates in the discharge. If we compare this tungsten dissociation pressure (the pressure resulting from dissociation of tungsten compounds) with the vapour pressure of tungsten, we obtain in a broad temperature range (1700- 5400 K) tungsten supersaturation and therefore the formation of metal clusters in the hot region of the discharge.

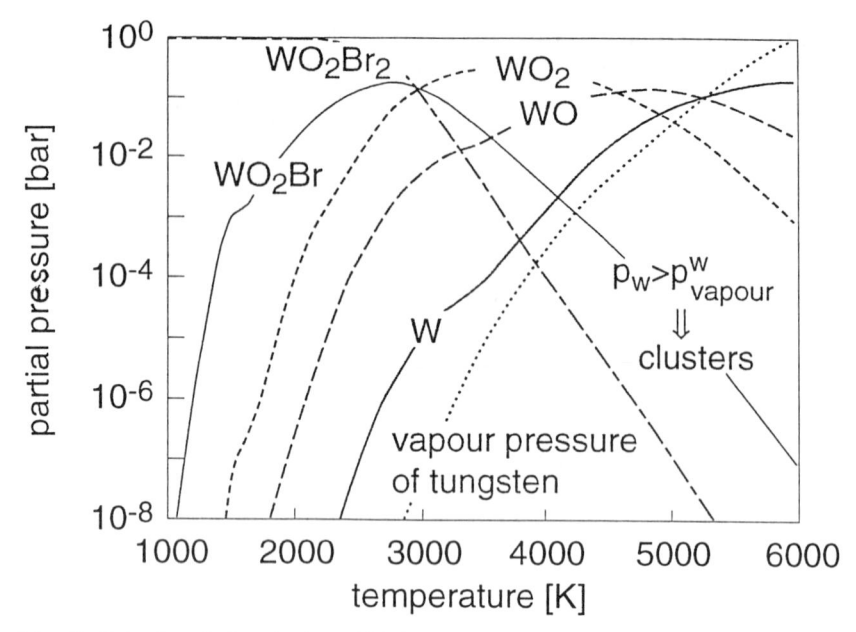

Fig. 5 Calculated chemical equilibrium composition in a WO2Br2 discharge.

Two main conditions have to be fulfilled to obtain a supersaturation of the metal vapour at high temperatures. Firstly, the metal has to be sufficiently evaporated in form of a compound mostly an oxide, halide or oxyhalide of the metal, to obtain sufficient dissociation pressure of the metal. Secondly in the hot region of the discharge the vapour pressure of the metal should be low, which is the case for metals with high boiling points like Re, W, Ta, Os, Nb, Mo, Hf and Ir.

The first condition - sufficient evaporation of a metal compound - is mostly met by using oxides, halides and/or oxyhalides in which the metal has its maximal oxidation state. An example is WO_2X_2 (X is a halogen atom), where the maximal oxidation state of 6 for tungsten is obtained.

Other candidates are Re_2O_7, OsO_4, MoO_2X_2, TaX_5, $TaOX_3$, NbX_5, $NbOX_3$, HfX_4, ZrX_4, TiX_4. These compounds filled in lamps show indeed a continuous emission spectrum (see fig.6a-d) and form clusters. But this is also the complete list of cluster forming substances. For other transition metals like Ir, Pt, Ru, Rh, V, Cr, Mn, Fe,... the maximal oxidation state can not be obtained and no compounds with sufficient vapour pressure exit. This has to do with an increasing noble metal character of these elements. For elements of the first, second and third group of the periodic table the maximal oxidation state can be obtained but the vapour pressure of the resulting halides, for example ScX_3, is too low. For all other elements of the periodic table the boiling points are too low to obtain metal supersaturation at high temperatures.

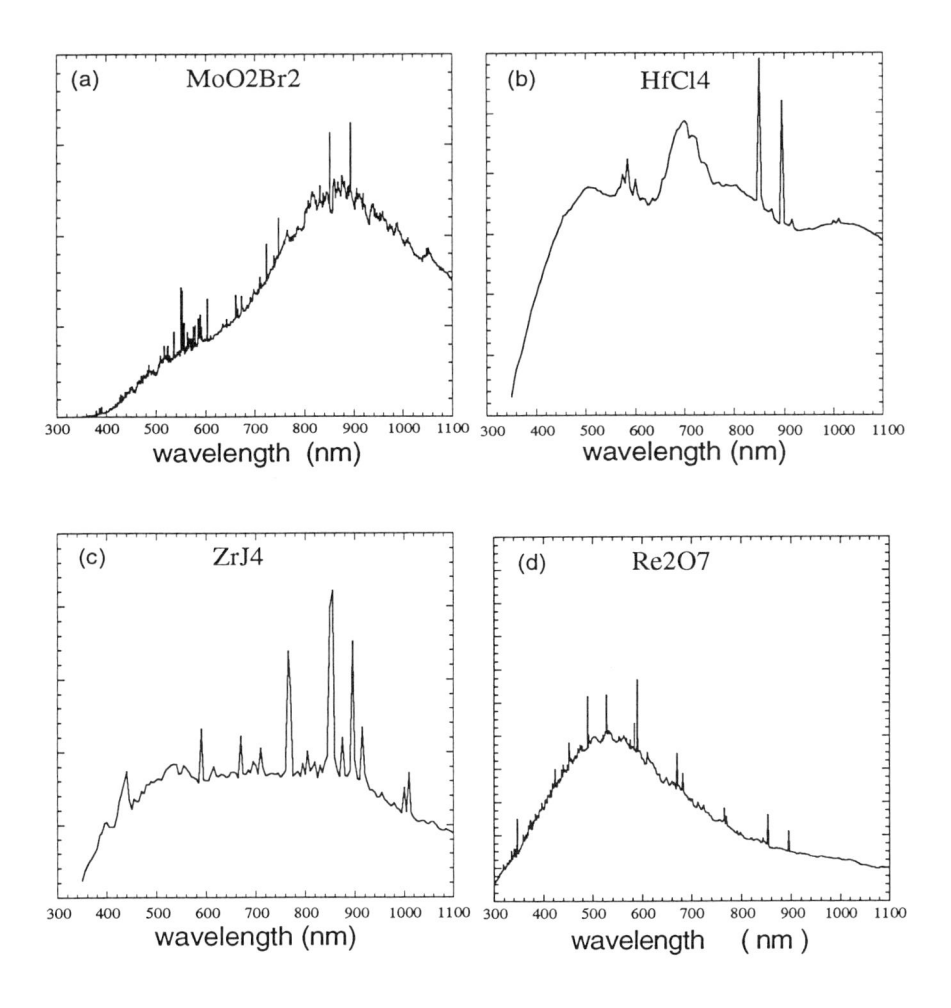

Fig.6 Measured emission spectra of (a) MoO$_2$Br$_2$, (b) HfCl$_4$, (c) ZrJ$_4$ and (d) Re$_2$O$_7$ discharges

A Theory of the Clusters' Emission Spectrum

In WO$_2$Br$_2$ cluster discharges a mean radius of the cluster of about 2,5 nm has been detected by laser scattering[5]. The clusters' emission spectrum can therefore theoretically be described by using the Rayleigh formula, an approximation of the Mie theory for objects much smaller than the wavelength of the emitted radiation[6].

$$\Phi_\lambda \sim \mathrm{Im}\left(3\frac{\varepsilon(\lambda)-1}{\varepsilon(\lambda)+1}\right) \cdot \frac{1}{\lambda} \cdot B_\lambda^{Planck}(T) \tag{1}$$

The above formula is valid for an ensemble of non-interacting, homogeneous spheres of temperature T. Φ_λ is the spectral radiant flux of the cluster ensemble, $B_\lambda^{Planck}(T)$ the spectral radiant exitance of a black body of temperature T and $\varepsilon(\lambda)$ the dielectric function. At high temperatures all resonant structures of $\varepsilon(\lambda)$ smear out [7] and the wavelength and temperature dependence of cluster radiation can be described as

$$\Phi_\lambda \sim \frac{1}{\lambda} \cdot B_\lambda^{Planck}(T) \tag{2}$$

For objects much smaller than the wavelength of the emitted radiation the black body radiation has therefore be corrected by a factor $1/\lambda$.

Fig.7 shows the measured emission spectrum of $HfBr_4$ cluster discharge together with the emission spectrum of a 5600 K black body and 4650 K Mie or Rayleigh radiator. Deviations of the measured spectrum from formula (2) can be explained by molecular band radiation of hafnium monohalides at 470, 620 and 730 nm [8] and absorption of cluster radiation in the UV and blue part of the spectrum by molecules in cooler parts of the discharge.

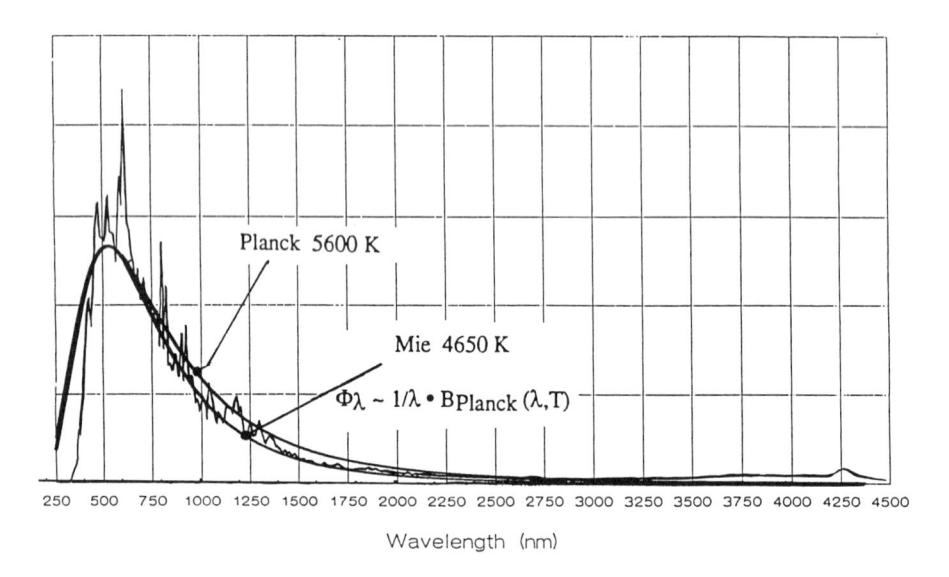

Fig.7 Emission spectrum of $HfBr_4$ discharge together with the emission spectrum of a 5600 K black body and 4650 K Mie or Rayleigh radiator

Lamp Properties

The most important feature of cluster lamps is the combination of a high efficiency and excellent colour rendering. Many of the metal compounds mentioned above have relatively low boiling points and are totally evaporated at a typical wall temperature of 1000 K. This leads to the following additional features:
- The lamps can be dimmed (reduction of operation power) without changing the colour rendering and efficiency; this makes them attractive for energy-saving applications.
- They have a good colour homogeneity, no salt shadows are seen when used for projection.
- The stability of the colour temperature and colour rendering is very high.
- Operation is independent of the burning position; the lamps can be operated both horizontally and vertically without colour change.
- The lamps allow low wall loading, which makes them suitable for low-wattage applications.
- The lamp filling react hardly with the quartz tube so that the wall is not attacked.

Conclusions

Cluster radiation represents a new development in the field of light generation. Via a chemical cycle it is possible to produce a stationary 'cloud' of continually regenerating hot clusters. Due to their high temperature the clusters emit a thermal emission spectrum which we use for light generation. The continuous spectra ensure perfect colour rendering and the characteristic cluster formation process leads to stable and reproducible colourmetric properties of a future cluster lamp.

REFERENCES

1. R. Scholl and B. Weber "Incandescent Radiation from 3500 K Hot Clusters" in Proc. Int. Sym. on the Physics ad Chemistry of Finite Systems: From Clusters to Crystals, Richmond, USA, 1991, edit. by P.Jena et al, Kluwer, Boston/London 1992

2. S. Offermanns "Electrodeless high pressure microwave discharges.", J. Appl. Phys. **67** (1) 115-123 (1990)

3. S. Offermanns "Resonance Characteristics of a Cavity Operated Electrodeless High Pressure Microwave Discharge System", IEEE Trans. Microwave Theory Tech. **38**, 904 (1990)

4. E.G. Zubler and F.A. Mosby " An iodine incandescent lamp with virtually 100 % lumen maintenance." Ill. Engng **54**, 734-740 (1959)

5. B. Weber and R. Scholl "A new kind of light-generation mechanism: Incandescent radiation from clusters.", J. Appl. Phys. **74** (1), 607-613 (1993)

6. C.F.Bohren, D.R.Huffman, "Absorption and Scattering of Light by Small Particles", Wiley, New York 1983

7. B.T.Barnes, "Optical Constants of Refractory Metals", J. Opt. Soc. Amer. **56**, 1546-1550 (1966)

8. T.Savithry, D.V.K.Rao and P.T.Rao, Current Science **42**, 533 (1973)

Author Index

A

Aleksić, N. B., 355
Alexandrovich, B., 166
Allen, J. E., 316
Allen, N. L., 247
Anderson, L. W., 1

B

Babich, L. P., 156
Bayer, R., 278
Becker, K. H., 75
Bergeson, S. D., 1
Brablec, A., 363
Brunetti, B., 234
Bruno, G., 146

C

Capezzuto, P., 146
Capitelli, M., 121
Černák, M., 136
Childs, M. A., 1

D

Dai, F., 176
Decomps, P., 306
Derouard, J., 204
Desoppere, E., 345
Deutsch, C., 12

F

Falcinelli, S., 234
Ferreira, C. M., 25
Fleurier, C., 12

G

Gadri, R. B., 306
Gamero, A., 257
Gardès, D., 12
Godyak, V., 166
Gorse, C., 121

H

Hubert, J., 25

I

Iasillo, D., 121
Ivanov, A. A., 41
Iwama, N., 289

J

Jastrabík, L., 299

K

Kapička, V., 363
Kasperczuk, A., 214
Klinger, T., 90
Kłos, Z., 105
Kogan, V. I., 267
Koidan, V. S., 192

L

Lawler, J. E., 1
Leys, C., 345
Li, C., 176
Longo, S., 121
Losurdo, M., 146

M

Margot, J., 25
Massines, F., 306
Maynard, G., 12
Mayoux, C., 306
Menningen, K. L., 1
Mentel, J., 278
Moisan, M., 25
Mullman, K. L., 1

N

Natour, G., 373
Novák, M., 299

P

Piejak, R., 166
Piel, A., 90
Pisarczyk, T., 214
Pitchford, L., 204

R

Rabehi, A., 306

S

Schein, J., 278
Scholl, R., 373
Schumann, M., 278

Ségur, P., 306
Šícha, L., 299
Šícha, M., 363
Sigmond, R. S., 332
Soukup, L., 299

T

Tammet, H., 224
Timotijević, D. V., 355
Tsai, J., 176

V

Vecchiocattivi, F., 234
Vuković, S. M., 355

W

Wołowski, J., 214
Wu, C.-H., 176

Y

Young, F. F., 176

Z

Zakrzewski, Z., 25

AIP Conference Proceedings

Title	L.C. Number	ISBN
No. 300 Discovery of Weak Neutral Currents: The Weak Interaction Before and After (Santa Monica, CA 1993)	94-70515	1-56396-306-X
No. 301 Eleventh Symposium Space Nuclear Power and Propulsion (3 Vols.) (Albuquerque, NM 1994)	92-75162	1-56396-305-1 (Set) 156396-301-9 (pbk. set)
No. 302 Lepton and Photon Interactions/ XVI International Symposium (Ithaca, NY 1993)	94-70079	1-56396-106-7
No. 303 Slow Positron Beam Techniques for Solids and Surfaces Fifth International Workshop (Jackson Hole, WY 1992)	94-71036	1-56396-267-5
No. 304 The Second Compton Symposium (College Park, MD 1993)	94-70742	1-56396-261-6
No. 305 Stress-Induced Phenomena in Metallization Second International Workshop (Austin, TX 1993)	94-70650	1-56396-251-9
No. 306 12th NREL Photovoltaic Program Review (Denver, CO 1993)	94-70748	1-56396-315-9
No. 307 Gamma-Ray Bursts Second Workshop (Huntsville, AL 1993)	94-71317	1-56396-336-1
No. 308 The Evolution of X-Ray Binaries (College Park, MD 1993)	94-76853	1-56396-329-9
No. 309 High-Pressure Science and Technology—1993 (Colorado Springs, CO 1993)	93-72821	1-56396-219-5 (Set)
No. 310 Analysis of Interplanetary Dust (Houston, TX 1993)	94-71292	1-56396-341-8
No. 311 Physics of High Energy Particles in Toroidal Systems (Irvine, CA 1993)	94-72098	1-56396-364-7
No. 312 Molecules and Grains in Space (Mont Sainte-Odile, France 1993)	94-72615	1-56396-355-8
No. 313 The Soft X-Ray Cosmos ROSAT Science Symposium (College Park, MD 1993)	94-72499	1-56396-327-2
No. 314 Advances in Plasma Physics Thomas H. Stix Symposium (Princeton, NJ 1992)	94-72721	1-56396-372-8
No. 315 Orbit Correction and Analysis in Circular Accelerators (Upton, NY 1993)	94-72257	1-56396-373-6

Title	L.C. Number	ISBN
No. 316 Thirteenth International Conference on Thermoelectrics (Kansas City, Missouri 1994)	95-75634	1-56396-444-9
No. 317 Fifth Mexican School of Particles and Fields (Guanajuato, Mexico 1992)	94-72720	1-56396-378-7
No. 318 Laser Interaction and Related Plasma Phenomena 11th International Workshop (Monterey, CA 1993)	94-78097	1-56396-324-8
No. 319 Beam Instrumentation Workshop (Santa Fe, NM 1993)	94-78279	1-56396-389-2
No. 320 Basic Space Science (Lagos, Nigeria 1993)	94-79350	1-56396-328-0
No. 321 The First NREL Conference on Thermophotovoltaic Generation of Electricity (Copper Mountain, CO 1994)	94-72792	1-56396-353-1
No. 322 Atomic Processes in Plasmas Ninth APS Topical Conference (San Antonio, TX)	94-72923	1-56396-411-2
No. 323 Atomic Physics 14 Fourteenth International Conference on Atomic Physics (Boulder, CO 1994)	94-73219	1-56396-348-5
No. 324 Twelfth Symposium on Space Nuclear Power and Propulsion (Albuquerque, NM 1995)	94-73603	1-56396-427-9
No. 325 Conference on NASA Centers for Commercial Development of Space (Albuquerque, NM 1995)	94-73604	1-56396-431-7
No. 326 Accelerator Physics at the Superconducting Super Collider (Dallas, TX 1992-1993)	94-73609	1-56396-354-X
No. 327 Nuclei in the Cosmos III Third International Symposium on Nuclear Astrophysics (Assergi, Italy 1994)	95-75492	1-56396-436-8
No. 328 Spectral Line Shapes, Volume 8 12th ICSLS (Toronto, Canada 1994)	94-74309	1-56396-326-4
No. 329 Resonance Ionization Spectroscopy 1994 Seventh International Symposium (Bernkastel-Kues, Germany 1994)	95-75077	1-56396-437-6
No. 330 E.C.C.C. 1 Computational Chemistry F.E.C.S. Conference (Nancy, France 1994)	95-75843	1-56396-457-0
No. 331 Non-Neutral Plasma Physics II (Berkeley, CA 1994)	95-79630	1-56396-441-4

	Title	L.C. Number	ISBN
No. 332	X-Ray Lasers 1994 Fourth International Colloquium (Williamsburg, VA 1994)	95-76067	1-56396-375-2
No. 333	Beam Instrumentation Workshop (Vancouver, B. C., Canada 1994)	95-79635	1-56396-352-3
No. 334	Few-Body Problems in Physics (Williamsburg, VA 1994)	95-76481	1-56396-325-6
No. 335	Advanced Accelerator Concepts (Fontana, WI 1994)	95-78225	1-56396-476-7 (Set) 1-56396-474-0 (Book) 1-56396-475-9 (CD-Rom)
No. 336	Dark Matter (College Park, MD 1994)	95-76538	1-56396-438-4
No. 337	Pulsed RF Sources for Linear Colliders (Montauk, NY 1994)	95-76814	1-56396-408-2
No. 338	Intersections Between Particle and Nuclear Physics 5th Conference (St. Petersburg, FL 1994)	95-77076	1-56396-335-3
No. 339	Polarization Phenomena in Nuclear Physics Eighth International Symposium (Bloomington, IN 1994)	95-77216	1-56396-482-1
No. 340	Strangeness in Hadronic Matter (Tucson, AZ 1995)	95-77477	1-56396-489-9
No. 341	Volatiles in the Earth and Solar System (Pasadena, CA 1994)	95-77911	1-56396-409-0
No. 342	CAM -94 Physics Meeting (Cacun, Mexico 1994)	95-77851	1-56396-491-0
No. 343	High Energy Spin Physics Eleventh International Symposium (Bloomington, IN 1994)	95-78431	1-56396-374-4
No. 344	Nonlinear Dynamics in Particle Accelerators: Theory and Experiments (Arcidosso, Italy 1994)	95-78135	1-56396-446-5
No. 345	International Conference on Plasma Physics ICPP 1994 (Foz do Iguaçu, Brazil 1994)	95-78438	1-56396-496-1
No. 346	International Conference on Accelerator-Driven Transmutation Technologies and Applications (Las Vegas, NV 1994)	95-78691	1-56396-505-4
No. 347	Atomic Collisions: A Symposium in Honor of Christopher Bottcher (1945-1993) (Oak Ridge, TN 1994)	95-78689	1-56396-322-1
No. 348	Unveiling the Cosmic Infrared Background (College Park, MD, 1995)	95-83477	1-56396-508-9

Title	L.C. Number	ISBN
No. 349 Workshop on the Tau/Charm Factory (Argonne, IL, 1995)	95-81467	1-56396-523-2
No. 350 International Symposium on Vector Boson Self-Interactions (Los Angeles, CA 1995)	95-79865	1-56396-520-8
No. 351 The Physics of Beams Andrew Sessler Symposium (Los Angeles, CA 1993)	95-80479	1-56396-376-0
No. 352 Physics Potential and Development of $\mu^+\mu^-$ Colliders: Second Workshop (Sausalito, CA 1994)	95-81413	1-56396-506-2
No. 353 13th NREL Photovoltaic Program Review (Lakewood, CO 1995)	95-80662	1-56396-510-0
No. 354 Organic Coatings (Paris, France, 1995)	96-83019	1-56396-535-6
No. 355 Eleventh Topical Conference on Radio Frequency Power in Plasmas (Palm Springs, CA 1995)	95-80867	1-56396-536-4
No. 356 The Future of Accelerator Physics (Austin, TX 1994)	96-83292	1-56396-541-0
No. 357 10th Topical Workshop on Proton-Antiproton Collider Physics (Batavia, IL 1995)	95-83078	1-56396-543-7
No. 358 The Second NREL Conference on Thermophotovoltaic Generation of Electricity	95-83335	1-56396-509-7
No. 360 The Physics of Electronic and Atomic Collisions XIX International Conference (Whistler, Canada, 1995)	95-83671	1-56396-440-6
No. 363 Phenomena in Ionized Gases XXII ICPIG (Hoboken, NJ, 1995)	96-83294	1-56396-550-X
No. 365 Latin-American School of Physics XXX ELAF Group Theory and Its Applications (México City, México, 1995)	96-83489	1-56396-567-4